Infinite Dimensional Analysis, Quantum Probability and Related Topics

Proceedings of the International Conference on
Infinite Dimensional Analysis, Quantum Probability and
Related Topics, QP38

QP–PQ: Quantum Probability and White Noise Analysis

Vol. 32: Infinite Dimensional Analysis, Quantum Probability and Related Topics
Proceedings of the International Conference on Infinite Dimensional Analysis, Quantum Probability and Related Topics, QP38
eds. L. Accardi, Si Si and N. Watanabe

Vol. 31: Quantum Bio-Informatics VI
From Quantum Information to Bio-Informatics
eds. L. Accardi, W. Freudenberg and N. Watanabe

Vol. 30: Quantum Bio-Informatics V
eds. L. Accardi, W. Freudenberg and M. Ohya

Vol. 29: Quantum Probability and Related Topics
eds. L. Accardi and F. Fagnola

Vol. 28: Quantum Bio-Informatics IV
From Quantum Information to Bio-Informatics
eds. L. Accardi, W. Freudenberg and M. Ohya

Vol. 27: Quantum Probability and Related Topics
eds. R. Rebolledo and M. Orszag

Vol. 26: Quantum Bio-Informatics III
From Quantum Information to Bio-Informatics
eds. L. Accardi, W. Freudenberg and M. Ohya

Vol. 25: Quantum Probability and Infinite Dimensional Analysis
Proceedings of the 29th Conference
eds. H. Ouerdiane and A. Barhoumi

Vol. 24: Quantum Bio-Informatics II
From Quantum Information to Bio-informatics
eds. L. Accardi, W. Freudenberg and M. Ohya

Vol. 23: Quantum Probability and Related Topics
eds. J. C. García, R. Quezada and S. B. Sontz

Vol. 22: Infinite Dimensional Stochastic Analysis
eds. A. N. Sengupta and P. Sundar

Vol. 21: Quantum Bio-Informatics
From Quantum Information to Bio-Informatics
eds. L. Accardi, W. Freudenberg and M. Ohya

*For the complete list of the published titles in this series, please visit:
www.worldscientific.com/series/qp-pq

QP–PQ
Quantum Probability and White Noise Analysis
Volume XXXII

Infinite Dimensional Analysis, Quantum Probability and Related Topics

Proceedings of the International Conference on
Infinite Dimensional Analysis, Quantum Probability and
Related Topics, QP38

Noda City, Chiba, Japan, 2 – 6 October 2017

Editors

Luigi Accardi
Università di Roma "Tor Vergata", Italy

Si Si
Aichi Prefectural University, Japan & Yangon University, Myanmar

Noboru Watanabe
Tokyo University of Science, Japan

World Scientific

NEW JERSEY • LONDON • SINGAPORE • BEIJING • SHANGHAI • HONG KONG • TAIPEI • CHENNAI • TOKYO

Published by

World Scientific Publishing Co. Pte. Ltd.
5 Toh Tuck Link, Singapore 596224
USA office: 27 Warren Street, Suite 401-402, Hackensack, NJ 07601
UK office: 57 Shelton Street, Covent Garden, London WC2H 9HE

British Library Cataloguing-in-Publication Data
A catalogue record for this book is available from the British Library.

QP-PQ: Quantum Probability and White Noise Analysis — Vol. 32
INFINITE DIMENSIONAL ANALYSIS, QUANTUM PROBABILITY AND RELATED TOPICS
Proceedings of the International Conference on Infinite Dimensional Analysis, Quantum Probability and Related Topics, QP38

ISBN 978-981-12-7598-2 (hardcover)
ISBN 978-981-12-7599-9 (ebook for institutions)
ISBN 978-981-12-7600-2 (ebook for individuals)

For any available supplementary material, please visit
https://www.worldscienti ic.com/worldscibooks/10.1142/13402#t=suppl

PREFACE

This volume collects papers discussed during the 38th international workshop of Infinite Dimensional Analysis and Quantum Probability and Related Topics QP38 held at Tokyo University of Sciences. This is a special issue dedicated to Prof. Takeyuki Hida, Prof. Luigi Accardi and Prof. Igor Volovich — initiators of this series of conferences.

The purpose of the workshop is to meet many eminent mathematicians who have contributed mainly in the areas of White Noise Analysis, Quantum Probability and Quantum Information Theory who made the workshop successful and cultivated new ideas. The goals of the workshop are to make a bridge among the fields on infinite dimensional analysis and quantum probability, to review recent research developments in these interdisciplinary fields, mainly focused on quantum probability and white noise analysis in line with IDAQP, and to accelerate the interaction of infinite dimensional analysis and quantum probability, and promoting the highly interdisciplinary research on new exciting areas in mathematical sciences like quantum information theory and white noise analysis.

In this workshop, particular attention was paid to quantum independence, Lévy process, quantum Markov semigroup, Brownian motion, nonpositive map, quantum-like modeling, non-commutative probability, holographic photosynthesis, entanglement entropy, and Hida distribution. Most speakers gave care to the relation between their own topics and the mystery of life.

All papers in this volume have been refereed. Their contents are related to one of the following subjects:

(1) Notions of quantum independence naturally arising from classical probability
(2) Lévy processes on the Lorentz-Lie algebra
(3) Basins of attraction of invariant states of quantum Markov semigroup
(4) Bimonotone Brownian motion
(5) Complex structures on three-point space
(6) Uncertainty relations and information loss for spin-1/2 measurements

(7) On effective methods of investigation of nonpositive maps
(8) Quantum-like modeling of the order effect in decision making
(9) Stability and perturbations of C_0-Markov semigroups on abstract states spaces
(10) Decohence in an excitation energy transport model
(11) Rough-paths and non-commutative probability
(12) A quick view on Professor Takeyuki Hida's work
(13) Representing the quantum operators in terms of multiplication and differentiation operators
(14) Holographic photosynthesis and entanglement entropy
(15) Conditional distribution of a random variable, conditioned by Hida distributions, on Euclidean quantum fields

This publication is supported by the Research Institute of Advanced Technology, SoftBank Corp., Japan

Luigi Accardi
Si Si
Noboru Watanabe

BIOGRAPHY: TAKEYUKI HIDA

Emeritus Professor, of Nagoya University and Meijo University, Takeyuki Hida was born on 12th November 1927. He was a visionary thinker. The greatest thing he contributed was the establishment of the White Noise Theory.

In 1970, Hida achieved "**The 33rd Chunichi Cultural Prize**" from Chunichi Shinbun and in 2007, "**The Order of the Sacred Treasure, Gold and Silver**" from the Emperor.

Hida was not only interested in Theory of mathematics but also in History of mathematics and in many other fields such as Physics, Biology and etc. His idea was based on that if one consider a random complex system, it should be dealt with the following steps:

Reductionism, Synthesis, Analysis.

With this idea, he established the White noise theory. In 2000, American Mathematical Society assigned **2000 AMS Subject Classification 60H40** to White Noise Theory.

Hida achieved the "Doctor of Science" under the supervision of Professor K. Ito with the dissertation: **"Canonical representation of Gaussian Processes and their Applications"**. In this paper the existence of generalized canonical representation concerned with multiplicity which is also obtained by H. Cramér at the same time and so it is called Hida-Cramér Theorem.

The use of reproducing kernel Hilbert spaces for research of Gaussian processes was introduced by Hida. It seems that he intended to use this idea establishing the analysis of Gaussian and Poisson white noises.

In 1970, the collection of his lectures on **"Stationary Stochastic processes"** was published by Princeton University Press. These lecture notes was given in Princeton University for Graduate course, during the period when he was invited there for one year by Professor William Feller.

In 1975, his famous book **"Brownian Motion"** was published in Japanese and it was translated to English in 1980. This book included Princeton University lecture notes mainly together with some background and developments. This book was also translated to Russian and Chinese in 1987.

We may say that **White Noise Theory** was newly born from his celebrated monograph **"Analysis of Brownian Functionals"**, (Carleton Mathematical Lecture Notes 13 (1975)), containing a series of his lectures delivered at Carleton University in 1975. He proposed the framework for the analysis of Brownian functionals from the white noise point of view by introducing the concept of "generalized functionals" of Brownian motion (white noise). He used the white noise $\dot{B}(t)$ as a basic variable of Brownian functionals.

Hida introduced the notion of generalized functionals of white noise being motivated by P. Lévy's functional analysis and various applications including quantum dynamics and nuclear biology. His essential idea is to use $\{\dot{B}(t); t \in R\}$ as a coordinate system. He first defined $\dot{B}(t)$ for each t and then consider non-linear functions of $\{\dot{B}(t); t \in R\}$ as a generalized functionals.

First Hida setup a triple $E \subset L^2(R^1) \subset E^*$ with a nuclear space E being dense in $L^2(R^1)$. Using Bochner-Minlos Theorem the probability measure μ for the probability distribution of white noise is obtained and then $L^2(E^*, \mu)$ is called a white noise space. He applied the Wiener-Ito theorem to decompose $L^2(E^*, \mu)$ into an orthogonal direct sum $L^2(E^*, \mu) = \sum_0^\infty \mathcal{H}_n$ of multiple Wiener integrals or homogeneous chaos. Since $\dot{B}(t) = \lim_{h\to 0} h^{-1}(B(t+h) - B(t))$ does not exist in the space $\mathcal{H}^1$, a weaker norm is introduced on $\mathcal{H}^1$ so that the limit exists. The variance of $\dot{B}(t)^2 = \lim_{h\to 0} \frac{1}{h}$ does not exist in the space $\mathcal{H}_2$. A weaker norm on $\mathcal{H}_2$ is introduced so that the limit exists with respect to the weaker norm. Thus, the renormalization $: \dot{B}(t)^2 :$ of $\dot{B}(t)^2$ is $\dot{B}(t)^2 - \frac{1}{h}$. Hida's idea to define $: \dot{B}(t)^2 :$ is a singular renormalization and thus $: \dot{B}(t)^2 :$ can be regarded as the generalized Wiener integral of $\delta_t \otimes \delta_t$ which is not in L^2. The same idea is carried out on each n and then take the orthogonal direct sum of the new spaces to obtain a generalized white noise functional space $(L^2)^-$.

White noise has long been regarded as an informal, mysterious, and imaginary object. Owing to tremendous effort of Hida, there is a mathematical theory of white noise.

Hida also contributed in many other fields and wrote many papers and books (more or less 200 papers and 20 books) in the fields: White noise analysis, Infinite dimensional harmonic analysis, Rotation group, Stochastic variational calculus, Quantum Theory, Feynman Integrals and Random fields, Variational Calculus and Random fields, Application to Biology and etc.

Hida served as a Professor of Nagoya University and after his retirement from there he was appointed as a Professor at Meijo University. He was a dean of Faculty of Science, Nagoya University and also in Meijo University. When he was in Nagoya University, he was invited to Indiana University, Stanford University and Princeton University to be a visiting Professor. He was assigned as a fellow of International Institute of Advanced Studies in 1999–2000. He served as a councillor of "The Operation Research Society of Japan", a director of "Mathematical Society of Japan", chairman of "Bernoulli Society" and "Stochastic Processes and their Applications", and Managing Editor of the Journal "Infinite Dimensional Analysis, Quantum Probability and Related Topics".

Hida organized the International Conferences on "Stochastic Processes and their Applications" in 1985, "Mathematics of Quantum Analysis and Quantum Information Theory" in 2000 and International Workshops on various topics. He achieved the Research Project grant for the Academic Frontier of Science promoted by the Ministry of Education in Japan, and so he organized a series of International Conferences on "Quantum Information"(Five times).

Hida used to look almost everything from the white noise viewpoint. He often used the term "idealized elementary random variables" to express the system $\{\dot{B}(t)\}$ since they are the most important mathematical elements to study random phenomena.

Hida continued his research until his last days. He has passed away on 29th December 2017. It is a great loss to mathematical society.

Si Si

BIOGRAPHY: LUIGI ACCARDI

After the emergence of quantum mechanics at the beginning of the last century, the role of probability in new subjects has been formulated in many aspects. However, it has grown significantly in the past few decades and it came to be called "quantum probability."

Luigi Accardi, professor of the University of Rome Tor Vergata, was born on August 13, 1947 in Naples. He is a pioneering thinker and one of the leaders in the development of quantum probability theory (quantum probability). He has produced a variety of excellent research results, and his contribution to quantum probability is immense.

He received the Vittorio Nobile Award for the highest degree in mathematics when he was awarded an honors degree in mathematics at the University of Naples on October 31, 1970.

In 1974, he received his doctorate at the Moscow State University in Moscow with a thesis on noncommutative Markov theory and square roots of measures under the supervision of Professor I. M. Gelfand and Professor Y. G. Sinai.

In 1982, he became Professor of Probability Theory at the University of Rome Tor Vergata and was invited as Visiting Professor worldwide. In 1996, he was also appointed as a Professor at the Graduate School of Polymathematics, Nagoya University for three years. From 1998 up to the present, he has been Professor at the Faculty of Economics, University of Rome Tor Vergata.

He has been a member of the Organizing, Scientific or Program Committee of many national and international conferences.

In 1983, he was a visiting scholar in New Delhi, along with K. R. Parthasarathy, and studied at Princeton (with J. S. Watson) and Warwick University (with D. Elworthy) in 1985 and 1986. In 1988/89, he was a visiting researcher with Paul-André Meyer at the University of Strasbourg and a visiting researcher with Rudolf Kalman at the University of Gainesville, Florida. He was also a visiting researcher at ETH Zurich.

In 2002, he published a book entitled "Quantum Theory and Its

Stochastic Limit" with Prof. Yun Gang Lu and Prof. Igor Volovich. This includes new mathematical methods, stochastic limit to solve nonlinear problems in quantum theory involving systems with infinitely many degrees of freedom, rules that can choose the dominant contribution to the dynamical evolution of systems in regimes with long time and small effects. In the stochastic limit, the original Hamiltonian theory is approximated using singular Hamiltonian theory. In particular, it can explicitly compute correlations or coherent vectors beyond the range of typical asymptotic methods, and can similarly estimate stochastic Schrodinger equation and master equation widely used in the Hamiltonian framework.

From 2002 to 2006, Prof. Accardi participated the joint research project supported by IIAS entitled "Study on mathematical theory of quantum information" initiated by Prof. Hida and continued by Prof. Ohya, which played an important role in several joint Italy-Japan scientific activities for example, the Italian-Japanese meeting of mathematicians (1997), the Japan-Italy joint workshop on quantum information theory (2001), within the wider frame of the Japan-Italy year (2001–2002) and the ICQI03IIAS (2003), which provided a testimony of the many new ideas and results in quantum information, white noise theory, infinite dimensional analysis and quantum probability. Under the research program of the joint TUS-Volterra group, Prof. Accardi studied the following topics: (1) adaptive systems, (2) simulation of quantum phenomena by classical computers, (3) quantum and classical cryptography, (4) complexity theory for classical and quantum algorithms, (5) quantum teleportation, (6) entanglement, (7) white noise and data analysis.

In addition, he became a visiting professor at Tokyo University of Science and conducted various joint researches with Prof. Masanori Ohya. In 2003, he co-hosted an international conference on quantum information, followed by the establishment of the QBIC international quantum information research conference. One of the members has been dedicated to holding QBIC and developing research in this field. In particular, Prof. Accardi was interested in the philosophy in many other fields, such as physics and biology, as well as mathematical theory. His idea of chameleon theory was based on the need to deal with these phenomena adaptively when considering random complex systems.

He has contributed to the development of the International Conference on Quantum Probability since the establishment of the International Conference on Quantum Probability QP "Quantum Probability". He is also one of the founding members of AQPIDA, "Association of Quantum Probability and Infinite Dimensional Analysis", and has contributed to the future direction and great progress of quantum probability research.

He also co-founded Centro V. Volterra with Tokyo University of Science and contributed to the establishment of the Tokyo University of Science's "Information Security" laboratory, which is currently involved in several relevant industrial projects. He also organized workshops on information security, etc., and lead industry-academia R & D in this field at Centro V. Volterra.

Prof. Accardi has collaborated with various researchers. For example, he has studied the weak coupling limit, Quantum Stochastic Processes, Quantum Markov Processes, Teleportation of General Quantum States, Compound Channels, Transition Expectations, and Liftings, a Stochastic Limit Approach to the SAT Problem, Dynamical Entropy Through Quantum Markov Chains, Markov States on the CAR Algebra, and so on.

Noboru Watanabe

BIOGRAPHY: IGOR VOLOVICH

Igor Vassilevich Volovich was born in Vochktomskoe (Archkangelskaja region, USSR) on August 9, 1946. His mathematical education started in 1970 at Moscow State University. In 1973, he entered the Graduate School of Steklov Mathematical Institute from which he received his Ph.D. in 1980 and the title of Doctor of Science together with the nomination as Professor of Mathematical Physics.

Since the beginning, his main research field has been mathematical physics, in particular mathematical problems in quantum field theory, and gradually was enlarged to include classical analysis, p-adic analysis, string theory, theory of classical and quantum gravity, quantum probability, quantum information, quantum optics, partial differential equations, differential geometry, integrable systems.

In the theory of nonlinear elliptic equations, he proved an existence and regularity theorem for solutions of equations of the form $\Delta\phi + P(\phi) = f(x)$ (P polynomial, f smooth) which arise in Euclidean quantum field theory and in the theory of integrable systems.

In 1982–1983 together with V. S. Vladimirov, he wrote mathematical papers putting in rigorous level the notion of super–space, used by physicists, with the help of analysis on Banach algebras.

In the theory of integrable systems, he suggested in 1983, a Lax-type representation for differential equations of super–symmetric Yang–Mills theory which leads to higher conservation laws. In this framework, he introduced super–self–duality equations and constructed solutions based on the theory of super–twistors. Later, with V. S. Vladimirov, he developed a general method for the construction of local and non-local conservation laws for integrable systems.

In 1985–1986 with I. Ya. Aref'eva, he found solutions of supergravity equations with non-standard signature in higher dimensional spaces using the theory of hyperbolic compact manifolds. In 1987 he introduced p-adic string theory and p-adic quantum mechanics and begun their systematic investigation. In 1989 with V. S. Vladimirov, he developed a formulation of p-adic quantum harmonic oscillator and performed spectral harmonic analysis of this system. The main results of this line of research were synthesized in the book "p-adic Analysis and Mathematical Physics" (with V. S. Vladimirov, E. I. Zelenov), World Sci. 1994.

In 1994 together with L. Accardi and P. I. Gibilisco he proved that the Yang–Mills equations for a connection on Euclidean space are equivalent to the fact that the associated parallel transport, considered as a function of the curve along which the transport takes place, is an harmonic function for the Levy-Laplacian. This result was extended, with R. Leandre, to general Riemannian manifolds.

In 1990 with I. Aref'eva, he investigated models of field theories with quantum group playing the role of the gauge group. They developed a mathematical approach for investigation of equations for functions taking values in quantum groups. For the quantum group chiral fields infinite numbers of conservation laws were constructed. With V. S. Vladimirov, he proved the equivalence between general theory of orthogonal polynomials and the Gaussian model of statistical mechanics on semi-axis. A generalized solution to the Wiener-Hopf problem was constructed.

He proposed a model of two-dimensional gravity with dynamical torsion and a string theory with torsion on the world sheet. The Cauchy problem for the corresponding system of nonlinear differential equations was considered and stationary solutions were constructed.

In the theory of elasticity, together with M. Katanaev, he extended in a geometrical way Volterra's theory of defects in elastic bodies introducing the notions of continuous distributions of dislocations and disclinations and applied it to the 3-dimensional theory of gravity with torsion.

In 1993 with M. Francaviglia and M. Ferraris, he found a new action for Einstein equations and considered its applications to quantum gravity, topological control and construction of new 2-dimensional models of strings and gravity.

With L. Accardi and Y. G. Lu, he developed the stochastic limit of quantum theory, which gives a mathematical deduction of the physical origins of quantum noises and quantum transport equations. From this theory several new results in physics and new fundamental structures in mathematics emerged. In physics, the stochastic bosonization of Fermi fields in 4-dimensional space-time, a new kind of semi-classical limit in quantum theory with quantum white noise as leading term, a natural notion of local equilibrium states, different from Zubarev's one, generalizing the usual equilibrium states and leading, for example, to a nonlinear extension of Einstein's coefficients in radiation theory, to the notion of dynamical detailed balance extending non-equilibrium regimes the usual (equilibrium) notion of detailed balance and, more generally to the first examples of non-equilibrium quantum field theories, the emergence of quasi-particles (like triphons) from nonlinear interactions, the emergence of dynamical q-deformations of the CCR from quantum electro-dynamics, as well as a multiplicity of applications in quantum optics (in particular to the dark states), the fact that, in strongly nonlinear regimes the usual Gaussian diagrams of Bose and Fermi noises are replaced by the free diagrams, and the quasi-free Bose or Fermi states, by a new class of operator valued states providing the first concrete example of module gaussianity (a systematic theory of module gaussian states is still missing nowadays), ...

In mathematics, with the notion of interacting Fock space which has allowed a purely algebraic approach to the classical theory of orthogonal polynomials and which is at the basis of the quantum decomposition of classical random variables and processes. Another, far reaching, mathematical fall out of the stochastic limit of quantum theory is the identification of usual stochastic differential equations (including the classical ones) with white noise Hamiltonian equations. This provides in particular a microscopic interpretation of the unitarity conditions of Hudson and Parthasarathy, in terms of the coefficients of the underlying Hamiltonian equation and suggests a natural extension of stochastic calculus to fields (classical or quantum), in which the 1-dimensional time parameter is replaced by a multi-dimensional index (space or space-time), as required by the applications to quantum fields. The monograph "Quantum Theory and its Stochastic Limit" (with L. Accardi and Yu. G. Lu, 2002, Springer) synthesizes the main results of this line of research up to 2002 with the exception of the low density limit. In collaboration with I. Ya. Aref'eva he established the anisotropic and large-N limit in quantum field theory and its relation with high energy scattering.

Finally, with I. Ya. Aref'eva and K. Visnewanathan he proposed a mechanism of black holes creation in the Planckian energy scattering of light particles by using solutions of Einstein equations describing colliding plane gravitational waves.

In quantum information he has worked on the application of quantum decoherence to the theory of quantum algorithms. He and M. Ohya have proposed an approach (which goes beyond of the standard paradigm of quantum computation) to the solution of NP-complete problem in polynomial time. A new approach to problems of quantum non-locality, Bell's inequalities, quantum cryptography was developed based on the study of spatial dependence of entangled states. The results of this line of investigation have been synthesized in the monograph "Quantum Information and Computations and Its Applications to Nano- and Bio-Systems" (with M. Ohya, 2011, Springer).

In the past 50 years, very few mathematicians in the world have given important and sometimes ground-breaking contributions to such a wide spectrum of mathematical disciplines as Igor Volovich.

These achievements have been recognized by a multiplicity of honors and awards, like the prize of the Mathematical Branch of the Academy of Science of USSR in 1984, the prize of Mathematical Branch of Academy of Science of USSR in 1986 and the election as Corresponding Member of the Russian Academy of Sciences in 2003. Since 2017, he is member of the Strategic Committee of the Volterra Center. Actually he is head of Department of Mathematical Physics, Steklov Mathematical Institute, Russian Academy of Sciences.

He is Managing Editor of the journal 'p-Adic Numbers, Ultrametric Analysis and Applications' (Springer) and member of the Editor Boards of the journals:
– Izvestiya: Mathematics (Izvestiya Rossiiskoi Akademii Nauk, Seriya Matematicheskaya).
– Infinite Dimensional Analysis, Quantum Probability and Related Topics, World Scientific.
– International Journal of Geometric Methods in Modern Physics, World Scientific.
– Journal of Stochastic Analysis and applications, JOSA.
– Nanosystems and Nanotechnologies, Moscow.

Among his students, most of whom are now established and internationally known researches, we mention Sergei Kozyrev, Alexander Pechen, Roman Roshin, Anton Trucheskin.

Luigi Accardi

CONTENTS

Infinite Dimensional Analysis, Quantum Probability and Related Topics

pp. 1–21

NOTIONS OF QUANTUM INDEPENDENCE NATURALLY ARISING FROM CLASSICAL PROBABILITY

LUIGI ACCARDI

Centro Vito Volterra, Università di Roma Tor Vergata, Italy
E-mail: accardi@volterra.uniroma2.it

YUN GANG LU

Dipartimento di Matematica
Universita degli Studi di Bari, Italy
E-mail: yungang.lu@uniba.it

We show how the basic notions of quantum independence, classical, tensor, monotone, free, Boolean (and in fact many more), naturally arise combining classical probability with the notion of Interacting Fock Spaces (IFS). In particular we discuss the Poisson central limit theorems associated to the various independences.

1. The classical Possion type central limit theorem (CLT)

Theorem 1.1. *Let* $(\Omega, \mathcal{F}, \mathbf{P})$ *be a classical probability space and, for each* $n \in \mathbb{N}^* := \mathbb{N} \setminus \{0\}$, *let*

$$\{\xi_n \ : \ n \in \mathbb{N}^*\} \tag{1}$$

be a sequence of $\mathbb{N}$*-valued random variables such that for each* $n \in \mathbb{N}^*$, *the distribution of each* ξ_n *is Bernoulli with parameter* p_n. *In this case we will write*

$$\xi_{n,k} \sim B(1, p_n) = \text{binomial of order } 1 \text{ with parameter } p_n$$

Then, if

$$\lim_{n\to\infty} np_n = \lambda$$

it follows that for any $m \in \mathbb{N}$ *one has:*

$$\lim_{n\to\infty} \mathbf{P}(\xi_n = m) = \frac{e^{-\lambda}\lambda^m}{m!} =: P_m(\lambda)$$

where $P(\lambda)$ denotes the Poisson distribution with parameter λ and $P(0)$ is interpreted as δ_0.

Proof. By assumption, for any $m, n \in \mathbb{N}^*$

$$\mathbf{E}(\xi_n^m) = \sum_{k=0}^{n} k^m \binom{n}{k} p_n^k (1-p_n)^{n-k} \tag{2}$$

$$= \sum_{k=0}^{n} \frac{k^m}{k!} n(n-1)\dots(n-k+1)\, p_n^k (1-p_n)^{n-k}$$

$$= \sum_{k=0}^{n} \frac{k^m}{k!} \prod_{j=0}^{k} (1-j/n)\,(np_n)^k (1-p_n)^{n-k}$$

$$\sim \sum_{k=0}^{n} \frac{k^m}{k!} \lambda^k \left(1-\frac{\lambda}{n}\right)^{n-k} = \sum_{k=0}^{n} \frac{k^m}{k!} \lambda^k \left(\left(1-\frac{\lambda}{n}\right)^n\right)^{(1-\frac{k}{n})}$$

$$\to \sum_{k=0}^{\infty} \frac{k^m}{k!} \lambda^k e^{-\lambda} = m\text{-th moment of } \xi \sim P(\lambda)$$

In the Poisson case moment convergence implies convergence of distributions.

Theorem 1.2. *Let $(\Omega, \mathcal{F}, \mathbf{P})$ be a classical probability space and, for each $n \in \mathbb{N}^* := \mathbb{N} \setminus \{0\}$, let*

$$\{\eta_{n,k} \ : \ k \in \mathbb{N}^* \ , \ k \le n\} \tag{3}$$

be a family of $\mathbb{N}$-valued random variables. Suppose that:
(i) For each $n \in \mathbb{N}^*$, the family (3) is independent and identically distributed (IID).
(ii) The distribution of each $\eta_{n,k}$ is Bernoulli with parameter p_n independent of k:

$$\eta_{n,k} \sim B(1, p_n) \qquad ; \qquad \forall k \le n$$

Then, if

$$\lim_{n\to\infty} np_n = \lambda$$

it follows that for any $m \in \mathbb{N}$ one has:

$$\lim_{n\to\infty} \mathbf{P}\left(\sum_{k=1}^{n} \eta_{n,k} = m\right) = \frac{e^{-\lambda}\lambda^m}{m!} =: P_m(\lambda)$$

Proof. If the distribution of each $\xi_{n,k}$ is binomial of order 1 with parameter p_n, then the distribution of the $\sum_{k=1}^{n}\xi_{n,k}$ is multi-nomial of order n with parameter p_n. In this case we will write

$$\sum_{k=1}^{n}\xi_{n,k}\sim M\left(n,p_n\right)$$

So one has

$$\begin{aligned}&\mathbf{E}\left(\left(\eta_{n,1}+\ldots+\eta_{n,n}\right)^{m}\right)\\ &=\sum_{0\le j_1,\ldots,j_n\le m,\ j_1+\ldots+j_n=m}\frac{m!}{j_1!\ldots j_n!}\mathbf{E}\left(\eta_{n,1}^{j_1}\cdots\eta_{n,n}^{j_n}\right)\\ &=\sum_{0\le j_1,\ldots,j_n\le m,\ j_1+\ldots+j_n=m}\frac{m!}{j_1!\ldots j_n!}\mathbf{E}\left(\eta_{n,1}^{j_1}\right)\ldots\mathbf{E}\left(\eta_{n,n}^{j_n}\right)\end{aligned}$$

We use the convention $0^0:=1$, hence

$$\eta_{n,k}^{j}=\begin{cases}1, & \text{if } j=0\\ \eta_{n,k}, & \text{if } j\ge 1\end{cases}$$

and since $\eta_{n,k}\sim B\left(1,p_n\right)$, one has

$$\mathbf{E}\left(\eta_{n,k}^{j}\right)=\begin{cases}1, & \text{if } j=0\\ p_n, & j\ge 1\end{cases}$$

Moreover, using the decomposition

$$\begin{aligned}&\left\{\left(j_1,\ldots,j_n\right):0\le j_1,\ldots,j_n\le m,\ j_1+\ldots+j_n=m\right\}\\ &=\bigcup_{r=1}^{m}\bigcup_{1\le k_1<\ldots<k_r\le n}\left\{\left(i_1,\ldots,i_n\right):1\le i_{k_1},\ldots,i_{k_r},\ i_1+\ldots+i_n=m\right\}\end{aligned}$$

one has

$$\begin{aligned}&\mathbf{E}\left(\left(\eta_{n,1}+\ldots+\eta_{n,n}\right)^{m}\right)\\ &=\sum_{r=1}^{m}\sum_{1\le k_1<\ldots<k_r\le n}\sum_{1\le j_1,\ldots,j_r\le m,\ j_1+\ldots+j_r=m}\mathbf{E}\left(\eta_{n,k_1}^{j_1}\cdots\eta_{n,k_r}^{j_r}\right)\\ &=\sum_{r=1}^{m}\sum_{1\le k_1<\ldots<k_r\le n}\sum_{1\le j_1,\ldots,j_r\le m,\ j_1+\ldots+j_r=m}\mathbf{E}\left(\eta_{n,k_1}^{j_1}\right)\ldots\mathbf{E}\left(\eta_{n,k_r}^{j_r}\right)\end{aligned}$$

$$= \sum_{r=1}^{m} \sum_{1 \le k_1 < \ldots < k_r \le n} \sum_{1 \le j_1, \ldots, j_r \le m, \; j_1 + \ldots + j_r = m} \mathbf{E}\left(\eta_{n,k_1}\right) \ldots \mathbf{E}\left(\eta_{n,k_r}\right)$$

$$= \sum_{r=1}^{m} \sum_{1 \le k_1 < \ldots < k_r \le n} \sum_{1 \le j_1, \ldots, j_r \le m, \; j_1 + \ldots + j_r = m} p_n^r (1 - p_n)^{n-r}$$

$$= \sum_{r=1}^{m} \binom{n}{r} r^m p_n^r (1 - p_n)^{n-r} \tag{4}$$

Combining this with (2), one gets, since $np_n \to \lambda > 0$

$$\lim_{n \to \infty} \mathbf{E}\left((\eta_{n,1} + \ldots + \eta_{n,n})^m \right) = \sum_{k=0}^{\infty} \frac{k^m}{k!} \lambda^k e^{-\lambda}$$
$$= m\text{-th moment of } \xi \sim P(\lambda) \tag{5}$$

2. The quantum decomposition of a classical random variable

We know that any random variable X with all moments has a quantum decomposition

$$X = a_X^+ + a_X^0 + a_X^-$$

We denote $\mathcal{P}(X)$ the polynomial algebra in X and $\mathcal{P}_0(a_X^+, a_X^0, a_X^-)$ the associated minimal quantum polynomial algebra, i.e. the algebra generated by the CAP operators a_X^+, a_X^0, a_X^- of X. The vacuum vector

$$\Phi_0 = 1$$

induces a state φ_0 on $\mathcal{P}_0(a_X^+, a_X^0, a_X^-)$ given by

$$\varphi_0(Q) := \langle \Phi_0, Q(X)\Phi_0 \rangle \; ; \; Q \in \mathcal{P}_0(a_X^+, a_X^0, a_X^-)$$

The restriction of the state φ_0 on $\mathcal{P}(X)$ is called the **distribution** of X. Note that, by construction, the CAP operators a_X^ε ($\varepsilon \in \{-1, 0, +1\}$) map $\mathcal{P}(X)$ into itself. Therefore the restriction of φ on $\mathcal{P}(X)$ uniquely determines the state φ_0 on $\mathcal{P}_0(a_X^+, a_X^0, a_X^-)$. The φ_0-**mean value** of X is

$$\varphi_0(X) = \langle \Phi_0, a_X^0 \Phi_0 \rangle =: \alpha_0$$

The φ_0-**variance** of X is

$$\varphi_0(X^2) - \varphi_0(X)^2 = \varphi_0(a_X^{+2} + a_X^+ a_X^0 + a_X^+ a_X^- + a_X^0 a_X^+ + a_X^{02} + a_X^0 a_X^-$$

$$+a_X^- a_X^+ + a_X^- a_X^0 + a_X^{-2}) - \alpha_0^2$$

Because of the Fock-type property

$$\varphi_0(a_X^+ Q) = \varphi_0(Q a_X^-) = 0 \qquad ; \qquad \forall Q \in \mathcal{P}_0(a_X^+, a_X^0, a_X^-)$$

one has

$$\varphi_0(X^2) - \varphi_0(X)^2 = \varphi_0(a_X^- a_X^+) = \omega_1$$

2.1. *The quantum decomposition of a sum of IID classical random variables*

Consider an algebraic probability space $(\mathcal{A}, \varphi)$ and a sequence of embeddings (algebraic random variables)

$$j_n : \mathcal{P}_0(a_X^+, a_X^0, a_X^-) \to \mathcal{A} \tag{6}$$

Definition. If the sequence of embeddings (j_n) in (6) satisfies the condition

$$\varphi \circ j_n = \varphi_0 \quad ; \quad \forall n \in \mathbb{N}$$

the sequence of classical operator random variables

$$X_n := j_n(X) \quad ; \quad n \in \mathbb{N}$$

is called φ-identically distributed. Now consider the sums

$$S_N(X) := \sum_{n=1}^{N} X_n \qquad ; \qquad N \in \mathbb{N}$$

The second moment of $S_N(X)$ is

$$\varphi(S_N(X)^2) = \sum_{n_1, n_2} \varphi(j_{n_1}(X) j_{n_2}(X)) \tag{7}$$

Denoting

$$\tilde{a}_{n,X}^{\varepsilon} := j_n(a_X^{\varepsilon}) \quad ; \quad \varepsilon \in \{-1, 0, +1\}, n \in \mathbb{N}$$

and assuming that φ is of Fock type, one has

$$\varphi(j_{n_1}(X) j_{n_2}(X)) = \varphi(\tilde{a}_{n_1,X}^- \tilde{a}_{n_2,X}^+) + \alpha_0^2$$

Assuming moreover that ε satisfies the **singleton condition**, which is shared by almost all known notions of stochastic independence, this becomes

$$\delta_{n_1,n_2} \varphi_0(a_X^- a_X^+) + \alpha_0^2 = \delta_{n_1,n_2} \omega_1 + \alpha_0^2$$

Therefore the second moment of $S_N(X)$ is

$$\varphi(S_N(X)^2) = N\omega_1 + N^2\alpha_0 \tag{8}$$

Note that, since the order of infinity are different in the two members on the right-hand side of (8), **there is no classical rescaling that allows to preserve non-trivial contributions from both limits.**
In fact the normalization

$$S_N(X) \to \frac{S_N(X)}{\sqrt{N}}$$

only preserves the contribution of the second moment, while the normalization

$$S_N(X) \to \frac{S_N(X)}{N}$$

only preserves the contribution of the mean. In classical CLT one subtracts the mean

$$X_n - \bar{X} = X_n - \alpha_0$$

which is equivalent to put $\alpha_0 = 0$. Thus the rescaling

$$\frac{1}{\sqrt{N}} S_N(X) = \sum_{n=1}^{N} \frac{1}{\sqrt{N}} j_n(X)$$

leads to a variance which, in the identically distributed case, is a multiple of the identity independent of N.
But subtracting the mean, under singleton condition, it is known that one can get only symmetric distributions in the moment CLT.

However taking advantage of the quantum decomposition one can look for a quantum rescaling that allows to preserve non-trivial contributions from both limits.
More precisely, rather than re-scaling the single classical random variable $X_n := j_n(X)$, one looks for a re-scaling of the form

$$X_{N,n} := q_N j_n(a_X^+) + p_N j_n(a_X^0) + \bar{q}_N j_n(a_X^-)$$

where q_N and p_N are positive real numbers. In this case the mean becomes

$$\varphi(S_N(X)) = N p_N \alpha_0$$

and the second moment

$$\varphi(S_N(X)^2) = q_N^2 N w_1 + (p_N N)^2 \alpha_0$$

it follows that, if q_N and p_N are chosen so that

$$Np_N \to c \quad ; \quad \sqrt{N}q_N \to c'$$

then both first and second moments of $S_N(X)$ converge.
We will see that, under additional conditions of independence, convergence of first and second moments is sufficient to obtain convergence not only of the classical, but also of the quantum moments. More precisely, it is known that given a notion $\mathcal{I}$ of stochastic independence, the moments of

$$\frac{1}{\sqrt{n}}\sum_{k=1}^{n}\left(a_k^{(-1)}+a_k^{(+1)}\right)=S_n^{(-1)}+S_n^{(+1)}$$

tend to the moments of the $\mathcal{I}$–Gaussian distribution denoted $\gamma_{\mathcal{I}}$.
For example:
– $\mathcal{I}$:= tensor independence $\Rightarrow \gamma_{\mathcal{I}}$ = the Normal distribution.

– $\mathcal{I}$:= Boolean independence $\Rightarrow \gamma_{\mathcal{I}}$ = the Boolean Gaussian distribution:

$$\frac{1}{2}\left(\delta_{-1}+\delta_1\right)$$

– $\mathcal{I}$:= free independence $\Rightarrow$
$\gamma_{\mathcal{I}}$ = the free Gaussian distribution
:= the Wigner distribution;

– $\mathcal{I}$:= monotone (anti-monotone) independence $\Rightarrow$
$\gamma_{\mathcal{I}}$ = the monotone (anti-monotone) Gaussian distribution := the arc–sine distribution.

We expect a similar correspondence for the quantized version of Poisson type CLT. That is, we expect to find in these CLT a **combination** of the (classical, Boolean, free monotone, anti-monotone, ...) Poisson distribution and of the corresponding Gaussian distribution.
We will prove that this expectation is correct.
The quantum decomposition of the classical random variables $X_{N,n}$ allows to associate, **in a canonical way**, a quantum limit theorem to any classical limit theorem. In fact, introducing the notations

$$a_k^{(\varepsilon)} := \begin{cases} a_k, & \text{if } \varepsilon = 1 \\ a_k^+, & \text{if } \varepsilon = 1 \\ a_k a_k^+, & \text{if } \varepsilon = 0 \\ a_k^+ a_k, & \text{if } \varepsilon = 2 \end{cases} \tag{9}$$

in correspondence with (14) we introduce the normalized sums

$$\begin{cases} B_n^{(\pm 1)} := \sqrt{p_n\,(1-p_n)}\sum_{k=1}^n a_k^{(\pm 1)} \\ B_n^{(0)} := p_n \sum_{k=1}^n a_k^{(0)} \\ B_n^{(2)} := (1-p_n)\sum_{k=1}^n a_k^{(2)} \end{cases} \tag{10}$$

and we study the moment central quantum limit problem for the operator random variables (10), namely:

$$\lim_{n\to\infty} \varphi\left(B_n^{(\varepsilon(1))} \ldots B_n^{(\varepsilon(m))}\right) \tag{11}$$

for $m \in \mathbb{N}$ and $\varepsilon \in \{-1,0,1,2\}^m$. Since

$$B_n := B_n^{(-1)} + B_n^{(+1)} + B_n^{(0)} + B_n^{(2)} = \sum_{k=1}^n X_{n,k}$$

is a sum of n $\mathcal{I}$-independent identically distributed classical Bernoulli random variables, i.e. all with distribution $B\,(1, p_n)$ ($\mathcal{I}$ = tensor, Boolean, free monotone, anti-monotone), it is natural to say that B_n has (classical, Boolean, free monotone, anti-monotone) **multinomial distribution** of order n and with parameter p_n.

3. Program

Consider an IFS $(\Gamma(\mathcal{H}), \langle \,\cdot\, , \,\cdot\, \rangle)$ over a test function space $\mathcal{H}$ with vacuum vector Φ and creator map $a^+ \ : \ v \in V \mapsto a_v^+$. Fix an ortho-normal basis $\{e_k\}_{k=1}^\infty$ of $\mathcal{H} \equiv (\mathcal{H}, \langle \,\cdot\, , \,\cdot\, \rangle_1)$ and for any $k \in \mathbb{N}^*$ and denote

$$a_k^+ := a_{e_k}^+ \qquad ; \qquad a_k := (a_{e_k}^+)^*$$

the creator (annihilator) with test function e_k,

$$\varphi := \langle \Phi \ , \ \cdot\ \Phi \rangle$$

the vacuum state and

$$\mathcal{A}_k := \left\{\text{polynomials of degree} \geq 1 \text{ in } a_k, a_k^+\right\}$$

we obtain an algebraic probability space $(\mathcal{A}, \varphi)$. We then consider the array

$$\{X_{n,k} \in \mathcal{A} \ : \ n, k \in \mathbb{N}^* \ , \ k \leq n\} \tag{12}$$

of classical operator random variables on $(\mathcal{A}, \varphi)$ defined by:

$$X_{n,k} := \sqrt{p_n\,(1-p_n)}\,\left(a_k + a_k^+\right) + p_n a_k a_k^+ + (1-p_n)\, a_k^+ a_k \tag{13}$$

for any $n \in \mathbb{N}^*$ and $k \leq n$ where

$$\lim_{n\to\infty} np_n = \lambda \quad ; \quad \{p_n\}_{n=1}^{\infty} \subset [0,1]$$

The right-hand side of (13) is therefore a quantum decomposition of the Bernoulli random variables $X_{n,k}$. The operator random variables

$$\sqrt{p_n\,(1-p_n)}a_k^{(-1)} \quad \text{and} \quad \sqrt{p_n\,(1-p_n)}a_k^{(+1)}$$

are mutually adjoint and are called **off-diagonal components** of $X_{n,k}$. The operator random variables

$$p_n a_k^{(0)} \quad \text{and} \quad (1-p_n)\,a_k^{(2)}$$

are Hermitean and are called **diagonal components** of $X_{n,k}$. The IFS $(\Gamma(\mathcal{H}), \langle\,\cdot\,,\,\cdot\,\rangle)$ will be chosen so that:
(H1) $X_{n,k} \overset{\varphi}{\sim} B\,(1,p_n)$ for any $n \in \mathbb{N}^*$ and $k \leq n$;
(H2) The family $\{\mathcal{A}_k\}_{k=1}^{\infty}$ is φ-$\mathcal{I}$-independent, where $\mathcal{I}$ is one of the following notions of stochastic independence:
– classical and tensor,
– Boolean,
– free,
– monotone,
– anti–monotone.
We will use the notations

$$a_k^{(\varepsilon)} := \begin{cases} a_k, & \text{if } \varepsilon = 1 \\ a_k^+, & \text{if } \varepsilon = 1 \\ a_k a_k^+, & \text{if } \varepsilon = 0 \\ a_k^+ a_k, & \text{if } \varepsilon = 2 \end{cases} \tag{14}$$

Condition (H1) above implies that (13) is **not the canonical quantum decomposition** of $X_{n,k}$ in the sense of x.

As a consequence of condition (H2) above, for any $n \in \mathbb{N}^*$ the operator random variables

$$\left\{a_k^{(\varepsilon_k)} \;:\; k \leq n\;,\; \varepsilon_k \in \{-1,0,1,2\}\right\}$$

are φ-$\mathcal{I}$-independent.
In particular the same is true for each $\{X_{n,k} \;:\; k \leq n\}$.

Since $\sum_{k=1}^{n} X_{n,k}$ is a sum of n φ-$\mathcal{I}$-independent classical Bernoulli random variables with parameter p_n, it is natural to call its distribution the **$\mathcal{I}$-multinomial distribution** of order n and with parameter p_n and to denote it $\mathcal{I}$-$M(n, p_n)$, where $\mathcal{I}$ stands for classical or Boolean or free or monotone or anti-monotone, independence.

In view of Theorem 1.2 it is then natural to call an $\mathcal{I}$-Poisson type moment CLT the study of the limit, for any $m \in \mathbb{N}$,

$$\lim_{n\to\infty} \varphi\left(\left(\sum_{k=1}^{n} X_{n,k}\right)^m\right) \tag{15}$$

Introducing the normalized sums

$$\begin{cases} B_n^{(\pm 1)} := \sqrt{p_n(1-p_n)}\sum_{k=1}^n a_k^{(\pm 1)} \\ B_n^{(0)} := p_n \sum_{k=1}^n a_k^{(0)} \\ B_n^{(2)} := (1-p_n)\sum_{k=1}^n a_k^{(2)} \end{cases} \tag{16}$$

The quantum decomposition (12) of the classical random variable $X_{n,k} \sim B(p_n)$ induces a quantum decomposition of the classical random variable

$$B_n := \sum_{k=1}^{n} X_{n,k} := B_n^{(-1)} + B_n^{(+1)} + B_n^{(0)} + B_n^{(2)}$$

$$\sim \mathcal{I} - M(n, p_n)$$

Consequently the Poisson type CLT (15) can be obtained as a consequence of the moment quantum CLT for the operator random variables (16), namely:

$$\lim_{n\to\infty} \varphi\left(B_n^{(\varepsilon(1))} \dots B_n^{(\varepsilon(m))}\right) \tag{17}$$

for $m \in \mathbb{N}$ and $\varepsilon \in \{-1, 0, 1, 2\}^m$.

The existence of the limit (17) and its explicit determination, together with the explicit description of the limit space, will be called a quantum Poisson type CLT.

3.1. *Preliminary results*

In order to prove the quantum Poisson type CLTs corresponding to the above listed notions of independence, we prove some results which hold in all these types of independence. In this section

- $(\Gamma(\mathcal{H}), \langle \cdot , \cdot \rangle , \Phi)$
- $\mathcal{H}$,

- $\{e_k\}_{k=1}^{\infty}$,
- a_k^+, a_k,
- $\mathcal{A}_k$,
- $\mathcal{A}$,

will be as described in section 3. In particular conditions (H1), (H2) are satisfied.

Lemma 3.1. *Given:*
– an algebraic probability space $(\mathcal{A}, \varphi)$,
– $p \in [0,1]$,
– $x_-, x_+ \in \mathcal{A}$,
define

$$x := \sqrt{p(1-p)}\,(x_- + x_+) + px_-x_+ + (1-p)\,x_+x_-$$

1) If $\forall X, Y \in \mathcal{A}$

$$\begin{cases} \varphi(Xx_-x_+) = \varphi(X) \\ \varphi(Xx_-) = 0 \\ \varphi\left(Xx_-^2Y\right) = 0 = \varphi\left(Xx_+^2Y\right) \end{cases} \tag{18}$$

then

$$\varphi(x^n) = p \qquad ; \qquad \forall n \in \mathbb{N}^* := \mathbb{N} \setminus \{0\} \tag{19}$$

consequently $x \overset{\mathcal{L}}{\sim} b(1,p)$.
2) If:

$$\begin{cases} x_+x_-x_+ = x_+ \\ x_-x_+x_- = x_- \\ x_-^2 = x_+^2 = 0 \end{cases} \tag{20}$$

then x is a projector (an Hermitean projector if $x_- = (x_+)^*$*).*

Proof. 1) From (18) it follows that

$$\varphi(x) = \varphi(px_-x_+) = p$$

and that the following identities hold for any $n \geq 2$:

$$\varphi\left(x^{n-2}(x_- + x_+)^2\right) = \varphi\left(x^{n-2}x_-x_+\right) = \varphi\left(x^{n-2}\right)$$

$$\varphi\left(x^{n-2}(x_- + x_+)\,x_-x_+\right) = \varphi\left(x^{n-2}x_+\right)$$

$$\varphi\left(x^{n-2}(x_- + x_+)\,x_+x_-\right) = 0$$

$$\varphi\left(x^{n-2}x_-x_+\left(x_-+x_+\right)\right)=0$$

$$\varphi\left(x^{n-2}\left(x_-x_+\right)^2\right)=\varphi\left(x^{n-2}x_-x_+\right)=\varphi\left(x^{n-2}\right)$$

$$\varphi\left(x^{n-2}x_-x_+x_+x_-\right)=0$$

$$\varphi\left(x^{n-2}x_+x_-\left(x_-+x_+\right)\right)=\varphi\left(x^{n-2}x_+\right)$$

$$\varphi\left(x^{n-2}x_+x_-x_-x_+\right)=0$$

$$\varphi\left(x^{n-2}\left(x_+x_-\right)^2\right)=0$$

Since

$$x^2=p\left(1-p\right)\left(x_-+x_+\right)^2+p\sqrt{p\left(1-p\right)}\left(x_-+x_+\right)x_-x_+ \tag{21}$$

$$+\left(1-p\right)\sqrt{p\left(1-p\right)}\left(x_-+x_+\right)x_+x_-+p\sqrt{p\left(1-p\right)}x_-x_+\left(x_-+x_+\right)$$

$$+\left(px_-x_+\right)^2+p\left(1-p\right)x_-x_+x_+x_-$$

$$+\left(1-p\right)\sqrt{p\left(1-p\right)}x_+x_-\left(x_-+x_+\right)+p\left(1-p\right)x_+x_-x_-x_+$$

$$+\left(\left(1-p\right)x_+x_-\right)^2$$

a consequence of the above identities is that, for any $n\geq 2$:

$$\varphi\left(x^n\right)=\varphi\left(x^{n-2}x^2\right) \tag{22}$$

$$=p\left(1-p\right)\varphi\left(x^{n-2}x_-x_+\right)+p\sqrt{p\left(1-p\right)}\varphi\left(x^{n-2}x_+\right)$$

$$+p^2\varphi\left(x^{n-2}x_-x_+\right)+\left(1-p\right)\sqrt{p\left(1-p\right)}\varphi\left(x^{n-2}x_+\right)$$

$$=\sqrt{p\left(1-p\right)}\varphi\left(x^{n-2}x_+\right)+p\varphi\left(x^{n-2}x_-x_+\right)$$

$$=\varphi\left(x^{n-2}\left(\sqrt{p\left(1-p\right)}\left(x_++x_-\right)\ +px_-x_++\left(1-p\right)x_+x_-\right)\right)$$
$$=\varphi\left(x^{n-1}\right)$$

and this proves (19).

2) If (20) holds, then

$$\left(x_- + x_+\right)^2 = x_-x_+ + x_+x_- \tag{23}$$

$$\left(x_- + x_+\right)x_-x_+ = x_+x_-\left(x_- + x_+\right) = x_+$$

$$\left(x_- + x_+\right)x_+x_- = x_-x_+\left(x_- + x_+\right) = x_-$$

$$\left(x_-x_+\right)^2 = x_-x_+$$

$$\left(x_+x_-\right)^2 = x_+x_-$$

$$x_-x_+x_+x_- = x_+x_-x_-x_+ = 0$$

and so (21) becomes

$$x^2 = p\left(1-p\right)\left(x_-x_+ + x_+x_-\right) + p\sqrt{p\left(1-p\right)}x_+$$

$$+\left(1-p\right)\sqrt{p\left(1-p\right)}x_-$$

$$+p\sqrt{p\left(1-p\right)}x_- + p^2x_-x_+ + \left(1-p\right)\sqrt{p\left(1-p\right)}x_+$$

$$+\left(1-p\right)^2 x_+x_- = x$$

□

4. Quantum realizations of the $\mathcal{I}$-Bernoulli processes in IFSs corresponding to different independence notions $\mathcal{I}$

In the following, given a notion $\mathcal{I}$ of stochastic independence, we will construct:
– a semi-Hilbert space $\mathcal{H}$,
– an Interacting Fock Space $\Gamma\left(\mathcal{H}\right)$ over $\mathcal{H}$,
such that, denoting:
– $\{e_k\}_k$ an onb of $\mathcal{H}$,
– $P_\Phi := \Phi\Phi^* : \Gamma\left(\mathcal{H}\right) \to \mathbb{C}\cdot\Phi$
the vacuum projector on $\Gamma\left(\mathcal{H}\right)$ and

$$\varphi := \text{ vacuum state } := \langle\Phi,\ \cdot\,\Phi\rangle$$

– $\mathbf{1}_\Gamma$ the identity operator on $\Gamma\left(\mathcal{H}\right)$,
– For any $x \in \mathcal{H}\setminus\{0\}$, the projector

$$P_x : \Gamma\left(\mathcal{H}\right) \to \mathbb{C}\cdot x := \{cx : c \in \mathbb{C}\}$$

$$\mathcal{A}_x^{(\geq 1)} :=:= \{\text{polynomials in } a(x) \text{ and } a^+(x) \text{ of degree} \geq 1\}$$

in particular for any k,

$$\mathcal{A}_k := \{\text{polynomials of degree} \geq 1 \text{ in } a_k := a_{e_k} \text{ and } a_k^+ := a_{e_k}^+\}$$

$$\mathcal{A}_{\mathbf{1}_\Gamma;x}^{(\geq 1)} := \text{ algebra generated by } \mathcal{A}_x^{(\geq 1)} \text{ and } \mathbf{1}_\Gamma$$

the above objects will be constructed so that conditions 1) and 2) of Lemma 3.1 hold whenever $x^{\pm}$ are replaced by $a^{\pm}k$ for any $k \in \mathbb{N}$. As a consequence of this, for any $k \in \mathbb{N}$, the random variable

$$X_{p,k} := \sqrt{p(1-p)}\left(a_k^- + a_k^+\right) + pa_k^- a_k^+ + (1-p)\, a_k^+ a_k^-$$

is φ-Bernoulli with parameter p. We will express this with the notation

$$X_{p,k} \overset{\mathcal{L}}{\sim} b(1,p)$$

(iii) the family of random variables $(a^{\pm}k)_{k\in\mathbb{N}}$ is $\mathcal{I}$-independent.

4.1. *Tensor independence*

The tensor independence case is simple and can be dealt in a purely algebraic way, that is we can use for the single Bernoulli random variable the standard quantum representation in terms of Pauli matrices.
The IFS is the obtained from the GNS construction.
The Pauli matrices are:

$$x_+ := \sigma^{(+1)} := \begin{pmatrix} 0 & 0 \\ 1 & 0 \end{pmatrix} \;;\; x_- := \sigma^{(-1)} := \begin{pmatrix} 0 & 1 \\ 0 & 0 \end{pmatrix} \tag{24}$$

and satisfy

$$\sigma^{(0)} \;: = \sigma^{(-1)}\sigma^{(+1)} = \begin{pmatrix} 1 & 0 \\ 0 & 0 \end{pmatrix} \;;\; \sigma^{(2)} := \sigma^{(+1)}\sigma^{(-1)} = \begin{pmatrix} 0 & 0 \\ 0 & 1 \end{pmatrix} \;;$$
$$\sigma^{(+1)} = \left(\sigma^{(-1)}\right)^*$$

They satisfy condition (2) of Lemma 3.1 and condition (1) is satisfied with respect to the state

$$\varphi_1(\,\cdot\,) := \left\langle \begin{pmatrix} 1 \\ 0 \end{pmatrix}, (\,\cdot\,) \begin{pmatrix} 1 \\ 0 \end{pmatrix} \right\rangle$$

Therefore for any $p \in [0,1]$,

$$x_p := \sqrt{p(1-p)}\left(\sigma^{(-1)} + \sigma^{(+1)}\right) + p\sigma^{(0)} + (1-p)\,\sigma^{(2)}$$

is an orthogonal projector satisfying $\varphi_1(x_p) = p$. Hence its φ_1-distribution is $b(1,p)$. Therefore, introducing algebraic probability space

$$\left(\bigotimes_{\mathbb{N}} \mathcal{M}_2 \ , \ \varphi := \bigotimes_{\mathbb{N}} \varphi_1\right)$$

the algebraic random variables

$$j_k\left(\sigma^{(\varepsilon)}\right) =: a_k^{(\varepsilon)} := \mathbf{1}_{(k-1)]} \otimes \sigma^{(\varepsilon)} \otimes \mathbf{1}_{[(k+1)} \quad ; \quad \forall \varepsilon \in \{-1,1\} \tag{25}$$

are φ-tensor independent when k varies hence, for any $p \in [0,1]$, the classical random variables

$$X_{p,k} := j_k(x_p) = \sqrt{p(1-p)}\left(a_k^{(-1)} + a_k^{(+1)}\right) + pa_k^{(0)} + (1-p)\,a_k^{(2)}$$

are φ-tensor independent orthogonal projectors with φ-distribution $b(1,p)$, i.e. $(a_k^{(\varepsilon)})$ is a φ-tensor Bernoulli process.

4.2. *Boolean case*

In the Boolean case the (semi-)Hilbert space $\mathcal{H}$ and its ortho-normal basis $\{e_k\}_{k=1}^{\infty}$ are arbitrary. The Boolean Fock space over $\mathcal{H}$ is the 1M-type-IFS $\Gamma(\mathcal{H}, \{\omega_n\}_n)$ with $\omega_1 = 1$ and $\omega_n = 0$ for any $n \geq 2$, i.e.

$$\Gamma_{Boolean}(\mathcal{H}) = \mathbb{C} \oplus \mathcal{H}$$

Lemma 4.1. *For $p \in [0,1]$ and $k \in \mathbb{N}$, define*

$$s_{k,p} := \sqrt{p(1-p)}\left(a_k^{(-1)} + a_k^{(+1)}\right) + pa_k^{(0)} + (1-p)\,a_k^{(2)}$$

Then:
– (20) holds for $x_\pm := a_k^{(\pm 1)}$ (the Boolean annihilation-creation operators). Consequently $s_{k,p}$ is a projector.
– $\varphi(s_{k,p}) = p$ in particular, $s_{k,p} \overset{\mathcal{L}}{\sim} b(1,p)$.

Proof. Denoting $P_\Phi := \Phi\Phi^*$ the vacuum projector, the definition of Boolean creator implies that

$$\begin{cases} a_k^+ a_k a_k^+ = a_k^+ P_\Phi = a_k^+ \\ a_k a_k^+ a_k = a_k P_\Phi = a_k \\ a_k^2 = (a_k^+)^2 = 0 \end{cases} \tag{26}$$

$\square$

Lemma 4.2. *If $\{e_k\}_k$ is an onb of $\mathcal{H}$, $\{\Phi, e_k\}_k$ is an onb of $\Gamma_{Boolean}(\mathcal{H})$. Moreover the following statements hold:*

1) for any k,

$$\mathcal{A}_k := \ alg.\ span\left\{a_k, a_k^+\right\}$$

$$= lin - sp.\left\{a_k, a_k^+, P_\Phi, P_{e_k}, a_k^+ P_\Phi, P_\Phi a_k, a_k P_{e_k}, P_{e_k} a_k^+\right\}$$

2) Each $\mathcal{A}_k$ is unital and in fact for any $x \in \mathcal{A}_k$,

$$\left(P_\Phi + P_{e_k}\right) x = x\left(P_\Phi + P_{e_k}\right) = x$$

i.e. $P_\Phi + P_{e_k}$ is the identity of $\mathcal{A}_k$ (but not necessarily of $\Gamma_{Boolean}(\mathcal{H})$);

3) the identity of $\Gamma_{Boolean}(\mathcal{H})$ belongs to $\mathcal{A}_k$ if and only if $\mathcal{H} = \{ce_k : c \in \mathbb{C}\}$ (so 1-dimensional);

4) the family $\{\mathcal{A}_k\}_{k=1}^{\infty}$ is Boolean independent with respect to the vacuum state $\varphi := \langle\Phi, \cdot\Phi\rangle$, hereinafter, $\Phi :=$ the vacuum vector.

4.3. *Free independence: The Fermi-free IFS*

One can prove that, on the standard full Fock space,
– for any annihilation operator a_f (resp. creation a_f^+) with unit-norm test function f (e.g. $a_{e_k} =: a_k$ and $a_{e_k}^+ := a_k^+$)
– and for any $p \in (0,1)$, **the vacuum distribution of**

$$Xp, f := \sqrt{p(1-p)}\left(a_f + a_f^+\right) + p a_f a_f^+ + (1-p)\, a_f^+ a_f \tag{27}$$

cannot be $b(1,p)$. For this reason we cannot realize our program using the standard full Fock space and the corresponding creation-annihilation operators, even if one can prove that the sequence of random variables $\{X_{n,k} : k \le n\}$ obtained replacing in (27) $a^\pm$ by $a_k^\pm$ is free independent for any $n \ge 2$. Our goal can be achieved using the **Fermi-free IFS**, introduced in [1]. Here we recall some of the basic properties of this IFS. Let $\mathcal{H} \equiv \mathcal{H}_1$ be an arbitrary (semi-)Hilbert space and

$$\Gamma_{FF}(\mathcal{H}) := \bigoplus_{n=0}^{\infty} \mathcal{H}_n$$

$$\mathcal{H}_0 \equiv \mathbb{C} \cdot \Phi \quad ; \quad \|\Phi\| = 1 \quad ; \quad \mathcal{H} \equiv \mathcal{H}_1$$

and, for $n \geq 2$, $\mathcal{H}_n$ is the (semi-)Hilbert space obtained by equipping the space

$$\text{lin–sp.}\left\{e_{\mathbf{k}(n)} \otimes \ldots \otimes e_{\mathbf{k}(1)} : \mathbf{k} \in \mathbb{F}_n^0\right\} \subset \mathcal{H}^{\otimes n}$$

$$\mathbb{F}_n^0 := \{\mathbf{k} \in \mathbb{F}_n : \mathbf{k}(i) \neq \mathbf{k}(j) \text{ for any } 1 \leq i \neq j \leq n\}$$

with scalar product obtained as restriction of the scalar product of the full Fock space. The (free) creation operator a_k^+ ($k \in \mathbb{N}$) is defined by linear extension of

$$a_k^+\left(e_{\mathbf{k}(n)} \otimes \ldots \otimes e_{\mathbf{k}(1)}\right) := \begin{cases} e_k \otimes e_{\mathbf{k}(n)} \otimes \ldots \otimes e_{\mathbf{k}(1)} \ , \ \text{if } k \notin range(\mathbf{k}) \\ 0 \ , \ \in range(\mathbf{k}) \\ \forall \mathbf{k} \in \mathbb{F}_n^0 \end{cases} \tag{28}$$

$$\begin{cases} e_k \otimes e_{\mathbf{k}(n)} \otimes \ldots \otimes e_{\mathbf{k}(1)}, & \text{if } k \notin range(\mathbf{k}) \\ 0, & \text{if } k \in range(\mathbf{k}) \end{cases}, \quad \forall \mathbf{k} \in \mathbb{F}_n^0$$

The following properties hold.

Γ_{FF}–**Property 1**

For any $k \in \mathbb{N}$, $n \in \mathbb{N}^*$, $\mathbf{k} \in \mathbb{F}_n^0$

$$a_k^+\left(e_{\mathbf{k}(n)} \otimes \ldots \otimes e_{\mathbf{k}(1)}\right) = e_k \otimes e_{\mathbf{k}(n)} \otimes \ldots \otimes e_{\mathbf{k}(1)} \prod_{j=1}^{n} \chi_{\{k\}^c}(\mathbf{k}(j))$$

$$= e_k \otimes e_{\mathbf{k}(n)} \otimes \ldots \otimes e_{\mathbf{k}(1)} \chi_{(range(\mathbf{k}))^c}(k))$$

In particular $\left\|a_k^+\right\| = 1$. The adjoint of a_k^+, the annihilation operator a_k, exists and is uniquely defined by the properties:

$$a_k \Phi = 0$$

$$a_k\left(e_{\mathbf{k}(n)} \otimes \ldots \otimes e_{\mathbf{k}(1)}\right) = \delta_{k,\mathbf{k}(n)} e_{\mathbf{k}(n-1)} \otimes \ldots \otimes e_{\mathbf{k}(1)}$$

$\forall n \in \mathbb{N}^*$, $\mathbf{k} \in \mathbb{F}_n^0$.

Γ_{FF}–**Property 2** One has the Fermi-like property

$$\left(a_k^+\right)^2 = a_k^2 = 0$$

and for any $m \in \mathbb{N}$, defining for any n, k, $\mathbf{k} \in \mathbb{F}_n^0$

$$\mathcal{H}_{n,k}^{(s)} := lin - sp.\left\{e_{\mathbf{k}(n)} \otimes \ldots \otimes e_{\mathbf{k}(1)} \ : \ k = \mathbf{k}(n)\right\} \tag{29}$$

$$\mathcal{H}_{n,k}^{(d)} := lin - sp.\left\{e_{\mathbf{k}(n)} \otimes \ldots \otimes e_{\mathbf{k}(1)} : k \notin range(\mathbf{k})\right\} \tag{30}$$

the following identities hold:

$$\left(a_k^+ a_k\right)^m = (\omega_1)^m \cdot \left(\text{the projector onto } \bigoplus_{n=1}^{\infty} \mathcal{H}_{n,k}^{(s)}\right) \tag{31}$$

$$\left(a_k a_k^+\right)^m = (\omega_1)^m \cdot \left(\text{the projector onto } \mathbb{C} \oplus \bigoplus_{n=1}^{\infty} \mathcal{H}_{n,k}^{(d)}\right) \tag{32}$$

<u>Γ_{FF}–**Property 3**</u> For any k,

$$\mathcal{A}_k = lin - sp.\{ \left(a_k a_k^+\right)^{m_1}, a_k^+ \left(a_k a_k^+\right)^{m_2}, \left(a_k^+ a_k\right)^{m_3}, a_k \left(a_k^+ a_k\right)^{m_4} : \tag{33}$$

$$m_1, m_2, m_3, m_4 \in \mathbb{N}\}$$

<u>Γ_{FF}–**Property 4**</u> For any k and $x_k \in \mathcal{A}_k$

$$x_k \Phi = \begin{cases} e_k, \text{ if } x_k = a_k^+ \left(a_k a_k^+\right)^m \\ \Phi, \text{ if } x_k = \left(a_k a_k^+\right)^m \\ 0, \text{ if } x_k \in \left\{\left(a_k^+ a_k\right)^m, a_k \left(a_k^+ a_k\right)^m\right\} \end{cases} \tag{34}$$

and for any $n \in \mathbb{N}^*$ and $\mathbf{k} \in \mathbb{F}_n^0$

$$x_k \left(e_{\mathbf{k}(n)} \otimes \ldots \otimes e_{\mathbf{k}(1)}\right) \tag{35}$$

$$= \begin{cases} \delta_{k,\mathbf{k}(n)} e_{\mathbf{k}(n-1)} \otimes \ldots \otimes e_{\mathbf{k}(1)}, & \text{if } x_k = a_k \left(a_k^+ a_k\right)^m \\ \delta_{k,\mathbf{k}(n)} e_{\mathbf{k}(n)} \otimes \ldots \otimes e_{\mathbf{k}(1)}, & \text{if } x_k = \left(a_k^+ a_k\right)^m \end{cases}$$

<u>Γ_{FF}–**Property 5**</u>

For any:

– k, $n \in \mathbb{N}^*$, $\mathbf{k} \in \mathbb{F}_n$

(now, $\mathbf{k}$ belongs to $\mathbb{F}_n$ but not necessarily to $\mathbb{F}_n^0$)

– $\{m_1, \ldots, m_n\} \subset \mathbb{N}$,

– $x_{\mathbf{k}(j)} \in \left\{a_{\mathbf{k}(j)}^+ \left(a_{\mathbf{k}(j)} a_{\mathbf{k}(j)}^+\right)^{m_j}, \left(a_{\mathbf{k}(j)}^+ a_{\mathbf{k}(j)}\right)^{m_j}, a_{\mathbf{k}(j)} \left(a_{\mathbf{k}(j)}^+ a_{\mathbf{k}(j)}\right)^{m_j}\right\}$

(or equivalently, $x_{\mathbf{k}(j)} \in \mathcal{A}_{\mathbf{k}(j)}$, $\varphi(x_{\mathbf{k}(j)}) = 0$ for $j \in \{1, \ldots, n\}$,

$$x_{\mathbf{k}(n)} \cdots x_{\mathbf{k}(1)} \Phi$$

$$= \begin{cases} e_{\mathbf{k}(n)} \otimes \ldots \otimes e_{\mathbf{k}(1)}, & \text{if } \mathbf{k} \in \mathbb{F}_n^0 \\ \text{and } x_{\mathbf{k}(j)} = a_{\mathbf{k}(j)}^+ \left(a_{\mathbf{k}(j)} a_{\mathbf{k}(j)}^+\right)^{m_j} & \\ \text{for any } j \in \{1, \ldots, n\} & \\ 0, & \text{otherwise} \end{cases} \tag{36}$$

Γ_{FF}–Property 6
With respect to the vacuum state the algebras $\mathcal{A}^{(\geq 1)}_{\mathbf{1}_\Gamma;e_k}$ are free independent and consequently the algebras $\{\mathcal{A}_k\}_k$ are free independent.
Recall that $\mathcal{A}^{(\geq 1)}_{\mathbf{1}_\Gamma;e_k}$ is the algebra generated by $\mathcal{A}^{(\geq 1)}_{e_k}$ and $\mathbf{1}_\Gamma$.

4.4. *Monotone and anti-monotone independences*

In the monotone case, $\mathcal{H}_0$, $\mathcal{H}$ and $\mathcal{H}_1$ are as in the free case and one constructs the IFS space as

$$\Gamma_{mon}(\mathcal{H}) := \bigoplus_{n=0}^{\infty} \mathcal{H}_n$$

where for $n \geq 2$, $\mathcal{H}_n$ is the (semi-)Hilbert space obtained by equipping the space

$$lin - sp. \left\{ e_{\mathbf{k}(n)} \otimes \ldots \otimes e_{\mathbf{k}(1)} : \mathbf{k} \in \mathbb{F}_n^{\uparrow} \right\} \subset \mathcal{H}^{\otimes n} \tag{37}$$

$$\mathbb{F}_n^{\uparrow} := \{\mathbf{k} \in \mathbb{F}_n : \mathbf{k}(n) > \ldots > \mathbf{k}(1)\}$$

with the scalar product obtained by restriction to $\mathcal{H}_n$ of the usual tensor product on $\mathcal{H}^{\otimes n}$.
Note that $\mathbb{F}_n^{\uparrow} \subset \mathbb{F}_n^0$.
The creation operator a_k^+ is defined $\forall \mathbf{k} \in \mathbb{F}_n^{\uparrow}$ as

$$a_k^+ e_{\mathbf{k}(n)} \otimes \ldots \otimes e_{\mathbf{k}(1)} := \chi_{(\mathbf{k}(n)}(k)\, e_k \otimes e_{\mathbf{k}(n)} \otimes \ldots \otimes e_{\mathbf{k}(1)} \tag{38}$$

where for any $h, k \in \mathbb{N}$

$$\chi_{(h}(k) := \begin{cases} 1, \text{ if } k > h \\ 0, \text{ if } k \leq h \end{cases}$$

The following properties hold.

Γ_{mon}–Property 1
For any k, $\|a_k^+\| = 1$ hence its adjoint, the annihilation operator a_k, is well defined and bounded. Moreover

$$\left(a_k^+\right)^2 = a_k^2 = 0$$

and $\forall n, k \in \mathbb{N}^*$, $\mathbf{k} \in \mathbb{F}_n^0$:

$$a_k \Phi = 0$$

$$a_k \left(e_{\mathbf{k}(n)} \otimes \ldots \otimes e_{\mathbf{k}(1)}\right) = \delta_{k,\mathbf{k}(n)} e_{\mathbf{k}(n-1)} \otimes \ldots \otimes e_{\mathbf{k}(1)}$$

Γ_{mon}–**Property 2**
Denoting, for any $n, k \in \mathbb{N}^*$,

$$\mathcal{H}_{n,k} :=$$

$$:= lin - sp. \left\{ e_{\mathbf{k}(n)} \otimes \ldots \otimes e_{\mathbf{k}(1)} : \mathbf{k} \in \mathbb{F}_n^{\uparrow} \text{ and } k = \mathbf{k}(n) \right\}$$

then

$$a_h|_{\mathcal{H}_{n,k}} = 0 \quad ; \quad \forall n \in \mathbb{N}^* \text{ and } h \neq k$$

$$a_h^+|_{\mathcal{H}_{n,k}} = 0 \quad ; \quad \forall n \in \mathbb{N}^* \text{ and } h < k$$

Consequently for any $x_h \in \mathcal{A}_h$

$$x_h|_{\mathcal{H}_{n,k}} = 0 \quad ; \quad \forall n \in \mathbb{N}^*$$

Γ_{mon}–**Property 3** For any $m, k \in \mathbb{N}_0$,

$$\left(a_k a_k^+\right)^m = \text{the projector onto } \mathbb{C} \oplus \bigoplus_{h<k} \bigoplus_{n=1}^{\infty} \mathcal{H}_{n,h} \tag{39}$$

$$\left(a_k^+ a_k\right)^m = \text{the projector onto } \bigoplus_{n=1}^{\infty} \mathcal{H}_{n,k} \tag{40}$$

and an element of $\mathcal{A}_k$ has either of the following forms:

$$\left(a_k a_k^+\right)^m, a_k^+ \left(a_k a_k^+\right)^m \quad , \quad \left(a_k^+ a_k\right)^m, a_k \left(a_k^+ a_k\right)^m$$

Γ_{mon}–**Property 4**
The sequence of algebras $\{\mathcal{A}_k\}_{k=1}^{\infty}$ is φ-monotone-independent.
Remark One has similar results for the anti-monotone case just by reversing all inequalities and exchanging $<$ and $>$.

Theorem 4.1. *If $p = p_N$ verifies*

$$\lim_{N \to \infty} N p_N = \lambda,$$

then for any $n \in \mathbb{N}$ and $\varepsilon \in \{-1, 0, 1, 2\}^n$, the limit, as $N \to \infty$ of the moments

$$\left\langle \Phi_N, B_N^{(\varepsilon(1))} \ldots B_N^{(\varepsilon(n))} \Phi_N \right\rangle \tag{41}$$

exists and is equal to

$$\left\langle \Phi, b^{(\varepsilon(1))} \ldots b^{(\varepsilon(n))} \Phi \right\rangle \lambda^{\sum_{k=1}^{n}(1-|\varepsilon(k)|/2)} \tag{42}$$

Where, for any $\varepsilon \in \{-1, 0, 1, 2\}$

$$b^{(\varepsilon)} := \begin{cases} b, & if\ \varepsilon = -1 \\ b^{+}, & if\ \varepsilon = 1 \\ \mathbf{1}, & if\ \varepsilon = 0 \\ \Lambda, & if\ \varepsilon = 2 \end{cases} \tag{43}$$

b (b^{+} and Λ respectively) is annihilation operator (creation operator and the number operator respectively) of the 1M–IFS $\Gamma\left(\mathbb{C}, \{n\}_n\right)$.

Before to prove Theorem 4.1, we give its useful consequence:

<u>**Corollary**</u> If $p = p_N$ verifies $\lim_{N\to\infty} Np_N = \lambda$, then for any $n \in \mathbb{N}$

$$\begin{aligned} &\lim_{N\to\infty} \left\langle \Phi_N, \left(\sum_{\varepsilon\in\{-1,0,1,2\}} c_\varepsilon B_N^{(\varepsilon)} \right)^n \Phi_N \right\rangle \\ &= \left\langle \Phi, \left(\sqrt{\lambda}\left(c_{-1}b + c_1 b^{+}\right) + c_0\lambda + c_2\Lambda \right)^n \Phi \right\rangle \end{aligned} \tag{44}$$

In particular, by taking $c_{-1} = c_1 = 1$ and $c_0 = c_2 = 0$

$$\lim_{N\to\infty} \left\langle \Phi_N, \left(B_N^{(-1)} + B_N^{(+1)} \right)^n \Phi_N \right\rangle = \lambda^{n/2} \left\langle \Phi, \left(a + a^{+}\right)^n \Phi \right\rangle \tag{45}$$

Reference

1. Y.G. Lu: Interacting Fock spaces and subproduct systems, Infinite Dimensional Analysis and Quantum Probability, and Related Topics, Vol. 23, No. 03, 2050017 (2020).

Infinite Dimensional Analysis, Quantum Probability and Related Topics

pp. 23–32

COVARIANT QUANTUM FIELDS VIA LORENTZ GROUP REPRESENTATION OF WEYL OPERATORS

RADHAKRISHNAN BALU

Army Research Laboratory Adelphi, MD, 21005-5069, USA
E-mail: radhakrishnan.balu.civ@mail.mil

Department of Mathematics & Norbert Wiener Center for Harmonic Analysis and Applications, University of Maryland, College Park, MD 20742, USA
E-mail: rbalu@umd.edu

The building blocks of Hudson-Parthasarathy quantum stochastic calculus start with Weyl operators on a symmetric Fock space. To realize a relativistically covariant version of the calculus we construct representations of Poincaré group in terms of Weyl operators on suitably constructed, Bosonic or Fermionic based on the mass and spin of the fundamental particle, Fock spaces. We proceed by describing the orbits of homogeneous Lorentz group on $\mathcal{R}^4$ and build fiber bundle representations of Poincaré group induced from the stabilizer subgroups (little groups) and build the Boson Fock space of the Hilbert space formed from the sections of the bundle. Our Weyl operators are constructed on symmetric Fock space of this space and the corresponding annihilation, creation, and conservation operators are synthesized in the usual fashion in relativistic theories for space-like, time-like, and light-like fields. We achieve this by constructing transitive systems of imprimitivity (second-quantized SI), which are dynamical systems with trajectories dense in the configuration space, by induced representations. We provide the details of the field operators for the case of massive Bosons as the rest are similar in construction and indicate the ways to construct adapted processes paving way for building covariant quantum stochastic calculus.

1. Introduction

Quantum stochastic calculus (QSC) [2] is a mathematically precise description of microscopic systems interacting with an environment that has been successfully applied to quantum optics based processes rigorously established by Accardi [1], quantum filters framework of Belavkin [3], and quantum information processing framework of Mabuchi [7] to name a few. A relativistically covariant version of the program was developed by Applebaum [4] for the Fermionic case and Frigerio for the Bosonic noise [5]. Relativistic effects are an important consideration in Fermionic systems of topological quantum systems [6]. Here, we present a systematic formulation of covariant QSC for all fundamental particles by extending the theory for relativistic quantum particles, that is based on systems of imprimitivity, to field theoretic context. Systems of imprimitivity is a covariant relationship between a pair of conjugate observables (can be thought of as "position" described by a projection measure and "momentum" operators by a unitary representation of the group) that can be used to describe a class of systems within unitary equivalence in a concise manner when there is a group representing the kinematics or dynamics of the system. Notions related relativistic quantum systems such as SI are based on orthodox quantum logic framework developed by Birkhoff-von Neumann and Mackey among others. Accordingly, our setup is separable Hilbert spaces that form standard quantum logics that are complemented orthomodular lattices [9]. The simplest example is to describe the Weyl commutation relation of a class of systems undergoing *Shrödinger* evolution. The SI in this case is defined with respect to the additive group of the real line to account for the time translation symmetry of the dynamics where the position operator is a canonical observable. Hahn-Hellinger theorem [2] enables decomposition of an arbitrary observable in terms canonical ones and so the SI approach is applicable to arbitrary quantum system and can be thought of as a quantization scheme. Some early examples for finite groups appeared in the works of von Neumann [13] that are ergodic systems. Wigner used these notions informally in his study on the theory of representations of the inhomogeneous Lorentz (Poincaré) group [14] that are transitive systems and Mackey [8] provided the systematic development of SI for systems with infinite degrees of freedom.

Systems of imprimitivity are a very useful characterization of dynamical systems, when the configuration space of a quantum system is described by a group, from which infinitesimal forms in terms of differential equations (*Shrödinger*, Heisenberg, and Dirac etc.), and the canonical commutation relations can be derived. For example, using SI arguments it can be shown [10] that massless elementary particles with spin less than or equal to 1 can't have well defined position operators, that is photons are not localizable. The concept of localization, where the position operator is properly defined in a manifold, and covariance in relativistic sense of systems can be shown to be a consequence of systems of imprimitivity. Now, we have a procedure called Mackey's machinery to synthesize SI, a class of unitarily equivalent systems, using representations induced by subgroups which in our primary example will be the little groups of inhomogeneous

Lorentz group. Our previous work in SI involved application to infinite unitary representations of the Poincaré group, that are induced by stabilizer subgroups or little groups in physics literature, in the context of quantum walks [11] and [12]. We follow the notions and notations from the work of Varadarajan [9] where the Dirac equation is derived in the most rigorous fashion as a consequence of SI. Varadarajan set up this program first by establishing standard quantum logics and geometry on orthomodular lattices and further developed the configuration space, of a Dirac particle, endowed with a topology, measure space, and a G-space of a locally compact second countable (lcsc) group G (for example the Poincaré group) providing a comprehensive view of the subject that exploits symmetry and is formulated in geometric terms. We extend this program by building the respective Fock spaces and then identifying Weyl operators indexed by elements of Poincaré group leading to covariant field operators (creation, annihilation, and preservation).

Let us first define the notion of SI and an important theorem by Mackey that characterizes such systems in terms of induced representations.

Definition 1.1. [9] A G-space of a Borel group G is a Borel space X with a Borel automorphism $\forall g \in G, t_g : x \to g.x, x \in X$ such that

$$t_e \text{ is an identity} \tag{1}$$

$$t_{g_1,g_2} = t_{g_1} t_{g_2}. \tag{2}$$

The group G acts on X transitively if $\forall x, y \in X, \exists g \in G \ni x = g.y$.

Definition 1.2. [9] A system of imprimitivity for a group G acting on a Hilbert space $\mathcal{H}$ is a pair (U, P) where $P : E \to P_E$ is a projection valued measure defined on the Borel space X with projections defined on the Hilbert space and U is a representation of G satisfying

$$U_g P_E U_g^{-1} = P_{g.E}. \tag{3}$$

The next step is to consider semidirect product of groups, that naturally describe the dynamics system of relativistic quantum particles, and use the representation of the subgroup to induce a representation in such a way that it is an SI.

Definition 1.3. Let A and H be two groups and for each $h \in H$ let $t_h : a \to h[a]$ be an automorphism (defined below) of the group A. Further, we assume that $h \to t_h$ is a homomorphism of H into the group of automorphisms of A so that

$$h[a] = hah^{-1}, \forall a \in A. \tag{4}$$

$$h = e_H, \text{ the identity element of } H. \tag{5}$$

$$t_{h_1 h_2} = t_{h_1} t_{h_2}. \tag{6}$$

Now, $G = H \rtimes A$ is a group with the multiplication rule of $(h_1, a_1)(h_2, a_2) = (h_1 h_2, a_1 t_{h_1}[a_2])$. The identity element is (e_H, e_A) and the inverse is given by $(h, a)^{-1} = (h^{-1}, h^{-1}[a^{-1}])$. When H is the homogeneous Lorentz group and A is R^4 we get the Poincaré group via this construction.

Definition 1.4. Let group G be lcsc and X be a standard Borel G-space, and M be a standard Borel group. A map f is a (G, X, M)-cocycle relative to α which is an invariant measure class if (1) f is a Borel map of $G \times X$ into M. (2) $f(e, x) = 1$ for all $x \in X$. (3) $f(g_1 g_2, x) = f(g_1, g_2 x) f(g_2, x)$ for almost all $(g_1, g_2, x) \in G \times G \times X$. If instead the map f satisfies conditions (2) and (3) for all x it is called a strict cocycle.

As an example let us construct a pair of observables obeying the SI relation.

Example 1.1. [9] Let G be lcsc and X is the G-space which is Borel and endowed with a σ-finite measure α. Then, we can define another σ-finite measure $\alpha^{g^{-1}}$ which is absolutely continuous with respect to α giving rise to a Radon-Nikodym derivative $r_g = \frac{d\alpha}{d\alpha^{g^{-1}}}$. Unitary operators that satisfy the SI relation as follows:

$$\mathcal{H} = \mathcal{L}^2(X, k, \alpha), \text{ where, k is a separable Hilbert space.}$$

$$U_g f(x) = \{r_g(g^{-1}x)\}^{\frac{1}{2}} f(g^{-1}x), f \in \mathcal{H} \text{ Momentum like operator.}$$

$$P_E F = \chi_E f \text{ Position like operator.}$$

Next, let us recollect few results [9] on systems of imprimivity that will help us establish our main result in this work.

Theorem 1.1. *(Theorem 5.19) There are quasi-invariant Borel measures on $X = G/G_0$, where G is a lcsc group and G_0 is a closed subgroup with the map $\beta : G \to X = g.x, x \in X$. If λ is a finite quasi-invariant measure on G, and $\tilde{\lambda}$ is the finite measure on X defined by the equation $\tilde{\lambda}(A) = \lambda(\beta^{-1}(A))$, then $\tilde{\lambda}$ is quasi-invariant. Any two quasi-invariant σ-finite measures on X are mutually absolutely continuous. If $E \subset X$ is any Borel set, the quasi-invariant quasi-invariant σ-finite measures on X vanish for E if and only* *if $\beta^{-1}(E)$ is a null set of G. If α is any Borel measure on X, α is quasi-invariant (invariant) if and only if α^0 is a quasi-invariant (left-invariant) measure on G.*

Theorem 1.2. *(Theorem 6.19) Let G be a connected, simply connected complex semisimple Lie group, and G_0 a maximal compact subgroup of G. Let α be an invariant Borel measure on $X = G/G_0$. Let m be an irreducible representation of G_0 in a Hilbert space $\mathcal{K}$, and let $m'(g \to m'(g))$ be the unique complex-analytic homomorphism of G into the group of invertible linear endomorphisms of $\mathcal{K}$ such that $m'(h) = m(h)$ for all $h \in G_0$. Then there exists exactly one map $x \to \langle .,. \rangle_x$ of X into the set of positive definite linear products on $\mathcal{K} \times \mathcal{K}$ such that $\langle .,. \rangle_{x_0} = \langle .,. \rangle$ and $\langle u, v \rangle_x = \langle m'(g)u, m'(g)v \rangle_{g.x}$ for all $(g, x) \in G \times X$. If $\mathcal{H}$ is the Hilbert space of Borel maps f of X into $\mathcal{K}$ such that $\|f\|^2 = \int_X \langle f(x), f(x) \rangle_x d\alpha(x) < \infty$, and we define P and U by*

$$U_g f(x) = m'(g) f(g^{-1}x), f \in \mathcal{H} \textit{ Momentum operator,}$$
$$P_E F = \chi_E f \textit{ Position operator,}$$

then (U, P) is a system of imprimitivity equivalent to the system induced by m.

2. Little groups (stabilizer subgroups)

Some of the different systems of imprimitivity that live on the orbits of the stabilizer subgroups are described below. It is good to keep in mind the picture that SI is an irreducible unitary representation of Poincaré group $\mathcal{P}^+$ induced from the representation of a subgroup such as SO_3, which is a subgroup of homogeneous Lorentz, as $(U_m(g)\psi)(k) = e^{i\{k,g\}}\psi(R_m^{-1}k)$ where g belongs to the $\mathcal{R}^4$ portion of the Poincaré group, m is a member of the rotation group, $\{k, g\} = k_0g_0 - k_1g_1 - k_2g_2 - k_3g_3$, and the expression is in momentum space.

The stabilizer subgroups of the Poincaré group $\mathcal{P}^+$ can be described as follows: [15]. Time-like particle: There is a reference frame in which the 4-component momentum is proportional to (0,0,0,1) and the stabilizer subgroup is SO_3 that describes the spin. The walker, a massive particle, is rest in this frame. Let us denote the eigen vector of the Casimir operator $P_\mu P^\mu$ as $|0\lambda\rangle$. Then, we can describe the invariant space under the Poincaré group $\mathcal{P}^+$ as $\Lambda_p |0\lambda\rangle = |p\lambda\rangle$ by applying the Lorentz boost. Any Lorentz operator operating on this space can be shown to be by a rotation (spin in our case).

Space-like particle: The Lorentz frame in which the walker is at rest has momentum proportional to (1,0,0,0) and the little group is again SO_3 and this time the rotations will change the helicity. In this imaginary mass case the little group is rotations around the space axis and the analysis above carries through.

Light-like particle: There is no frame in which the relativistic quantum particle is at rest but the frame where the momentum is proportional to $(1,1,0,0)$ has the stabilizer subgroup with elements of the form J_1, that is a rotation around the first component of momentum and the boost Λ_p in the spatial direction [15]. These two operators commute and the induced representation can be constructed as above.

3. Fiber bundle representation of relativistic quantum particles

The states of a freely evolving relativistic quantum particles are described by unitary irreducible representations of Poincaré group that has a geometric interpretation in terms of fiber bundles.

The 3+1 spacetime discrete Lorentz group $\hat{O}(3,1)$-orbits of the momentum space $\mathcal{R}^4$, where the systems of imprimitivity established will live, described by the symmetry $\hat{O}(3,1) \rtimes \mathbb{R}^4$. The orbits have an invariant measure α_m^+ whose existence is guaranteed as the groups and the stabilizer groups concerned are unimodular and in fact it is the Lorentz invariant measure $\frac{dp}{p_0}$ for the case of forward mass hyperboloid. The orbits are defined as:

$$\begin{aligned} X_m^{+,2} &= \{p : p_0^2 - p_1^2 - p_2^2 - p_3^2 = m^2, p_0 > 0\}, \\ &\quad \text{forward mass hyperboloid.} \qquad (7) \\ X_m^{+,2} &= \{p : p_0^2 - p_1^2 - p_2^2 - p_3^2 = m^2, p_0 < 0\}, \\ &\quad \text{backward mass hyperboloid.} \qquad (8) \\ X_{00} &= \{0\}, \text{origin.} \qquad (9) \end{aligned}$$

Each of these orbits are invariant with respect to $\hat{O}(3,1)$ and let us consider the stabilizer subgroup of the first orbit at p = (0,0,0,1). Now, assuming that the spin of the particle is 1 (massive Boson) let us define the corresponding fiber bundles (vector) for the positive mass hyperboloid that corresponds to the positive-energy states by building the total space as a product of the orbits and $\mathbb{C}^4$.

$$\hat{B}_m^{+,2} = \{(p,v)\ p \in \hat{X}_m^+,\ v \in \mathcal{C}^2, \sum_k p_k \gamma_k v = mv\}. \tag{10}$$

$$\begin{aligned} \hat{\pi}\ :\ (p,v) \to p.\ &\text{Projection from the total space} \\ \hat{B}_m^{+,2}\ &\text{to the base}\ \hat{X}_m^{+,2}. \end{aligned} \tag{11}$$

The states of the particles are defined on the Hilbert space $\hat{\mathcal{H}}_m^{+,2}$, square integrable functions on Borel sections of the bundle $\hat{B}_m^{+,2}$ with respect to the invariant measure α_m^+, whose norm induced by the inner product is given below:

$$\|\phi\|^2 = \int_{X_m^+} p_0^{-1} \langle \phi p, \phi p \rangle . d\alpha^+(p). \tag{12}$$

To be more technical, we need Schwartz spaces, and their duals tempered distributions to guarantee Fourier transforms, to move with ease between configuration and momentum space descriptions. However, this is needed when setting up differential equations and so we will deal with them when we work with Fermionic examples in the future.

Let us now state and establish the main result for the case of massive Boson with time-like momentum by constructing a strict cocycle from the representation of a subgroup following the prescription (Lemma 5.24) in Varadarajan' text. The SI is a consequence of strict cocycle property and as he noted the construction is not canonical.

Lemma 5.24 [9]: Let m be a Borel homomorphism of G_0 into M. Then there exists a Borel map b of G into M such that

$$B(e) = 1. \tag{13}$$

$$b(gh) = b(g)m(h), \forall (g,h) \in G \times G_0. \tag{14}$$

Corresponding to any such map b, there is a unique strictly (G, X)-cocycle f such that

$$f(g, g^1) = b(gg^1)b(g^1)^{-1}. \tag{15}$$

$\forall (g, g^{-1}) \in G \times G$. f defines m at x_0. Conversely, f is a strict (G, X)-cocycle and b is a Borel map such that it satisfies (13) equation pair, then the restriction of b to G_0 coincides with the homomorphism m defined of at x_0 and b satisfies (15).

Theorem 3.1. *Time-like Weyl representation of Poincaré group $\mathcal{P}$ is a transitive system of imprimitivity.*

Proof. Let $g : \mathcal{G} \to U_g(\hat{\mathcal{H}}_m^{+,2})$ be a homomorphism from the rotation group $\mathcal{G} = SO_3$, which is a subgroup of $\mathcal{P}$, to the unitary representation of the group in $\hat{\mathcal{H}}_m^{+,2}$. We note that it is a stabilizer subgroup which is also closed at the momentum $(m,0,0,0)$ and so $\mathcal{P}/\mathcal{G}$ is a transitive space.

Consider a map $v(g) : \mathcal{G} \to \hat{\mathcal{H}}_m^{+,2}$ satisfying the first order cocycle relation $v(gh) = v(g) + U_g v(h), g, h \in \mathcal{G}$. For an example of such a map consider the following: [2]

$$\begin{aligned}
\mathcal{H} &= \oplus_{j=0}^{\infty}\mathcal{H}_j.\\
H &= 1 \oplus \oplus_{j=1}^{\infty} H_j.\\
U_t &= e^{-itH}, t \in \mathcal{G}.\\
v(t) &= tu_0 \oplus \oplus_{j=1}^{\infty}(e^{-itH_j}u_j - u_j).
\end{aligned}$$

Now, we can define the Weyl operator $V_g = W_g(v(g), U_g)$ where $g \in \mathcal{G}$ in the Fock space $\Gamma_s(\hat{\mathcal{H}}_m^{+,2})$.

This is a projective unitary representation satisfying the commutator relation $V_g V_h = e^{iIm\langle v_g, U_g v_h\rangle} V_h V_g$ and let us denote the homomorphism from $\mathcal{G}$ to V_g by m. This guarantees a map b that satisfies $b(gh) = b(g)m(h)$, $g \in \mathcal{P}$, $h \in \mathcal{G}$ and such map can be constructed by considering the map $c(x \to c(x)$ as Borel section of $\mathcal{P}/\mathcal{G}$ (not a canonical choice) with $c(x_0) = e$. The map β maps $g \in \mathcal{P} \to g\mathcal{G}$

$$a(g) = c(\beta(g))^{-1}. \tag{16}$$

$$b(g) = m(a(g)). \tag{17}$$

Then the strict cocycle ϕ satisfies $\phi(g_1, g_2) = b(g_1 g_2) b(g_2)^{-1}$.

We can now set up the SI relation as follows: Let us denote the unique invariant measure class of X, noting that it is lcsc, by G and a member of this class as α and the Radon-Nikodym derivative $r_g = \frac{d\alpha}{d\alpha^{g^{-1}}}$.

$$U_g f(x) = \{r_g(g^{-1}x)\}^{\frac{1}{2}}\phi(g, g^{-1}x)f(g^{-1}x),$$

$$f \in \mathcal{H} \text{ Momentum operator.} \tag{18}$$

$$P_E F = \chi_E f \text{ Position operator.} \tag{19}$$

□

We can construct the conjugate pair of field operators for the Fock space $\Gamma_s(\hat{\mathcal{H}}_m^{+,2}))$ as follows:

Let p_g be the stone generator for the family of operators $P_{gt,p}, g \in \mathcal{P}, t \in \mathbb{R}$ and $q(g) = p(ig)$ and we get the creation and annihilation operators as $a(g)^\dagger = \frac{1}{2}(q(g) - ip(g)$ and $a(g) = \frac{1}{2}(q(g) + ip(g)$.

4. Summary and conclusions

We derived the covariant field operators using induced representations of groups and expressed them in terms of systems of imprimitivity. We established the results for the massive Boson case by inducing a representation of Poincaré group from the subgroup that is a stabilizer at the momentum (1,0,0,0). In the next step we will develop the Fermionic QSC and detail the basic adopted processes in this case. This opens up the ways for treating topological quantum materials with relativistic electrons that are of importance to a specific form of quantum computation in a rigorous fashion.

References

1. L. Accardi, A. Frigerio, and Y. Lu, The weak coupling limit as a quantum functional central limit, Commun. Math. Phys., 131 (1990), pp. 537–570.
2. K. R. Parthasarathy, An Introduction to Quantum Stochastic Calculus, Birkhauser, Basel (1992).
3. J. Gough, V. P. Belavkin, and O. G. Smolyanov, Hamilton-Jacobi-Bellman equations for quantum filtering and control, J. Opt. B: Quantum Semiclass. Opt., 7 (2005), pp. 237–244.
4. D. Applebaum, Fermion stochastic calculus in Dirac-Fockspace, J. Phys. A, 28 (1995), 257–270.
5. A. Frigerio and M. Ruzzier, Relativistic transformation properties of quantum stochastic calculus, Ann. Inst. Henri Poincare 51, 67 (1989).
6. E. Witten, Three Lectures on Topological Phases of Matter, Nuovo Cim. Riv. Ser. 39, 313 (2016).
7. N. Tezak, A. Niederberger, D. S. Pavlichin, G. Sarma, and H. Mabuchi, Specification of photonic circuits using quantum hardware description language. Philosophical transactions A, 370, 5270 (2012).
8. Infinite dimensional group representations, Bulletin of the American Mathematical Society, 69, 628 (1963).
9. V. S. Varadarajan, Geometry of quantum theory, Springer (1985)
10. T. D. Newton and E. P. Wigner, Localized states for elementary systems, Rev. Mod. Phys. 21 400 (1949).
11. N. Pradeep Kumar, Radhakrishna Balu, Raymond Laflamme, C. M. Chandrashekar, Bounds on the dynamics and entanglement in a periodic quantum walks, Phys. Rev. A (2018).
12. Radhakrishnan Balu, Kinematics and Dynamics of Quantum Walk in terms of Systems of Imprimitivity, J. Phys. A: Mathematical and Theoretical (2019).
13. Von Neumann, On rings of operators. Reduction theory, Annals of Mathematics, 50, 401 (1949).
14. Wigner, Unitary representations of the inhomogeneous Lorentz group, Annals of Mathematics, 40, 149 (1939).

15. Y. S. Kim and E. S. Noz, Phase Space Picture of Quantum Mechanics: Group Theoretical Approach, Lecture Notes in Physics Series, World Scientific Pub Co Inc (1991).
16. Chaobin Liu and Radhakrishnan Balu, Steady States of Continuous-time open quantum walks, Quant. Inf. 16, 173 (2017).
17. Wigner, Unitary representations of the inhomogeneous Lorentz group: Ann. Math., 40, 845 (1939). Mico Durdevic, Quantum principal bundles and corresponding gauge theories.
18. J. Wolf, Spaces of constant curvature, Univ. California, Berkeley (1972).
19. Akihito Hora, Nobuaki Obata, Quantum Probability and Spectral Analysis of Graphs, springer (2007).
20. E. Schrödinger, Naturwiss. 14 (1926) 664.

Infinite Dimensional Analysis, Quantum Probability and Related Topics

pp. 33–46

LÉVY PROCESSES ON THE LORENTZ-LIE ALGEBRA

AMEUR DHAHRI

Dipartimento di Matematica,
Politecnico di Milano,
Piazza Leonardo da Vinci 32,
I-20133 Milano, Italy
Email: ameur.dhahri@polimi.it

UWE FRANZ

Laboratoire de mathématiques de Besançon,
Université de Bourgogne Franche-Comté,
16, route de Gray, 25 030 Besangn cedex, France
Email: uwe.franz@univ-fcomte.fr

Lévy processes in the sense of Schürmann on the Lie algebra of the Lorentz group are studied. It is known that only one of the irreducible unitary representations of the Lorentz group admits a non-trivial one-cocycle. A Schürmann triple is constructed for this cocycle and the properties of the associated Lévy process are investigated. The decompositions of the restrictions of this triple to the Lie subalgebras $so(3)$ and $so(2,1)$ are described.

Introduction

Factorisable representations of current groups and current algebras have a long history, cf. [4,17,12], and they played an important role in the development of quantum stochastic calculus, see [21] and the references therein.

Factorisable representations of current groups of a Lie group G can be viewed as Lévy processes in the sense of Schürmann [19] on the level of the associated universal enveloping algebra $U(\mathfrak{g})$ or the Lie algebra $\mathfrak{g}$, see [8,7]. We followed this approach in [3], where we associated classical Lévy processes to the representation introduced in [23] and studied, e.g., their marginal distributions. Furthermore, in [2,1], this approach was used to define "quadratic" exponential vectors and a "quadratic" second quantization functor.

Interesting factorisable representations of current groups of a Lie group exist only if the Lie group has a representation which admit a non-trivial cocycle, see [22,25,10]. If we restrict our attention to unitary representations and simple Lie groups, then this leaves only the two series. $G = SO(n,1)$ and $G = SU(n,1)$. These are the simple Lie groups which do not have Kazhdan's property (T). According to Graev and Vershik [10,24], this fact was first observed in 1973–74 by Gelfand, Graev, and Vershik [22] (but we where not able to confirm this claim from the paper's English translation), according to Shalom [20] this was first proved by Delorme [6] and Hotta and Wallach [13].

In [3], we focussed on the lowest-dimensional case, i.e. the unique non-compact form $so(2,1) \cong su(1,1) \cong sl(2,\mathbb{R})$ of the unique simple Lie algebra of rank one. In the present paper we study the Lie algebra $so(3,1)$ of the Lorentz group.

In Section 1, we recall the definition of the Lorentz group and its Lie algebra and we introduce some notations which we shall use in this paper. For the purpose of self-containedness we also recall several facts about their unitary representations.

In Section 2, we recall the definitions of Lévy processes and Schürmann triples on real Lie algebras. We also construct a Schürmann triple on $so(3,1)$, which has a non-trivial 1-cocycle. From the general theory mentioned above, it is clear that this Schürmann triple is unique up to rescaling and adding a coboundary. Therefore the Lévy process associated to this Schürmann triple must be the infinitesimal version of the factorisable representations on $SO(3,1)$ studied in [10,24].

In the remaining sections we study restrictions of this Schürmann triple to Lie subalgebras of $so(3,1)$. In Section 3, we obtain the decomposition of the restriction to $so(3)$. In Section 4, we study the restriction to $so(2,1)$.

1. The Lorentz group and its Lie algebra

The *Lorentz group* $O(3,1)$ is the group of all isometries of Minkowski space-time, i.e., it is the group of all 4×4 matrices that leave invariant the Minkowksi inner product

$$\left\langle \begin{pmatrix} t \\ x \\ y \\ z \end{pmatrix}, \begin{pmatrix} t' \\ x' \\ y' \\ z' \end{pmatrix} \right\rangle = tt' - xx' - yy' - zz'.$$

The identity component $SO(3,1)^+$ of $O(3,1)$ is called the *restricted Lorentz group*. It consists of the 4×4 matrices $A = (a_{jk}) \in O(3,1)$ with $\det(A) = +1$ and $a_{11} \geq 1$.

The restricted Lorentz group $SO(3,1)^+$ is isomorphic to the projective special linear group (or Möbius group) $PSL(2,\mathbb{C})$.

The Lie algebra $so(3,1) \cong sl(2,\mathbb{C})$ of the Lorentz group $O(3,1)$ has as basis $\{H_1, H_2, H_3, F_1, F_2, F_3\}$ with the relations

$$[H_j, H_k] = i\epsilon_{jk\ell}H_\ell. \qquad [F_j, F_k] = -i\epsilon_{jk\ell}F_\ell,$$
$$[F_j, H_k] = i\epsilon_{jk\ell}H_\ell,$$

for $j, k, \ell \in \{1,2,3\}$, where $\epsilon_{jk\ell}$ is the Levi-Civita symbol,

$$\epsilon_{j,k,\ell} = \begin{cases} +1 & \text{if} \quad (jk\ell) = (1,2,3), (2,3,1) \text{ or } (3,1,2), \\ -1 & \text{if} \quad (jk\ell) = (2,1,3), (3,2,1) \text{ or } (1,3,2), \\ 0 & \text{else.} \end{cases}$$

We shall consider $so(3,1)$ with the involution which makes these six elements hermitian. Then the infinitesimals of unitary representations of $SO(3,1)^+$ are *-representations of $so(3,1)$. For our computations we will also use the bases $\{A_1, A_2, A_3, B_1, B_2, B_3\}$ and $\{H_3, H_-, H_+, K_3, K_-, K_+\}$ with

$$A_j = \frac{1}{2}(H_j + iF_j), \quad B_j = \frac{1}{2}(H_j - iF_j), \quad j = 1,2,3,$$
$$H_3, \quad H_\pm = H_1 \pm iH_2, \quad F_3, \quad K_\pm = K_1 \pm iK_2.$$

In terms of these bases the relations become

$$[A_j, A_k] = i\epsilon_{jk\ell}A_\ell, \quad [B_j, B_k] = i\epsilon_{jk\ell}B_\ell, \quad [A_j, B_k] = 0$$
$$[H_+, H_-] = 2H_3, \quad [F_+, F_-] = -2H_3,$$
$$[H_3, H_\pm] = \pm H_\pm, \quad [F_3, F_\pm] = \mp H_\pm,$$
$$[H_3, F_\pm] = \pm F_\pm, \quad [F_3, H_\pm] = \pm F_\pm,$$
$$[H_\pm, F_\mp] = \pm F_3, \quad [H_\pm, F_\pm] = 0, \quad [F_3, H_3] = 0,$$

and

$$A_j^* = B_j, \quad H_\pm^* = H_\mp, \quad F_\pm^* = F_\mp.$$

The representation theory of $SO(3,1)^+$ can be found in the monographs [9,15,18]. Here we are only interested in unitary representations. On the level of the Lie algebra $so(3,1)$ they can be described as follows. For $\ell_0 \in \frac{1}{2}\mathbb{Z}$, $\ell_0 \geq 0$, and $\ell_1 \in \mathbb{C}$, let

$$D_{\ell_0\ell_1} = \text{span}\{\xi_{\ell m}; \ell = \ell_0, \ell_0+1, \ldots, m = -\ell, -\ell+1, \ldots, \ell\}$$

and set

$$C_\ell = \begin{cases} i\frac{\sqrt{(\ell^2-\ell_0^2)(\ell^2-\ell_1^2)}}{\ell\sqrt{4\ell^2-1}} & \text{if } \ell \geq 1, \\ 0 & \text{if } \ell = 0, \frac{1}{2}, \end{cases}$$

$$A_\ell = \begin{cases} \frac{i\ell_0\ell_1}{\ell(\ell+1)} & \text{if } \ell > 0, \\ 0 & \text{if } \ell = 0, \end{cases}$$

(note that $\ell = 0$ or $\ell = \frac{1}{2}$ can only occur if $\ell_0 = 0$ or $\ell_0 = \frac{1}{2}$, resp.).

We define an action of $so(3,1)$ on $D_{\ell_0\ell_1}$ by

$$\rho_{\ell_0\ell_1}(H_3)\xi_{\ell m} = m\xi_{\ell m}, \quad \rho_{\ell_0\ell_1}(H_\pm)\xi_{\ell m} = \sqrt{(\ell \mp m)(\ell \pm m + 1)}\xi_{\ell, m\pm 1},$$

$$\rho_{\ell_0\ell_1}(F_3)\xi_{\ell m} = C_\ell\sqrt{\ell^2 - m^2}\xi_{\ell-1,m} - mA_\ell\xi_{\ell m}$$
$$-C_{\ell+1}\sqrt{(\ell+1)^2 - m^2}\xi_{\ell+1,m},$$

$$\rho_{\ell_0\ell_1}(F_\pm)\xi_{\ell m} = \pm C_\ell\sqrt{(\ell \mp m)(\ell \mp m - 1)}\xi_{\ell-1,m\pm 1}$$
$$-A_\ell\sqrt{(\ell \mp m)(\ell \pm m + 1)}\xi_{\ell,m\pm 1} \pm C_{\ell+1}\sqrt{(\ell \pm m + 1)(\ell \pm m + 2)}\xi_{\ell+1,m\pm 1}$$

(we use the same basis as [9] and [15], note that [5] uses $\psi_{\ell m} = i^{m-\ell}\xi_{\ell m}$ instead).

As was observed in [5], "formally" there exists also a basis $\{\phi_{m_1,m_2}\}$ on which A_3, $A_\pm = A_1 \pm iA_2$, B_3, and $B_\pm = B_1 \pm iB_2$ act as

$$A_3\phi_{m_1,m_2} = m_1\phi_{m_1,m_2}, \quad A_\pm = \sqrt{(j_1 \mp m_1)(j_1 \pm m_1 + 1)}\phi_{m_1\pm 1,m_2},$$
$$B_3\phi_{m_1,m_2} = m_2\phi_{m_1,m_2}, \quad B_\pm = \sqrt{(j_2 \mp m_2)(j_2 \pm m_2 + 1)}\phi_{m_1,m_2\pm 1},$$

where $j_1 = \overline{j_2} = \frac{1}{2}(\ell_0 + \ell_1 - 1)$.

Define an inner product on $D_{\ell_0\ell_1}$ s.t. the family $\{\xi_{\ell m}\}$ is an orthonormal system and denote by $H_{\ell_0\ell_1}$ the completion of $D_{\ell_0\ell_1}$.

The representation $\rho_{\ell_0\ell_1}$ is the infinitesimal representation of an irreducible unitary representation of $SO(3,1)^+$ on $H_{\ell_0\ell_1}$ in the following two cases:

a) ℓ_1 is purely imaginary (and no restriction on $\ell_0 \in \frac{1}{2}\mathbb{Z}_+$), this is the *principal series.*
b) $\ell_0 = 0$ and $0 \leq \ell_1 < 1$, this is the *supplementary series.*

The representations ρ_{0,ℓ_1} and $\rho_{0,-\ell_1}$ with ℓ_1 purely imaginary are easily seen to be equal, since only the square of ℓ_1 occurs in their definition. The remaining representations are all inequivalent. Together with the trivial representation ε, which sends $so(3,1)$ identically to 0, these two families exhaust all irreducible unitary representations of $SO(3,1)^+$. Note that the representation ρ_{01} is not irreducible. It can be decomposed as $\rho_{01} \cong \varepsilon \oplus \rho_{10}$ on

$$D_{01} = \text{span}\{\xi_{00}\} \oplus \text{span}\{\xi_{\ell m}; \ell \geq 1, m = -\ell, \ldots, \ell\}.$$

The elements $C_A = 2A_3^2 + A_+A_- + A_-A_+$ and $C_B = 2B_3^2 + B_+B_- + B_-B_+$ generate the center of the universal enveloping algebra $U\big(so(3,1)\big)$ and satisfy $C_A^* = C_B$. The Casimir invariants

$$J_1 = C_A + C_B = 2H_3^2 + H_+H_- + H_-H_+ - 2F_3^2 - F_+F_- - F_-F_+,$$
$$J_2 = -i(C_A - C_B)/2 = H_+F_- + H_-F_+ + F_+H_- + F_-H_+ + 4F_3H_3,$$

are hermitian and also generate the center of $U\big(so(3,1)\big)$. In the irreducible unitary representations defined above they act as

$$\rho_{\ell_0\ell_1}(J_1) = (\ell_0^2 + \ell_1^2 - 1)\text{id}_{D_{\ell_0\ell_1}} \quad \text{and} \quad \rho_{\ell_0\ell_1}(J_2) = i\ell_0\ell_1\text{id}_{D_{\ell_0\ell_1}}.$$

In this paper we will work with (in general) unbounded involutive representations of involutive complex Lie algebras $(\mathfrak{g}, *)$ or their universal enveloping algebras $U(\mathfrak{g})$ acting on some pre-Hilbert space D. See [16] for necessary and sufficient conditions for such a representation to be the infinitesimal representation associated to a unitary representation $H = \overline{D}$ of the connected simply connected Lig group G associated to the real Lie algebra $\mathfrak{g}_{\mathbb{R}} = \{X \in \mathfrak{g}; X^* = -X\}$.

2. Schürmann triples on $so(3,1)$ and their Lévy processes

We start by recalling the definition of Schürmann triples and Lévy processes on real Lie algebras. Let $\mathfrak{g}$ be a real Lie algebra, $(\mathfrak{g}_{\mathbb{C}}, *)$ its complexification equipped with the involution that makes the elements of $\mathfrak{g}$ anti-hermitian, and $U_0(\mathfrak{g})$ its universal enveloping algebra, without unit, but with the involution induced from $(\mathfrak{g}_{\mathbb{C}}, *)$.

For D a complex pre-Hilbert space with, we let $\mathcal{L}(D)$ be algebra of linear operators on D having an adjoint defined everywhere on D, and $\mathcal{L}_{\text{AH}}(D)$ the anti-Hermitian linear operators on D.

Definition 2.1. [[7], Definition 8.1.1] Let D be a pre-Hilbert space and $\omega \in D$ a unit vector. A family

$$\big(j_{st} : \mathfrak{g} \to \mathcal{L}_{\text{AH}}(D)\big)_{0 \leq s \leq t}$$

of representations of $\mathfrak{g}$ is a *Lévy process on* $\mathfrak{g}$ *over* (D, ω) if

a) (increment property) we have
$$j_{st}(X) + j_{tu}(X) = j_{su}(X)$$
for all $0 \leq s \leq t \leq u$ and all $X \in \mathfrak{g}$;

b) (independence) we have
$$[j_{st}(X), j_{s't'}(Y)] = 0, \qquad X, Y \in \mathfrak{g},$$
$0 \leq s \leq t \leq s' \leq t'$, and
$$\langle \omega, j_{s_1t_1}(X_1)^{k_1} \cdots j_{s_nt_n}(X_n)^{k_n}\omega\rangle$$
$$= \langle \omega, j_{s_1t_1}(X_1)^{k_1}\omega\rangle \cdots \langle \omega, j_{s_nt_n}(X_n)^{k_n}\omega\rangle,$$
for $n, k_1, \ldots, k_n \in \mathbb{N}$, $0 \leq s_1 \leq t_1 \leq s_2 \leq \cdots \leq t_n$, $X_1, \ldots, X_n \in \mathfrak{g}$;

c) (stationarity) for $n \in \mathbb{N}$, $X \in \mathfrak{g}$, $0 \leq s \leq t$,
$$\langle \omega, j_{st}(X)^n\omega\rangle = \langle \omega, j_{0,t-s}(X)^n\omega\rangle$$
i.e., the moments depend only on the difference $t - s$;

d) (pointwise continuity) we have
$$\lim_{t \searrow s} \langle \omega, j_{st}(X)^n\omega\rangle = 0, \qquad n \in \mathbb{N}, \quad X \in \mathfrak{g}.$$

To a Lévy process we can associate the states $\varphi_t = \langle \omega, j_{0,t}(a)^n\omega\rangle$ for $a \in U_0(\mathfrak{g})$, $t \geq 0$, and the functional
$$L(a) = \frac{\mathrm{d}}{\mathrm{d}t}\bigg|_{t=0} \varphi_t(a), \qquad a \in U_0(\mathfrak{g}).$$
The functional L is a generating functional in the sense of the following definition: A linear functional $L : U_0(\mathfrak{g}) \to \mathbb{C}$ on the non-unital *-algebra $U_0(\mathfrak{g})$ is called a *generating functional on* $\mathfrak{g}$ if

a) L is Hermitian, i.e., $L(u^*) = \overline{L(u)}$ for $u \in U_0(\mathfrak{g})$;

b) L is positive, i.e., $L(u^*u) \geq 0$ for $u \in U_0(\mathfrak{g})$.

Definition 2.2. [[7], Definition 8.2.1] Let D be a pre-Hilbert space. A *Schürmann triple on* $\mathfrak{g}$ *over* D is a triple (ρ, η, ψ), where

a) $\rho : \mathfrak{g} \to \mathcal{L}_{\mathrm{AH}}(D)$ is a representation of $\mathfrak{g}$ on D, i.e.,
$$\rho([X, Y]) = \rho(X)\rho(Y) - \rho(Y)\rho(X)$$
for $X, Y \in \mathfrak{g}$;

b) $\eta : \mathfrak{g} \to D$ is a ρ-1-cocycle, i.e. it satisfies
$$\eta([X, Y]) = \rho(X)\eta(Y) - \rho(X)\eta(Y), \qquad X, Y \in \mathfrak{g};$$

c) $\psi : \mathfrak{g} \to \mathbb{C}$ is a linear functional with imaginary values s.t. the bilinear map $(X, Y) \longmapsto \langle \eta(X), \eta(Y) \rangle$ is the 2-ε-ε-coboundary of ψ (where ε denotes the trivial representation), i.e.,

$$\psi([X, Y]) = \langle \eta(Y), \eta(X) \rangle - \langle \eta(X), \eta(Y) \rangle, \qquad X, Y \in \mathfrak{g}.$$

See [11] for more information on the cohomology of Lie algebras and Lie groups.

The Schürmann triple and in particular the linear functional ψ in a Schürmann triple have unique extensions to $\mathfrak{g}_{\mathbb{C}}$ (by linearity) and to $U_0(\mathfrak{g})$ (as representation, cocycle and coboundary, resp.) and the extension of the functional is a generating functional. Therefore it corresponds to a Lévy process on $\mathfrak{g}$ (which is unique in distribution). This Lévy process can be realised on the symmetric Fock space $\Gamma\big(L^2(\mathbb{R}_+, D)\big) = \bigoplus_{n=0}^{\infty} L^2(\mathbb{R}_+, D)^{\otimes_s n}$ over $L^2(\mathbb{R}_+, D)$ as

$$j_{st}(X) = \Lambda_{st}\big(\rho(X)\big) + A_{st}^{+}\big(\eta(X)\big) + A_{st}^{-}\big(\eta(X^*)\big) + \psi(X)(t-s)\mathrm{id}, \qquad X \in \mathfrak{g}_{\mathbb{C}},$$

cf. [19]. Here $\Lambda_{st}(M) = \Lambda(M \otimes \mathbf{1}_{[s,t]})$, $A_{st}^{+}(v) = A(v \otimes \mathbf{1}_{[s,t]})$, and $A_{st}^{-}(v) = A(v \otimes \mathbf{1}_{[s,t]})$, with $M \in \mathcal{L}(D)$, $v \in D$, denote the conservation, creation, and annihilation operators, see, e.g., [[7], Chapter 5].

If the cocycle η is a coboundary, then the associated Lévy process is unitarily equivalent to the second quantisation of ρ, cf. [[7], Proposition 8.2.7], and the Lévy processes of cocycles that differ only by a coboundary are also unitarily equivalent. Therefore it is most interesting to study the processes associated to non-trivial cocycles. The Lévy processes associated to reducible Schürmann triples can be constructed as tensor products of the Lévy processes of their irreducible components, for this reason we shall study only irreducible representations.

If the Casimir invariants are invertible in some representation, then all 1-cocyles are coboundaries, cf. [[3], Lemma 2.2]. Therefore the only non-trivial irreducible unitary representation that can have a non-trivial cocycle is ρ_{10}. Since $so(3,1)$ is simple, it is clear that the trivial representation has no non-zero cocycles at all, cf. [[3], Lemma 2.1].

From [6,13], we know that only one irreducible unitary representation of $SO(3,1)$ admits a non-trivial 1-cocycle, and this is ρ_{10}. We will describe a non-trivial 1-cocycle of this representation below, after recalling a useful lemma.

Lemma 2.1. *(Raabe-Duhamel test) Let* $(u_n)_{n\in\mathbb{N}}$ *be a sequence of strictly* *positive real numbers such that*

$$\frac{u_{n+1}}{u_n} = 1 - \frac{\alpha}{n} + o\left(\frac{1}{n}\right).$$

If $\alpha < 1$, then the series $\sum_{n\in\mathbb{N}} u_n$ diverges, if $\alpha > 1$, then the series $\sum_{n\in\mathbb{N}} u_n$ converges (nothing can be concluded in the case $\alpha = 1$).

We can now construct a non-trivial 1-cocycle for ρ_{10}.

Proposition 2.1. *There exists a 1-ρ_{10}-cocycle c with $c(F_+) = \xi_{11}$. The cocycle c is not a coboundary and every other non-trivial 1-ρ_{10}-cocycle is a linear combination of c and some 1-ρ_{10}-coboundary.*

Proof. The cocycle c can not be a coboundary, since the vector ξ_{11} is not in the image of $\rho_{10}(F_+)$. Indeed, assume there exists a vector $\zeta = \sum x_{\ell m}\xi_{\ell m} \in H_{10}$ s.t. $\rho_{10}(F_+)\zeta = \xi_{11}$. Then we have $x_{\ell m} = 0$ for all $m \neq 0$ and for pairs with $m = 0$ and ℓ odd. For $m = 0$ and even ℓ we find the recurrence relation

$$x_{\ell,0} = -\frac{C_{\ell-1}}{C_\ell} x_{\ell-2,0}$$

with the initial condition $x_{20} = -i\sqrt{\frac{5}{2}}$. We have

$$\begin{aligned} \frac{|x_{\ell,0}|^2}{|x_{\ell-2,0}|^2} &= \frac{((\ell-1)^2-1)(4\ell^2-1)}{(\ell^2-1)(4(\ell-1)^2-1)} = \frac{(\ell^2-2\ell)(4\ell^2-1)}{(\ell^2-1)(4\ell^2-8\ell+3)} \\ &= \frac{4\ell^4-8\ell^3-\ell^2+2\ell}{4\ell^4-8\ell^3-\ell^2+8\ell+3} = 1 + o\left(\frac{1}{\ell}\right) \end{aligned}$$

i.e., $\alpha = 0 < 1$. So the Raabe-Duhamel test implies that the series $\sum |x_{\ell m}|^2$ diverges and therefore there exists no such vector ζ in H_{10}.

It was shown in [6,13] that the first cohomology group of ρ_{10} has dimension one, this implies the uniqueness. To prove existence, one checks that

$$c(H_3) = c(H_\pm) = 0$$
$$c(F_3) = -\frac{1}{\sqrt{2}}\xi_{10}, \quad c(F_\pm) = \pm\xi_{1,\pm 1}$$

defines indeed a 1-ρ_{10}-cocycle. □

Remark 2.1. Let $\ell_0 = 0$. The 1-ρ_{0,ℓ_1}-coboundary of ξ_{00} is given by

$$\partial\xi_{00}(H_3) = 0 = \partial\xi_{00}(H_\pm),$$
$$\partial\xi_{00}(F_3) = -iC_1(0,\ell_1)\xi_{10}, \quad \partial\xi_{00}(F_\pm) = \pm iC_1(0,\ell_1)\sqrt{2}\xi_{1,\pm 1},$$

with $C_1(0,\ell_1) = \frac{\sqrt{1-\ell_1^2}}{\sqrt{3}}$. We see that c is formally the limit of the 1-ρ_{0,ℓ_1}-coboundaries $\frac{1}{iC_1(0,\ell_1)\sqrt{2}}\partial\xi_{00}$ as ℓ_1 tends to 1.

Proposition 2.2. *There exists a unique Schürmann triple* (ρ_{10}, c, ψ) *containing the representation* ρ_{10} *and the cocycle* c.

Proof. This is a consequence of the fact that $so(3,1)$ is simple and that therefore the second cohomology group of the trivial representation is trivial. Any element in $so(3,1)$ can be written as a commutator and so Condition c) in Definition 2.2 allows to deduce the values of ψ from the values of c. We have, e.g.,

$$2\psi(F_3) = \psi([F_+, H_-]) = \Big\langle c((F^+)^*), c(H_-)\Big\rangle - \Big\langle c((H^-)^*), c(F_+)\Big\rangle = 0.$$

It turns out that ψ is identically equal to 0 on $su(3,1)$. □

We would like to characterise the Lévy process associated to the Schürmann triple (ρ_{10}, c, ψ). For this purpose one could compute the action of the Casimir invariants on the vacuum vector $\Omega \in \Gamma\big(L^2(\mathbb{R}_+, D)\big)$.

Since ψ vanishes on $\mathfrak{g}$, we have a simple formula for the action of Lie algebra elements on the vacuum vector,

$$j_{st}(X)\Omega = c(X) \otimes \mathbf{1}_{[s,t]}, \qquad X \in \mathfrak{g}_{\mathbb{C}}, \quad 0 \le s \le t,$$

so we have

$$j_{st}(H_3)\Omega = 0 = j_{st}(H_\pm)\Omega,$$
$$j_{st}(F_3)\Omega = -\frac{1}{\sqrt{2}}\xi_{10} \otimes \mathbf{1}_{[s,t]}, \quad j_{st}(F_\pm)\Omega = \pm\xi_{1,\pm1} \otimes \mathbf{1}_{[s,t]}.$$

For the Casimir invariants we get

$$j_{st}(J_2)\Omega = j_{st}(H_+)\big(-\xi_{1,-1} \otimes \mathbf{1}_{[s,t]}\big) + j_{st}(H_-)\big(\xi_{1,1} \otimes \mathbf{1}_{[s,t]}\big) = 0$$

and

$$\begin{aligned} j_{st}(J_1)\Omega &= -2j_{st}(F_3)\left(-\frac{1}{\sqrt{2}}\xi_{10} \otimes \mathbf{1}_{[s,t]}\right) - j_{st}(F_+)\big(-\xi_{1,-1} \otimes \mathbf{1}_{[s,t]}\big) \\ &\qquad -j_{st}(F_-)\big(\xi_{11} \otimes \mathbf{1}_{[s,t]}\big) \\ &= -(t-s)\Omega + \sqrt{2}\big(\xi_{1,1} \otimes \mathbf{1}_{[s,t]}\big) \otimes \big(\xi_{1,-1} \otimes \mathbf{1}_{[s,t]}\big) \\ &\qquad +\sqrt{2}\big(\xi_{1,-1} \otimes \mathbf{1}_{[s,t]}\big) \otimes \big(\xi_{1,1} \otimes \mathbf{1}_{[s,t]}\big) \\ &\qquad -\sqrt{2}\big(\xi_{10} \otimes \mathbf{1}_{[s,t]}\big) \otimes \big(\xi_{10} \otimes \mathbf{1}_{[s,t]}\big). \end{aligned}$$

The action of $j_{st}(J_1)$ on the vacuum vector shows that $j_{st}(J_1)$ is not a multiple of the identity, which implies that the representation j_{st} restricted to the subspace generated from the vacuum vector can not be irreducible.

To get a better understanding of the representations j_{st}, $0 \leq s \leq t$, we will now consider the restrictions of ρ_{10} to Lie subalgebras of $so(3,1)$.

3. Restriction to the Lie sub algebra $so(3)$

The basis we used to describe the representations of $so(3,1)$ is already adapted to the subalgebra $so(3) = \text{span}\{H_3, H_+, H_-\}$, so it is easy to decompose the restriction of representations of $so(3,1)$ to its Lie subalgebra $so(3)$ into its irreducible parts.

The representation ρ_{10} restricted to $so(3) = \text{span}\{H_3, H_+, H_+\}$ decomposes into a direct sum of finite-dimensional irreducible representations. Recall that the irreducible representations of $so(3)$ are all unitarily equivalent to one of the following. Let $s \in \frac{1}{2}\mathbb{Z}$, set $E_s = \text{span}\{e_{-s}, e_{-s+1}, \ldots, e_s\}$, where $e_{-s}, \ldots, e_s$ form an orthonormal basis, and set

$$\pi_s(H_3)e_m = me_m, \quad \pi_s(H_\pm)e_m = \sqrt{(s \mp m)(s \pm s + 1)}e_{m\pm 1},$$

for $m = -s, \ldots, s$. It is not difficult to check that we have

$$(D_{10}, \rho_{10}|_{so(3)}) \cong \bigoplus_{s=3}^{\infty}(E_s, \pi_s).$$

4. Restriction to Lie sub algebra $so(2,1)$

The basis elements H_3, F_+, F_- span a Lie sub algebra of $so(3,1)$ that is isomorphic to the non-compact form $sl(2;\mathbb{R}) \cong su(1,1) \cong so(2,1)$ of the three-dimensional simple Lie algebra $sl(2)$. We will now describe the restriction of our Lévy processes on $so(3,1)$ to this Lie sub algebra.

Recall that $so(2,1)$ admits the highest and lowest weight representations π_t^+ and π_t^-, with $t > 0$, acting on $D_t^\pm = \text{span}\{e_n; n \in \mathbb{N}\}$ (where (e_n) are an orthonormal basis) as

$$\pi_t^\pm(H_3)e_n = \pm(n+t)e_n,$$
$$\pi_t^+(F_+)e_n = \sqrt{(n+1)(n+2t)}e_{n+1}, \quad \pi_t^-(F_+)e_n = \sqrt{n(n+2t-1)}e_{n-1}$$
$$\pi_t^+(F_-)e_n = \sqrt{n(n+2t-1)}e_{n-1}, \quad \pi_t^-(F_-)e_n = \sqrt{(n+1)(n+2t)}e_{n+1},$$

cf. [3]. There is also a third family $\pi_{c,\mu}$ acting on $D^0_{c,\mu} = \text{span}\{f_n, n \in \mathbb{Z}\}$ (where (f_n) is an orthonormal basis) as

$$\begin{aligned}\pi_{c,\mu}(H_3)f_n &= (n-\mu)f_n,\\ \pi_{c,\mu}(F_+)f_n &= \sqrt{n^2+(1-2\mu)n+\mu(\mu-1)-c}f_{n+1},\\ \pi_{c,\mu}(F_-)f_n &= \sqrt{(n-1)^2+(1-2\mu)(n-1)+\mu(\mu-1)-c}f_{n-1},\end{aligned}$$

with $0 \le \mu < 1$ and $c < \mu(\mu-1)$.

The families π_t^+ and π_t^- are called the positive and the negative discrete series. Our third family contains both the principal unitary series and the complementary unitary series. See, e.g., [[26], Section 6.4] for more information on the representation theory of $SU(1,1)$.

Denote by $K = H_3^2 - \frac{1}{2}(F_+F_- + F_-F_+) = H_3(H_3-1) - F_-F_+$ the Casimir element of $so(2,1)$. Then we have

$$\pi_t^{\pm}(K) = 2t(t-1)\,\text{id} \qquad \text{and} \qquad \pi_{c,\mu}(K) = 2c\,\text{id}.$$

The subrepresentations π_t^+ and π_t^- can be detected by their cyclic vector e_0 which is characterised by the equations

$$\pi_t^+(F_-)e_0 = 0,\ \pi_t^+(H_3)e_0 = te_0$$
$$(\text{or } \pi_t^-(F_+)e_0 = 0,\ \pi_t^+(H_3)e_0 = -te_0 \text{ resp.}).$$

Proposition 4.1. *If we restrict the representation ρ_{10} of $so(3,1)$ to the Lie subalgebra $so(2,1)$, then it decomposes as*

$$(D_{10}, \rho_{10}|_{so(2,1)}) \cong (D_1^+, \pi_1^+) \oplus (D_1^-, \pi_1^-) \oplus (D_R, \pi_R)$$

where the "rest" (π_R, D_R) is a direct sum of unitary irreducible representations $(\pi_{c,0}, D^0_{c,0})$ belonging to the third family.

Proof. We need to determine all eigenvectors of $\rho_{10}(H_3)$ that are annihilated by $\rho_{10}(F_-)$. A non-zero vector $\xi = \sum_{\ell=1}^{\infty}\sum_{m=-\ell}^{\ell} x_{\ell m}\xi_{\ell m}$ is an eigenvector of $\rho_{10}(H_3)$, if and only if there exists an integer $m_0 \in \mathbb{Z}$ s.t. $x_{\ell m} = 0$ for $m \neq m_0$. And this integer is then its eigenvalue.

Let us now study, when such a vector is annihilated by $\rho_{10}(F_-)$. We consider first $m_0 \in \{-1, 0, 1\}$. For $\xi = \sum_{\ell=1}^{\infty} x_\ell \xi_{\ell m_0} \in D_{10}$ we have

$$\rho_{10}(F_-)\xi = -\sum_{\ell=2}^{\infty}\Big(x_{\ell+1}C_{\ell+1}\sqrt{(\ell+m_0+1)(\ell+m_0)} \tag{1}$$

$$+ x_{\ell-1}C_\ell\sqrt{(\ell-m_0)(\ell-m_0+1)}\Big)\,\xi_{\ell m_0}. \tag{2}$$

If this vector vanishes, then the coefficients $x_{2\ell+1}$, $\ell \in \mathbb{N}$, are determined by x_1 via the recurrence relation

$$x_{\ell+1} = -\frac{C_\ell\sqrt{(\ell-m_0)(\ell-m_0+1)}}{C_{\ell+1}\sqrt{(\ell+m_0+1)(\ell+m_0)}}x_{\ell-1}. \tag{3}$$

We get

$$\begin{aligned}\frac{|x_{\ell+1}|}{|x_{\ell-1}|} &= \frac{(\ell^2-1)\big(4(\ell+1)^2-1\big)(\ell-m_0)(\ell-m_0+1)}{(4\ell^2-1)\big((\ell+1)^2-1\big)(\ell+m_0+1)(\ell+m_0)}\\ &= 1-\frac{8m_0}{2\ell}+o\left(\frac{1}{\ell}\right)\end{aligned}$$

and the Raabe-Duhamel test shows that for $x_1 \neq 0$ this series can only converge if $m_0 = 1$.

Furthermore, in the case $m_0 = 1$ we get $x_2 = 0$ from the coefficient of ξ_{10} in (1), and then $x_{2\ell} = 0$ for all $\ell > 1$ from the recurrence relations (3).

We set $x_1 = 1$, $x_2 = 0$ and let $(x_\ell)_{\ell\geq 1}$ denote the solution of the recurrence relation (3). Then $\xi^+ = \sum_{\ell=1}^\infty x_\ell \xi_{\ell 1}$ is a non-zero vector s.t.

$$\rho_{10}(H_3)\xi^+ = \xi^+, \quad \rho_{10}(F_-)\xi^+ = 0,$$

it is therefore the cyclic vector of a subrepresentation of $(D_{10}, \rho_{10}|_{so(2,1)})$ that is unitarily equivalent to (D_1^+, π_1^+).

A careful study of the equation $\rho_{10}(F_-)\xi = 0$ shows that all solutions are of the form $\lambda\xi^+$ for some $\lambda \in \mathbb{C}$.

Indeed, the discussion above shows this already for $m_0 \in \{-1, 0, 1\}$. Furthermore, there are no solution with $m_0 > 1$, because the condition $\rho_{10}(F_-)\xi = 0$ implies immediately $x_1 = 0 = x_2$. And the Raabe-Duhamel test allows to show that there are no solutions with $m_0 < 1$, either.

The discussion of the condition $\rho_{10}(F_+)\xi = 0$, which leads to subrepresentations that are unitarily equivalent to a representation of the form (D_t^-, π_t^-) is similarly. Set $y_1 = 1$, $y_2 = 0$ and let $(y,\ell)_{\ell\geq 1}$ be the sequence determined from these values via the recurrence relation

$$y_{\ell+1} = \frac{C_\ell\sqrt{(\ell+m)(\ell+m+1)}}{C_{\ell+1}\sqrt{(\ell-m)(\ell-m+1)}}$$

with $m = 1$. Set $\xi^- = \sum_{\ell=1}^\infty y_\ell \xi_{\ell,-1}$. Then we have

$$\{\xi \in D_{10}; \rho_{10}(F_+)\xi = 0\} = \mathbb{C}\xi^-,$$

which implies that $(D_{10}, \rho_{10}|_{so(2,1)})$ contains a unique subrepresentation that is unitarily equivalent to (D_{-1}^+, π_{-1}^+).

Since the spectrum of $\rho_{10}(H_3)$ is equal to $\mathbb{Z}$, it follows that the remaining subrepresentations have to belong to the family $(D_{c,0}^0, \pi_{c,0})$, $c < 0$. □

5. Conclusion

We have identified the Schürmann triple underlying the factorizable current representations of the Lorentz group in [10,24].

The decomposition in Proposition 4.1 can be used to compute the classical distribution of elements of $j_{st}(F_+ + F_- + \lambda H_3)$, since the distributions of elements of this form are known for the irreducible unitary representations of $so(2,1)$, cf. [14] and [3], and since the direct sums at the level of the Schürmann triple translate into tensor products for the associated Lévy processes.

It would be interesting to extend these results to the higher rank groups $O(n,1)$ and $U(n,1)$ in the future.

References

1. Luigi Accardi, Ameur Dhahri, Quadratic exponential vectors. J. Math. Phys. 51 (2010), no. 2, 122113.
2. Luigi Accardi, Ameur Dhahri, Quadratic exponential vectors. J. Math. Phys. 50 (2009), no. 12, 122103.
3. Luigi Accardi, Uwe Franz, and Michael Skeide, Renormalized squares of white noise and other non-Gaussian noises as Lévy processes on real Lie algebras. Comm. Math. Phys. 228 (2002), no. 1, 123–150.
4. Huzuhiro Araki, Factorizable representation of current algebra. Non commutative extension of the Léy-Kinchin formula and cohomology of a solvable group with values in a Hilbert space. Publ. Res. Inst. Math. Sci. 5 1969/1970 361–422.
5. N.M. Atakishiyev and S.K. Suslov, The Hahn and Meixner polynomials of an imaginary argument and some of their applications. J. Phys. A 18 (1985), no. 10, 1583–1596.
6. P. Delorme, 1-cohomologie des représentations unitaires des groupes de Lie semi-simples et rÃ©solubles, in Produits Tensoriels Continus et Représentations, Bull. Soc. Math. France 105 (1977), 281–336.
7. Uwe Franz and Nicolas Privault, Probability on real Lie algebras. Cambridge Tracts in Mathematics, 206. Cambridge University Press, New York, 2016.
8. Uwe Franz, Lévy process on real Lie algebras. Recent developments in stochastic analysis and related topics, 166–181, World Sci. Publ., Hackensack, NJ, 2004.
9. I.M. Gelfand, R.A. Minlos, and Z.Ya. Shapiro, Representations of the rotation and Lorentz group and their applications, Translated by C. Cummings and T. Boddington, translation edited by H.K. Farahat. A Pergamon Press Book The Macmillan Co., New York 1963.
10. I.M. Graev, A.M. Vershik, The basic representation of the current group $O(n,1)^X$ in the L^2 space over the generalized Lebesgue measure. Indag. Math. (N.S.) 16 (2005), no. 3-4, 499–529.

11. Alain Guichardet, Cohomologie des groupes topologiques et des algèbres de Lie. Textes Mathématiques, 2. CEDIC, Paris, 1980.
12. Alain Guichardet, Symmetric Hilbert spaces and related topics. Infinitely divisible positive definite functions. Continuous products and tensor products. Gaussian and Poissonian stochastic processes. Lecture Notes in Mathematics, Vol. 261. Springer-Verlag, Berlin-New York, 1972.
13. R. Hotta and N. Wallach, On Matsushima's formula for the Betti numbers of a locally symmetric space, Osaka J. Math. 12 (1975), 419–431.
14. H.T. Koelink, J. Van Der Jeugt, Convolutions for orthogonal polynomials from Lie and quantum algebra representations. SIAM J. Math. Anal. 29 (1998), no. 3, 794–822.
15. M.A. Naimark, Linear representations of the Lorentz group, Translated by Ann Swinfen and O.J. Marstrand, translation edited by H.K. Farahat. A Pergamon Press Book The Macmillan Co., New York 1964.
16. E. Nelson, Analytic vectors, Ann. Math. 70 (1959), no. 3, 572–615.
17. K.R. Parthasarathy and K. Schmidt, Positive definite kernels, continuous tensor products, and central limit theorems of probability theory. Lecture Notes in Mathematics Vol. 272. Springer-Verlag, Berlin-New York, 1972.
18. W. Rühl, The Lorentz group and harmonic analysis. W. A. Benjamin, Inc., New York, 1970.
19. Michael Schürmann, White noise on bialgebras. Lecture Notes in Mathematics, 1544. Springer-Verlag, Berlin, 1993.
20. Yehuda Shalom, Rigidity, unitary representations of semisimple groups, and fundamental groups of manifolds with rank one transformation group. Ann. of Math. (2) 152 (2000), no. 1, 113–182.
21. R.F. Streater, Classical and quantum probability. J. Math. Phys. 41 (2000), no. 6, 3556–3603.
22. A.M. Vershik, I.M. Gelfand, and M.I. Graev, Irreducible representations of the group G^X and cohomology. Funkcional. Anal. i Priložen. 8 (1974), no. 2, 67–69.
23. A.M. Vershik, I.M. Gelfand, M.I. Graev, Representations of the group $SL(2, R)$, where R is a ring of functions. Uspehi Mat. Nauk 28 (1973), no. 5(173), 83–128.
24. A.M. Vershik and I.M. Graev, The structure of complementary series and special representations of the groups $O(n, 1)$ and $U(n, 1)$. (Russian. Russian summary) Uspekhi Mat. Nauk 61 (2006), no. 5(371), 3–88; translation in Russian Math. Surveys 61 (2006), no. 5, 799–884.
25. A.M. Vershik, S.I. Karpushev, Cohomology of groups in unitary representations, neighborhood of the identity and conditionally positive definite functions. Mat. Sb. (N.S.) 119(161) (1982), no. 4, 521–533, 590.
26. N.Ja. Vilenkin, A.U. Klimyk, Representation of Lie groups and special functions. Vol. 1. Simplest Lie groups, special functions and integral transforms. Translated from the Russian by V.A. Groza and A.A. Groza. Mathematics and its Applications (Soviet Series), 72. Kluwer Academic Publishers Group, Dordrecht, 1991.

Infinite Dimensional Analysis, Quantum Probability and Related Topics

pp. 47–58

BASINS OF ATTRACTION OF INVARIANT STATES OF A QUANTUM MARKOV SEMIGROUP

F. FAGNOLA

Dipartimento di Matematica, Politecnico di Milano,
Piazza Leonardo da Vinci 33, I-20133 Milano, Italy
E-mail: franco.fagnola@polimi.it

E. SASSO* and V. UMANITA†

Dipartimento di Matematica, Università di Genova,
Via Dodecaneso 35, I-16146 Genova, Italy
*E-mails: *sasso@dima.unige.it, †umanita@dima.unige.it*

We determine the basins of attraction of invariant states of a quantum Markov semigroup with a faithful normal invariant state and atomic decoherence-free subalgebra.

Keywords: Quantum Markov semigroup, invariant state, basin of attraction.

1. Introduction

Quantum Markov semigroups (QMS) are the key structure for describing irreversible evolutions of open quantum systems. Their importance in the study of decoherence, relaxation to equilibrium, dephasing and control of quantum states can hardly be underestimated. Our understanding of their structure, however, is somewhat limited with the exception, perhaps, of some special case such as quasi-free and generic QMSs (see Ref.[2]).

In our papers Refs.[4,8] we provided a full description of the structure of uniformly continuous QMSs with a decoherence-free subalgebra which is atomic (this is always the case, in particular, for finite dimensional systems). We also established the general form of normal invariant states when the QMS itself admits a faithful normal invariant state.

In this note we study domains of attraction of invariant states namely, the sets initial states converging to a given invariant state.

These results are the key tool for the implementation of the control quantum states by exploiting the decoherence due to the coupling with environment (see e.g. Refs. [1,15] and the references therein). With this technique, called control through decoherence, one can get as a target state any invariant state in a stable way not only for by thermal or vacuum environments but also for more general non-equilibrium environments. Moreover, by appropriately choosing control parameters such as the system-environment coupling and initial environment state one can drive the system to a variety of pre-assigned quantum states.

Here we explicitly determine domains of attraction of extremal invariant states. From this starting point, we also find domains of attraction of arbitrary invariant states as convex combinations of extremal invariant states. In this way, we provide additional information on QMSs with an atomic decoherence-free subalgebra that can be useful in quantum control.

The paper is organized as follows. In Section 2 we recall the fundamental results on QMSs with atomic decoherence-free subalgebra. In Section 3, we characterize extremal invariant states and find their domains of attraction in Section 4. Finally, in Section 5 we consider general invariant states.

2. The structure of the decoherence-free algebra

Let h be a complex separable Hilbert space and let $\mathcal{B}(\mathsf{h})$ be the algebra of all bounded operators on h with unit $\mathbb{1}$. A QMS on $\mathcal{B}(\mathsf{h})$ is a semigroup $\mathcal{T} = (\mathcal{T}_t)_{t\geq 0}$ of completely positive, identity preserving, normal maps which is also weakly* continuous. We assume $\mathcal{T}$ uniformly continuous i.e.

$$\lim_{t\to 0^+} \sup_{\|x\|\leq 1} \|\mathcal{T}_t(x) - x\| = 0,$$

so that there exists a linear bounded operator $\mathcal{L}$ on $\mathcal{B}(\mathsf{h})$ such that $\mathcal{T}_t = e^{t\mathcal{L}}$. In this case, we know that (see Ref. [14] Th. 30.16) there exists a bounded selfadjoint operator H and a sequence $(L_\ell)_{\ell\geq 1}$ of elements in $\mathcal{B}(\mathsf{h})$ such that:

(1) $\sum_{\ell\geq 1} L_\ell^* L_\ell$ is strongly convergent,
(2) if $\sum_{\ell\geq 0} |c_\ell|^2 < \infty$ and $c_0\mathbb{1} + \sum_{\ell\geq 1} c_\ell L_\ell = 0$ for scalars $(c_\ell)_{\ell\geq 0}$ then $c_\ell = 0$ for every $\ell \geq 0$,
(3) $\mathcal{L}(x) = \mathrm{i}[H, x] - \frac{1}{2}\sum_{\ell\geq 1}(L_\ell^* L_\ell x - 2L_\ell^* x L_\ell + x L_\ell^* L_\ell)$ for all $x \in \mathcal{B}(\mathsf{h})$.

The representation of the generator $\mathcal{L}$ is called Gorini-Kossakowski-Sudarshan-Lindblad (GKSL) form. The choice of operators H and $(L_\ell)_{\ell\geq 1}$ is not unique, but this will not create any inconvenience in this paper.

The *decoherence-free (DF) subalgebra* of $\mathcal{T}$ is defined as the set of $x \in \mathcal{B}(\mathsf{h})$ such that

$$\mathcal{T}_t(x^*x) = \mathcal{T}_t(x)^*\mathcal{T}_t(x), \quad \mathcal{T}_t(xx^*) = \mathcal{T}_t(x)\mathcal{T}_t(x)^* \quad \forall t \geq 0. \tag{1}$$

It is a well known fact that $\mathcal{N}(\mathcal{T})$ is the biggest von Neumann subalgebra of $\mathcal{B}(\mathsf{h})$ on which every $\mathcal{T}_t$ acts as a $*$-automorphism (each map $\mathcal{T}_t$ is invertible and its inverse is $\mathcal{T}_{-t}$) by the following (see e.g. Evans[7] Theorem 3.1, Dhahri, Fagnola and Rebolledo[5] Proposition 2.1).

Proposition 2.1. *Let $\mathcal{T}$ be a quantum Markov semigroup on $\mathcal{B}(\mathsf{h})$ and let $\mathcal{N}(\mathcal{T})$ be the set defined by (1). Then*

(1) $\mathcal{N}(\mathcal{T})$ is $\mathcal{T}_t$-invariant for all $t \geq 0$,
(2) $\mathcal{N}(\mathcal{T})$ is a von Neumann subalgebra of $\mathcal{B}(\mathsf{h})$,
(3) $\mathcal{N}(\mathcal{T}) = \{\delta_H^{(n)}(L_k), \delta_H^{(n)}(L_k^*) \,:\, n \geq 0\}'$,
(4) $\mathcal{T}_t(x) = e^{\mathrm{i}tH} x e^{-\mathrm{i}tH}$ for all $x \in \mathcal{N}(\mathcal{T})$, $t \geq 0$,

As shown in Ref.[4], the structure of the decoherence-free subalgebra $\mathcal{N}(\mathcal{T})$ provides information on the whole QMS. We recall here these results. The algebra $\mathcal{N}(\mathcal{T})$ is atomic if and only if there exist two countable sequences $(\mathsf{k}_i)_i$ and $(\mathsf{m}_i)_i$ of Hilbert spaces such that $\mathsf{h} = \oplus_{i\in I}(\mathsf{k}_i \otimes \mathsf{m}_i)$ and

$$\mathcal{N}(\mathcal{T}) = \oplus_{i\in I}\left(\mathcal{B}(\mathsf{k}_i) \otimes \mathbb{1}_{\mathsf{m}_i}\right). \tag{2}$$

As a consequence, defining p_i as the orthogonal projection onto $\mathsf{k}_i \otimes \mathsf{m}_i$, we obtain a family $(p_i)_{i\in I}$ of mutually orthogonal non-zero projections, which are minimal projections in the center of $\mathcal{N}(\mathcal{T})$ and satisfying $\sum_{i\in I} p_i = \mathbb{1}$. Moreover we have that

$$\mathcal{T}_t(a \otimes b) = \mathcal{T}_t^{\mathsf{k}_i}(a) \otimes \mathcal{T}_t^{\mathsf{m}_i}(b) = e^{itK_i} a e^{-itK_i} \otimes \mathcal{T}_t^{\mathsf{m}_i}(b)$$

for all $a \in \mathcal{B}(\mathsf{k}_i)$, $b \in \mathcal{B}(\mathsf{m}_i)$, $i \in I$, $t \geq 0$.

Throughout this paper we assume $\mathcal{N}(\mathcal{T})$ atomic and the existence of a faithful invariant state ρ.

With these assumptions we have useful tools of ergodic theory at our disposal (see e.g. Refs.[10,11,12]); moreover, we know the structure of *rotating states*, i.e. states belonging to the vector space

$$\begin{aligned}\mathcal{R}(\mathcal{T}_*) :&= \overline{\mathrm{span}}\{\sigma \in \mathfrak{I}(\mathsf{h}) \,:\, \mathcal{T}_{*t}(\sigma) = e^{\mathrm{i}t\lambda}\sigma \text{ for some } \lambda \in \mathbb{R},\ \forall t \geq 0\} \\ &= \overline{\mathrm{span}}\{\sigma \in \mathfrak{I}(\mathsf{h}) \,:\, \mathcal{L}_*(\sigma) = \mathrm{i}\lambda\,\sigma \text{ for some } \lambda \in \mathbb{R}\},\end{aligned}$$

where $\mathfrak{I}(\mathsf{h})$ denotes the space of trace-class operators on h. In particular $\mathcal{R}(\mathcal{T}_*)$ is the predual of the decoherence-free algebra $\mathcal{N}(\mathcal{T})$ (see Ref.[9]). Moreover by Proposition 3.2 in Kümmerer[13] and Theorem 2.8 in Engel and Nagel[6] we can decompose $\mathfrak{I}(\mathsf{h})$ as

$$\mathfrak{I}(\mathsf{h}) = \mathcal{R}(\mathcal{T}_*) \oplus \{\sigma \in \mathfrak{I}(\mathsf{h}) \,:\, 0 \in \overline{\{\mathcal{T}_{*t}(\sigma)\}_{t\geq 0}}^{w}\}. \tag{3}$$

On the other hand, since $\mathcal{N}(\mathcal{T})$ is atomic the following facts hold:

(F1). $\mathcal{B}(\mathsf{h}) = \mathcal{N}(\mathcal{T}) \oplus \mathcal{M}_0$ with $\mathcal{N}(\mathcal{T}) = \mathfrak{M}_r = \operatorname{Ran}\mathcal{E}$ and $\mathcal{M}_0 = \mathfrak{M}_s = \operatorname{Ker}\mathcal{E}$, where $\mathcal{E} : \mathcal{B}(\mathsf{h}) \to \mathcal{N}(\mathcal{T})$ is a conditional expectation compatible with the faithful state ρ (see Theorem 10 in Ref.[9]);

(F2). $\mathfrak{I}(\mathsf{h}) = {}^{\perp}\mathcal{M}_0 \oplus {}^{\perp}\mathcal{N}(\mathcal{T})$ with

$${}^{\perp}\mathcal{M}_0 = \operatorname{Ran}\mathcal{E}_* \simeq \mathcal{N}(\mathcal{T})_*, \qquad {}^{\perp}\mathcal{N}(\mathcal{T}) = \operatorname{Ker}\mathcal{E}_* \simeq \mathcal{M}_{2*}.$$

Moreover each $\mathcal{T}_{*t}$ acts as a surjective isometry on ${}^{\perp}\mathcal{M}_0$, and $\lim_t \mathcal{T}_{*t}(\sigma) = 0$ for all $\sigma \in {}^{\perp}\mathcal{N}(\mathcal{T})$ (see Theorem 10 in Ref.[3]).

As a consequence, every state $\omega \in \mathcal{N}(\mathcal{T})_*$ is represented by a unique density σ in ${}^{\perp}\mathcal{M}_0$, and, in this case, we write $\omega = \omega_\sigma$ to mean that $\omega(x) = \operatorname{tr}(\sigma x)$ for all $x \in \mathcal{N}(\mathcal{T})$. Therefore, if we denote by $\mathcal{S} = (\mathcal{S}_t)_{t\geq 0}$ the restriction of $\mathcal{T}$ to $\mathcal{N}(\mathcal{T})$, we have

$$(\mathcal{S}_{*t}\omega_\sigma)(x) = \omega_\sigma(\mathcal{T}_t(x)) = \operatorname{tr}\left(\sigma\, e^{\mathrm{i}tH} x e^{-\mathrm{i}tH}\right) = \operatorname{tr}\left(\mathcal{E}_*(e^{-\mathrm{i}tH}\sigma e^{\mathrm{i}tH})x\right)$$

for all $x = \mathcal{E}(x) \in \mathcal{N}(\mathcal{T})$, concluding that $\mathcal{S}_{*t}\omega_\sigma$ is represented by the density $\mathcal{E}_*(e^{-\mathrm{i}tH}\sigma e^{\mathrm{i}tH}) \in {}^{\perp}\mathcal{M}_0$. In a equivalent way, we have

$$\mathcal{T}_{*t}(\sigma) = \mathcal{E}_*(e^{-\mathrm{i}tH}\sigma e^{\mathrm{i}tH}) \qquad \forall \sigma \in {}^{\perp}\mathcal{M}_0. \tag{4}$$

The $\mathcal{N}(\mathcal{T})$ structure allows us to characterize the form of normal invariant states. Indeed, by Theorem 4.3 in Ref.[4], we know that every invariant state $\eta \in \mathfrak{I}(\mathsf{h})$ can be written as

$$\eta = \sum_{i\in I} \alpha_i \eta_i \otimes \tau_i = \sum_{i\in I} \operatorname{tr}_{\mathsf{m}_i}(p_i \eta p_i) \otimes \tau_i, \tag{5}$$

$\operatorname{tr}_{\mathsf{m}_i}$ denoting the partial trace, for some η_i normal state on k_i which commutes with K_i, τ_i the unique invariant state of $\mathcal{T}^{\mathsf{m}_i}$, and $\alpha_i \geq 0$ such that $\sum_i \alpha_i = 1$.

We want to study the convergence of the predual semigroup to an invariant state. For this purpose we first note that the set of normal invariant states is the weak* closure of the convex hull of its extremal states which are described in the next section.

3. Extremal states

In this section we characterize extremal points of the set $\mathcal{F}(\mathcal{T}_*)_1$ of invariant states of $\mathcal{T}$.

We recall that, given a convex subset C of $\mathfrak{I}(\mathsf{h})_+$, a positive functional $\rho \in C$ is an extremal point of C if it does not lie in any open line segment joining two points of C. In particular, a normal invariant state ρ is extremal in $C = \mathcal{F}(\mathcal{T}_*)_1$ if and only if the only positive normal invariant functionals majorized by ρ are of the form $\lambda\rho$ with $\lambda \in [0,1]$.

Proposition 3.1. *An invariant state η is extremal in $\mathcal{F}(\mathcal{T}_*)_1$ if and only if $\eta = \eta_i \otimes \tau_i$ for some state η_i extremal in $\mathcal{F}(\mathcal{T}_*^{\mathsf{k}_i})_1$.*

Proof. Let η be as in (5). Since $\eta \geq \alpha_i \eta_i \otimes \tau_i$ for all $i \in I$, if η is extremal in $\mathcal{F}(\mathcal{T}_*)_1$ there exists a unique $i \in I$ such that $\eta_i \otimes \tau_i = \eta$, with η_i invariant state for $\mathcal{T}^{\mathsf{k}_i}$.
We have to prove that η_i is extremal in $\mathcal{F}(\mathcal{T}_*^{\mathsf{k}_i})_1$. Assume then $\eta_i = \lambda\omega_1 + (1-\lambda)\omega_2$ for some $\lambda \in [0,1]$ and $\omega_1, \omega_2 \in \mathcal{F}(\mathcal{T}_*^{\mathsf{k}_i})_1$, so that

$$\eta = \eta_i \otimes \tau_i = \lambda\,(\omega_1 \otimes \tau_i) + (1-\lambda)\,(\omega_2 \otimes \tau_i)$$

with $\omega_1 \otimes \tau_i$ and $\omega_2 \otimes \tau_i$ invariant states. The extremality of η in $\mathcal{F}(\mathcal{T}_*)_1$ implies either $\lambda = 0$ or $\lambda = 1$, and then η_i is extremal also in $\mathcal{F}(\mathcal{T}_*^{\mathsf{k}_i})_1$.

Conversely, assume η_i extremal in $\mathcal{F}(\mathcal{T}_*^{\mathsf{k}_i})_1$ and prove that $\eta := \eta_i \otimes \tau_i$ is extremal in $\mathcal{F}(\mathcal{T}_*)_1$. It is clear that $\eta_i \otimes \tau_i$ is invariant for the semigroup, since η_i is invariant for $\mathcal{T}^{\mathsf{k}_i}$. If $\eta_i \otimes \tau_i = \lambda\omega_1 + (1-\lambda)\omega_2$ with $\lambda \in [0,1]$ and $\omega_1, \omega_2 \in \mathcal{F}(\mathcal{T}_*)_1$,

$$\omega_k = \sum_{j \in J} \alpha_{k,j}\omega_{k,j} \otimes \tau_j, \qquad k = 1,2$$

(see equation (5)), then $\alpha_{k,j} = 0$ for all $j \neq i$ and $k = 1,2$, i.e. every $\omega_k = \omega_{k,i} \otimes \tau_i$. Since ω_k is in $\mathcal{F}(\mathcal{T}_*^{\mathsf{k}_i})_1$ and η_i is extremal in this set, we can conclude that $\eta_i \otimes \tau_i$ is extremal in $\mathcal{F}(\mathcal{T}_*)_1$. $\square$

Now we explicitly describe extremal states of $\mathcal{F}(\mathcal{T}_*^{\mathsf{k}_i})_1$.

Proposition 3.2. *Let $\mathcal{T}^{\mathsf{k}}$ be a QMS on a Hilbert space k, $\mathcal{T}_t^{\mathsf{k}}(x) = e^{\mathrm{i}tK} x e^{-\mathrm{i}tK}$ for all $x \in \mathcal{B}(\mathsf{k})$, with $K = K^* \in \mathcal{B}(\mathsf{k})$.*
An invariant state σ is extremal in $\mathcal{F}(\mathcal{T}_*^{\mathsf{k}})_1$ if and only if $\sigma = |e\rangle\langle e|$ for some eigenvector e of K.

Proof. Assume $K = \sum_j \lambda_j q_j$ with $\lambda_j \in \mathbb{R} \setminus \{0\}$ and q_j spectral projection of K corresponding to the non-zero eigenvalue λ_j. Since $[\sigma, K] = 0$, σ commutes with every q_j, and so it can be written as

$$\sigma = \sum_j \sigma_j + \omega, \tag{6}$$

where ω and σ_j are states acting on $\ker(K)$ and $q_j(\mathsf{k})$, respectively.

Now, σ can be extremal if and only if its range is one dimensional, i.e. $\sigma = |e\rangle\langle e|$ with or $e \in q_j(\mathsf{k})$ for some j, either $e \in \ker K$. In any case e is an eigenvector of K. □

As a consequence, we obtain the following characterization of normal invariant extremal states.

Theorem 3.1. *An invariant state η is extremal in $\mathcal{F}(\mathcal{T}_*)_1$ if and only if $\eta = |e^i\rangle\langle e^i| \otimes \tau_i$ for some eigenvector e^i of K_i, $e^i \in \mathsf{k}_i$.*

4. Basins of attraction

In the following we will denote by $\mathfrak{S}(\mathsf{k})$ density matrices associated with states on $\mathcal{B}(\mathsf{k})$ for an arbitrary Hilbert space k.

Definition 4.1. Let η be an invariant state. We call *basin of attraction* or *domain of attraction* of η the set $D(\eta)$ of normal states weakly convergent to η under the action of the semigroup $\mathcal{T}$. In other words

$$D(\eta) := \{\sigma \in \mathfrak{S}(\mathsf{h}) \ : \ w - \lim_t \mathcal{T}_{*t}(\sigma) = \eta\}.$$

Given an initial state σ, we can first remark that its off-diagonal parts weakly converge to 0 under the action of the predual semigroup. More precisely we have the following result

Lemma 4.1. *For every state $\sigma \in \mathfrak{S}(\mathsf{h})$ we have*

$$w - \lim_t \mathcal{T}_{*t}(\sigma) = w - \lim_t \mathcal{T}_{*t}\left(\sum_{i\in I} p_i \sigma p_i\right).$$

Proof. Let $x \in \mathcal{B}(\mathsf{h})$, $x = x_1 + x_2$ with $x_1 \in \mathcal{N}(\mathcal{T})$, $x_2 \in \mathcal{M}_0$ (by property (F1) above), and let σ be a state on h. By definition of $\mathcal{M}_0$ we have

$$\lim_t \operatorname{tr}(\mathcal{T}_{*t}(\sigma)x) = \lim_t \operatorname{tr}(\sigma\mathcal{T}_t(x_1)) + \lim_t \operatorname{tr}(\sigma\mathcal{T}_t(x_2)) = \lim_t \operatorname{tr}(\sigma\mathcal{T}_t(x_1)).$$

Now, since every $\mathcal{T}_t$ preserves $\mathcal{N}(\mathcal{T}) = \oplus_i (\mathcal{B}(\mathsf{k}_i) \otimes \mathbb{1}_{\mathsf{m}_i})$, we get

$$\lim_t \mathrm{tr}\,(\mathcal{T}_{*t}(\sigma)x) = \sum_i \lim_t \mathrm{tr}\,(p_i \sigma p_i \mathcal{T}_t(x_1)) = \sum_i \lim_t \mathrm{tr}\,(\mathcal{T}_{*t}(p_i \sigma p_i)x)$$
$$= \lim_t \mathrm{tr}\left(\mathcal{T}_{*t}(\sum_i p_i \sigma p_i)x\right),$$

giving the thesis. □

Remark 4.1. The existence of $w - \lim_t \mathcal{T}_{*t}(\sigma)$ is not guaranteed for all $\sigma \in \mathfrak{S}(\mathsf{h})$. Under our assumptions we only know that the Cesaro means

$$\frac{1}{t}\int_0^t \mathcal{T}_{*s}(\sigma)ds$$

converge to an invariant state. More precisely, if σ is an initial state

$$\mathcal{T}_{*t}(\sigma) = \sum_{i,j\in I} p_i \mathcal{T}_{*t}(\sigma)p_j$$

and $w{-}\lim_t \mathcal{T}_{*t}(p_i\sigma p_j) = w{-}\lim_t \mathcal{T}_{*t}(p_i\sigma p_j) = 0$ for all $i \neq j$ by Lemma 4.1.

Now, writing $p_i\sigma p_i$ as $\sum_{j,k} |e^i_j\rangle\langle e^i_k| \otimes \sigma^i_{jk}$ with $K_i e^i_j = \kappa^i_j e^i_j$ and $\sigma^i_{jk} \in \mathfrak{I}(\mathsf{m}_i)$, we have

$$\mathcal{T}_{*t}(p_i\sigma p_i) = \sum_{j,k\geq 1} e^{\mathrm{i}(\kappa^i_j - \kappa^i_k)t} |e^i_j\rangle\langle e^i_k| \otimes \mathcal{T}^{\mathsf{m}_i}_{*t}(\sigma^i_{jk}), \tag{7}$$

so that

$$w - \lim_{t\to\infty} \frac{1}{t}\int_0^t \mathcal{T}_{*s}(p_i\sigma p_i)ds = \sum_{\{j,k\,|\,\kappa^i_j = \kappa^i_k\}} tr\left(\sigma^i_{jk}\right) |e^i_j\rangle\langle e^i_k| \otimes \tau_i.$$

This implies that, given an invariant state $\eta = \sum_i \alpha_i \eta_i \otimes \tau_i$, the net $(\mathcal{T}_{*t}(\sigma))_{t\geq 0}$ weakly mean converges to η if and only if

$$\alpha_i\eta_i = \sum_{\{j,k\,|\,\kappa^i_j = \kappa^i_k\}} tr\left(\sigma^i_{jk}\right) |e^i_j\rangle\langle e^i_k|.$$

We want to try to extend this result by using the weak convergence of $(\mathcal{T}_{*t}(\sigma))_{t\geq 0}$.

First of all note that equation (7) immediately gives the following result.

Proposition 4.1. *Let $(e^i_n)_n$ be an orthonormal basis of k_i given by eigenvectors of the Hamiltonian K_i for all $i \in I$, with $K_i e^i_n = \kappa^i_n e^i_n$. If $\sigma \in \mathfrak{S}(\mathsf{h})$ with $p_i \sigma p_i = \sum_{j,k} |e^i_j\rangle\langle e^i_k| \otimes \sigma^i_{jk}$ and $\sigma^i_{jk} \in \mathfrak{I}(\mathsf{m}_i)$, the following conditions are equivalent:*

(1) the weak limit of $(\mathcal{T}_{*t}(\sigma))_t$ exists,
(2) $\operatorname{tr}\left(\sigma^i_{jk}\right) = 0$, when $\kappa^i_j \neq \kappa^i_k$.

Moreover, in this case, we have

$$w-\lim_t \mathcal{T}_{*t}(\sigma) = \sum_{i\in I}\left(\sum_{\{j,k\,|\,\kappa^i_j=\kappa^i_k\}} \operatorname{tr}\left(\sigma^i_{jk}\right) |e^i_j\rangle\langle e^i_k|\right)\otimes\tau_i = \sum_{i\in I} \operatorname{tr}_{\mathsf{m}_i}(p_i\sigma p_i)\otimes\tau_i$$

with $K_i e^i_j = \kappa^i_j e^i_j = K_i e^i_k$ for all $j \neq k$.

Proof. It is enough to show that $(\mathcal{T}_{*t}(\sigma))_t$ converges if and only if $(\mathcal{T}_{*t}(p_i\sigma p_i))_t$ is convergent for all $i \in I$. Moreover, since there always exists $w - \lim \mathcal{T}^{\mathsf{m}_i}_{*t}(\omega) = \tau_i$ for all $i \in I$ and ω state on m_i, by equation (7) the convergence of $(\mathcal{T}_{*t}(p_i\sigma p_i))_t$ is equivalent to that one of

$$\left(\sum_{j,k\geq 1} \operatorname{tr}\left(\sigma^i_{jk}\right) e^{\mathrm{i}(\kappa^i_j-\kappa^i_k)t} |e^i_j\rangle\langle e^i_k|\right)_t = \left(\operatorname{tr}_{\mathsf{m}_i}(\mathcal{T}_{*t}(p_i\sigma p_i))\right)_t,$$

i.e. $\operatorname{tr}\left(\sigma^i_{jk}\right) = 0$, when $\kappa^i_j \neq \kappa^i_k$. □

Proposition 4.1 characterizes the invariant state to which the evolution of an initial state σ weakly converges (when this limit exists).
Vice-versa we can determine the basin of attraction of an arbitrary invariant state, recalling equation (5).

Theorem 4.1. *Let η be a normal invariant state. Then the basin of attraction $D(\eta)$ of η is the set of states*

$$\sum_{i\in I} \operatorname{tr}_{\mathsf{m}_i}(p_i\eta p_i)\otimes\omega_i + \sigma_0$$

where $\omega_i \in \mathfrak{I}(\mathsf{m}_i)$, $\sigma_0 \in \mathfrak{I}_0$ for all $i \in I$ and

$$\mathfrak{I}_0 := \{\omega \in \mathfrak{I}(\mathsf{h}) \,:\, \operatorname{tr}(\omega) = 0,\ w - \lim_t \mathcal{T}_{*t}(\omega) = 0\}. \tag{8}$$

Unfortunately in general is not so trivial to determine the explicit form of σ_0. However, the situation simplifies whenever we consider the basin of attraction of extremal states and their convex combinations.

5. Domains of attraction of general invariant states

Let us consider η an extremal state in the set $\mathcal{F}(\mathcal{T}_*)_1$ of invariant states of $\mathcal{T}$, $\eta = |e\rangle\langle e| \otimes \tau_i$ with $e \in \mathsf{k}_i$ an eigenvector of K_i. We look for its domain of attraction.

Proposition 5.1. *Assume $\eta_i = |e^i\rangle\langle e^i| \otimes \tau_i$ for some eigenvector e^i of K_i, $e^i \in \mathsf{k}_i$ and $i \in J$. If $\sigma \in D(\eta_i)$, then $p_j \sigma\, p_j = 0$ for all $j \neq i$. Moreover, as consequence, $p_j \sigma\, p_n = p_n \sigma\, p_j = 0$, for every $n \in I$.*

Proof. Let $\sigma \in D(\eta_i)$. For all $j \in J$, $j \neq i$, we have

$$\mathrm{tr}\,(p_j \sigma p_j) = \mathrm{tr}\,(\sigma p_j) = \mathrm{tr}\,(\mathcal{T}_{*t}(\sigma) p_j) \to \mathrm{tr}\,\big((|e^i\rangle\langle e^i| \otimes \tau_i)\, p_j\big) = 0,$$

since p_j is a fixed point. Therefore, $p_j \sigma p_j = 0$ for all $j \neq i$. It remains to prove that $p_j \sigma\, p_n = p_n \sigma\, p_j = 0$, for every $n \in I$. Assuming that $p_j \sigma p_j = 0$ and $n \neq j$, we know that

$$0 \leq \langle (p_n + z p_j)(x), \sigma(p_n + z p_j)(x)\rangle = \langle p_n \sigma p_n(x), x\rangle + 2\Re\mathrm{e}(z\langle p_j \sigma p_n(x), x\rangle)$$

for each $z \in \mathbb{C}$ and for all $x \in \mathsf{h}$, so $p_j \sigma p_n = 0$. In a similar way we obtain that $p_n \sigma p_j = 0$. □

Theorem 5.1. *Assume $\mathcal{N}(\mathcal{T}) = \mathcal{B}(\mathsf{k}) \otimes \mathbb{1}_{\mathsf{m}}$ a type I factor, and let $e \in \mathsf{k}$ be a norm one eigenvector of K. Then the elements of $D(|e\rangle\langle e| \otimes \tau)$ are of the form*

$$|e\rangle\langle e| \otimes \omega$$

with ω a state on m.

Proof. It is clear that every state $|e\rangle\langle e| \otimes \omega$ belongs to $D(|e\rangle\langle e| \otimes \tau)$.

Vice-versa, let us consider an orthonormal basis $(e_n)_{n\geq 0}$ of k such that $K = \sum_{n\geq 0} \kappa_n |e_n\rangle\langle e_n|$, $\kappa_j \in \mathbb{R}$ and $e_0 = e$. Then if $\sigma = \sum_{i,j} |e_i\rangle\langle e_j| \otimes \sigma_{ij}$ is in $D(|e\rangle\langle e| \otimes \tau)$, with $\sigma_{ij} \in \mathfrak{I}(\mathsf{m})$, by Proposition 4.1 we have

$$w - \lim_t \mathcal{T}_{*t}(\sigma) = \sum_{\{i,j\,|\,\kappa_i = \kappa_j\}} \mathrm{tr}\,(\sigma_{ij})\, |e_i\rangle\langle e_j| \otimes \tau,$$

giving $\mathrm{tr}\,(\sigma_{ij}) = 0$ for all $i, j \neq 0$ and $\mathrm{tr}\,(\sigma_{00}) = 1$. Consequently we obtain $\sigma = |e\rangle\langle e| \otimes \sigma_{00}$ with σ_{00} a state on m. Indeed, by positivity of every σ_{ii}, we get $\sigma_{ii} = 0$ for all $i \neq 0$ so that $\sigma_{ij} = 0$ for all $(i,j) \neq (0,0)$ (see Proposition 5.1). □

We are now in a position to characterize the domain of attraction of an extremal invariant state.

Theorem 5.2. *Assume $\mathcal{N}(\mathcal{T})$ atomic. Given an arbitrary extremal invariant state $\eta_i = |e^i\rangle\langle e^i| \otimes \tau_i$ for some eigenvector e^i of K_i, $e^i \in \mathsf{k}_i$ and $i \in J$,*

$$D(\eta_i) = \{|e^i\rangle\langle e^i| \otimes w^i,\ w^i \in \mathfrak{S}(\mathsf{m}_i)\} = \{\mathrm{tr}_{\mathsf{m}_i}(\eta_i) \otimes w^i,\ w^i \in \mathfrak{S}(\mathsf{m}_i)\}.$$

Proof. Prove first equality, since the second one is trivial. Let σ be in $D(\eta_i)$. Proposition 5.1 immediately implies $p_h \sigma p_h = 0$ for all $h \neq i$. Therefore, by Lemma 4.1, we have $w-\lim_t \mathcal{T}_{*t}(\sigma) = w-\lim_t \mathcal{T}_{*t}(p_i \sigma p_i)$, i.e. $p_i \sigma p_i$ belongs to $D(\eta_i)$. We can then conclude thanks to Theorem 5.1.

The converse is clear. □

Now we can find the attractive domain of states given by convex combinations of extremal invariant states.

From Theorem 5.2 directly follows the characterization of $D(\rho)$ when ρ is a convex combination of extremal invariant states in the same factor of $\mathcal{N}(\mathcal{T})$.

Proposition 5.2. *Assume F to be a finite set, $(e^i_n)_{n\in F} \subseteq \mathsf{k}_i$ a family of orthonormal eigenvectors of K^i, and τ_i the unique (faithful) invariant state of $\mathcal{T}^{\mathsf{m}_i}$. Let*

$$\rho_i := \sum_{n\in F} \lambda_n \left(|e^i_n\rangle\langle e^i_n| \otimes \tau_i\right)$$

for some set $(\lambda_n)_{n\in F} \subseteq [0,1]$ such that $\sum_{n\in F} \lambda_n = 1$, and consider $\sigma \in \mathfrak{S}(\mathsf{h})$. Then the following conditions are equivalent:

(1) σ belongs to $D(\rho_i)$,
(2) σ is of the form

$$\sigma = \sum_{n\in F_i} \lambda_n (|e^i_n\rangle\langle e^i_n| \otimes \omega^i) = (\mathrm{tr}_{\mathsf{m}_i} \rho_i) \otimes \omega^i \tag{9}$$

for some $\omega^i \in \mathfrak{S}(\mathsf{m}_i)$.

Proof. It is enough to show that $\sum_{n\in F_i} \lambda_n |e^i_n\rangle\langle e^i_n| = \mathrm{tr}_{\mathsf{m}_i}(\rho_i)$. But this is clear since, by definition of partial trace, we have

$$\begin{aligned}\mathrm{tr}\left((\mathrm{tr}_{\mathsf{m}_i}(\rho_i))\, x\right) = \mathrm{tr}\left(\rho_i (x \otimes \mathbb{1}_{\mathsf{m}_i})\right) &= \mathrm{tr}\left(\left(\sum_{n\in F_i} \lambda_n |e^i_n\rangle\langle e^i_n| \otimes \tau_i\right)(x \otimes \mathbb{1}_{\mathsf{m}_i})\right)\\ &= \mathrm{tr}\left(\sum_{n\in F_i} \lambda_n |e^i_n\rangle\langle e^i_n| x\right)\end{aligned}$$

for all $x \in \mathcal{B}(\mathsf{k}_i)$. □

Finally, Proposition 5.1 allows us to extend the above result before to invariant states given as w^*-limit of convex combinations of arbitrary extremal states (not necessarily acting on the same factor).

Theorem 5.3. *For all $i \in I$ let F_i be a countable set, $(e_n^i)_{n\in F_i} \subseteq \mathsf{k}_i$ a family of orthonormal eigenvectors of K^i, and τ_i the unique (faithful) invariant state of $\mathcal{T}^{\mathsf{m}_i}$. Define*

$$\rho := \sum_{i\in I}\sum_{n\in F_i} \lambda_n^i \left(|e_n^i\rangle\langle e_n^i| \otimes \tau_i\right)$$

for some sequence $(\lambda_n^i)_{n\in F, i\in I} \subseteq [0,1]$ such that $\sum_{i\in I}\sum_{n\in F_i}\lambda_n^i = 1$. Then a state σ belongs to $D(\rho)$ if and only if

$$\sigma = \sum_{i\in I}\sum_{n\in F_i} \lambda_n^i(|e_n^i\rangle\langle e_n^i| \otimes \omega^i) + \sigma_0 \qquad (10)$$

for some $\omega^i \in \mathfrak{S}(\mathsf{m}_i)$ and $\sigma_0 \in \mathfrak{I}_0$.
In particular,

$$\sigma_0 = \sum_{i\in I}\sum_{\{n,m\in F_i, n\neq m\}} q_n^i \sigma q_m^i,$$

where $q_n^i = |e_n^i\rangle\langle e_n^i|$ for all $n \in F_i$ and for all $i \in I$.

Acknowledgement

FF and SE thank Professor Noboru Watanabe for hospitality and the enjoyable QP38 conference at Tokyo University of Science.
We would like to congratulate Professor Luigi Accardi and Igor Volovich for their remarkable career and scholarly influence; their work is a vivid source of inspiration for many researchers around the world.
We deeply regret that Professor Takeyuki Hida is no longer with us. He has been a guide and a great master for all of us by his scientific contributions to infinite dimensional analysis and quantum probability and his example.

References

1. L. Accardi and K. Imafuku, Control of quantum states by decoherence, in: *Quantum Information: Mathematical, Physical Engineering and Industrial Aspects* (Tokio, 2003), QP–PQ: Quantum Probab. White Noise Anal., Vol. 19 (World Scientific, 2005).
2. R. Carbone, E. Sasso and V. Umanità, On the asymptotic behavior of generic quantum Markov semigroups, *Infin. Dimens. Anal. Quantum Probab. Relat. Top.* **17**, No. 1, (2014) 1450001.

3. R. Carbone, E. Sasso and V. Umanità, Environment induced decoherence for markovian evolutions, *J. Math. Phys.* **56**, 092704 (2015).
4. J. Deschamps, F. Fagnola, E. Sasso, V. Umanità, Structure of Uniformly Continuous Quantum Markov Semigroups, *Rev. Math. Phys.* **28**, N. 1 (2016), 165003.
5. A. Dhahri, F. Fagnola and R. Rebolledo, The decoherence-free subalgebra of a quantum Markov semigroup with unbounded generator, *Infin. Dimens. Anal. Quantum Probab. Relat. Top.* **13** (2010) 413–433.
6. K. J. Engel, R. Nagel, *One-parameter semigroups for linear evolution equations*, Springer-Verlag, New York, 2000.
7. D.E. Evans, Irreducible quantum dynamical semigroups, *Commun. Math. Phys.* **54** (1977) 293–297.
8. F. Fagnola, E. Sasso, V. Umanità, Structure of Uniformly Continuous Quantum Markov Semigroups with Atomic Decoherence-free Subalgebra, *Open Syst. Inf. Dyn.* **24** (2017) 1740005.
9. F. Fagnola, E. Sasso, V. Umanità, The role of the atomic decoherence-free subalgebra in the study of Quantum Markov Semigroups, preprint 2018.
10. A. Frigerio, Stationary States of Quantum Dynamical Semigroups, *Commun. Math. Phys.* **63** (1978) 269–276.
11. A. Frigerio and M. Verri, Long-Time Asymptotic Properties of Dynamical Semigroups on W^*-algebras, *Math. Z.* **180** (1982), 275–286.
12. M. Hellmich, Quantum dynamical semigroups and decoherence, *Advances in Math. Phys.*, 625978 p. 16 (2011).
13. B. Kummerer, and R. Nagel, Mean ergodic semigroups on W^*-algebras, *Acta Sci. Math.* (Szeged) **41** (1979), no. 1-2, 151–159.
14. K. R. Parthasarathy, *An introduction to quantum stochastic calculus*, *Monographs in Mathematics* 85, Birkhäuser-Verlag, Basel 1992.
15. S. Wüster, Quantum Zeno suppression of dipole-dipole forces, arXiv:1611.07697v1.

Infinite Dimensional Analysis, Quantum Probability and Related Topics

pp. 59–76

BI-MONOTONE BROWNIAN MOTION

MALTE GERHOLD

Institut für Mathematik und Informatik,
Ernst Moritz Arndt Universität Greifswald,
Walther-Rathenau-Straße 47,
17487 Greifswald,
Germany
E-mail: mgerhold@uni-greifswald.de
Website: www.math-inf.uni-greifswald.de/index.php/mitarbeiter/282-malte-gerhold

We define bi-monotone independence, prove a bi-monotone central limit theorem and use it to study the distribution of bi-monotone Brownian motion, which is defined as the two-dimensional operator process with monotone and antimonotone Brownian motion as components.

Keywords: Noncommutative probability, bi-monotone independence, Lévy processes, central limit theorem, bi-monotone partitions.

1. Introduction

It is a key feature of noncommutative probability that there is not a unique notion of independence, but a number of different ones, each of which allows to a distinct theory with its own limit theorems and limit distributions, Lévy processes, de Finetti theorems, etc. Although weaker concepts have been considered, a full theory comparable to classical probability theory is only achieved with a *universal product independence*, i.e. independence for a family of noncommutative random variables is defined as factorization of the joint distribution with respect to a universal product construction for distributions of noncommutative random variables; this product construction replaces the tensor product of probability measures in the classical case. One possibility is to use the tensor product of states, and this leads to the notion of independence most frequently used in quantum mechanics. The most famous example of an alternative product is the free product of states which leads to the notion of *freeness*, and which has been studied extensively, exhibiting many connections to random matrix theory and the theory of operator algebras, cf. [9,21]. Based on work of Speicher [20] and

Ben Ghorbal and Schürmann [2], Muraki showed that besides the tensor product and the free product, there are only three more universal products of states, namely the Boolean, the monotone and the antimonotone product [17,18]. Extending Muraki's result to a purely algebraic setting, Gerhold and Lachs classified all universal products of linear functionals on associative algebras; these are all deformations of Muraki's five, the Boolean products admits a two-parameter deformation, while the others admit a one-parameter deformation [6]. Results on cummulants and Lévy processes for general universal product independences can be found in [4,14,7].

In 2014, Voiculescu introduced the notion of bi-freeness in a series of papers [22,23,24]. The main difference between bi-freeness and the universal product independences mentioned above is that Voiculescu does not consider a notion of independence for single random variables, but for pairs of random variables, or *two-faced* random variables. This corresponds to considering only such algebras which are structured as a free product of subalgebras. Since universality only has to hold for algebra homomorphisms which respect the two-faced structure, the possibility arises to find universal products of states besides Muraki's five. Bi-freeness is built on the fact that there are two distinct natural realizations of the GNS-representation of a free product state, a left and a right one. The monotone product was introduced by Muraki [15,16] and Lu [12]. In contrast to the free case, the monotone product is not symmetric, i.e. the monotone product $\triangleright$ and its opposite, the anti-monotone product $\varphi_1 \triangleleft \varphi_2 := \varphi_2 \triangleright \varphi_1$, do not coincide. There is a "left" and a "right" product representation, which corresponds to the monotone product and the anti-monotone product, respectively. In this paper we will introduce a new universal product for states on two-faced $*$-algebras, the *bi-monotone product*, and study its associated notion of independence for two-faced random variables (Section 3), construct a bi-monotone Brownian motion on monotone Fock space (Section 4), introduce bi-monotone partitions (Section 5), and use these to give a combinatorial description of the distribution of bi-monotone Brownian motion via a bi-monotone central limit theorem (Sections 6 and 7).

During the work on this manuscript, bi-monotone independence was independently discovered by Gu, Hasebe and Skoufranis, where it is called bi-monotone independence of type II [5]. Their main interest, however, lies in bi-monotone independence of type I which is derived from conditionally bi-free independence introduced by Gu and Skoufranis[8]. Recently, Liu constructed a (Free-[11])Free-Boolean[10] independence.

2. Preliminaries and notation

For any natural number n, we denote by $[n]$ the set $\{1,\ldots,n\}$. If the set $[n]$ appears as an argument inside usual brackets, we omit the inner square bracket. By an *algebra*, we always mean a not necessarily unital, complex, associative algebra. The free product of algebras $A_1,\ldots,A_n$ is denoted by $A_1 \sqcup \cdots \sqcup A_n$. We write $A = A^{(1)} \sqcup \cdots \sqcup A^{(n)}$ if $A^{(1)},\ldots,A^{(n)}$ are all subalgebras of A and the canonical homomorphism $A^{(1)} \sqcup \cdots \sqcup A^{(n)} \to A$ (i.e. the free product of the embeddings $\iota^{(i)}\colon A^{(i)} \hookrightarrow A$) is an isomorphism. An algebra A with subalgebras $A^{(1)},\ldots,A^{(n)}$ such that $A = A^{(1)}\sqcup\cdots\sqcup A^{(n)}$ is called an *n-faced algebra*. Let B_1, B_2 be n-faced algebras. Then the free product $B_1 \sqcup B_2$ is again an n-faced algebra with respect to the subalgebras $B^{(i)} := B_1^{(i)} \sqcup B_2^{(i)}$. Recall that an *augmented algebra* is a unital algebra A with a character, i.e. a non-zero homomorphism to $\mathbb{C}$, whose kernel is called *augmentation ideal* (cf. for example [13]). For any algebra B, its the unitization $\widetilde{B}$ is an augmented algebra with augmentation ideal B and, conversely, every augmented algebra is isomorphic to the unitization of its augmentation ideal. Therefore, we will always denote an augmented algebra as $\widetilde{A}$ where A denotes its augmentation ideal. We say that $\widetilde{A}$ is an *augmented n-faced algebra* if it is the unitization of an n-faced algebra A. In this case $\widetilde{A} = (A^{(1)}\sqcup\cdots\sqcup A^{(n)})^{\sim} = \widetilde{A}^{(1)}\sqcup_1\cdots\sqcup_1 \widetilde{A}^{(n)}$, where $\sqcup_1$ denotes the free product of unital algebras. If A, B are $*$-algebras, then we consider $\widetilde{A}$ and $A \sqcup B$ as $*$-algebras in the obvious way.

A *noncommutative probability space* is a pair (A, Φ), where A is a unital $*$-algebra and Φ is a state on A. Let B be an n-faced $*$-algebra and (A, Φ) a noncommutative probability space. A $*$-homomorphism $j\colon B \to A$ is called *n-faced random variable.* We call j augmented if B is an augmented algebra and j is unital, that is if $B = \widetilde{B'}$ and $j = \widetilde{j \upharpoonright B'}$. Just as for algebras, we will write augmented random variables as $\widetilde{\jmath}$ from the start with j its restriction to the augmentation ideal. An element of $(b_1,\ldots,b_n) \in A^n$ gives rise to the $*$-algebra homomorphism $j_{b_1,\ldots,b_n}\colon * - \mathrm{alg}(b_1) \sqcup \cdots \sqcup * - \mathrm{alg}(b_n) \to A$ determined by $j_{b_1,\ldots,b_n}(b_i) = b_i$. In this sense, we will consider elements of A^n as n-faced random variables. For an n faced random variable $b = (b_1,\ldots b_n) \in A^n$ and an m tuple $\delta \in [n]^{[m]}$ we define $b^\delta := b_{\delta_1}\cdots b_{\delta_m}$. The numbers $\Phi(b^\delta)$ are called *moments* of b and the collection of all moments is called *distribution* of b.

In this paper we will only consider two-faced algebras. We write a two-faced algebra as $A = A^{\updownarrow} \sqcup A^{\nabla}$. $A^{\updownarrow}$ is called *left face* and A^{∇} is called *right face* of A.

A *pointed representation* of an algebra A consists of a pre-Hilbert space H, an algebra homomorphism $\pi\colon A \to L(H)$ and a unit vector $\Omega \in H$. For every pointed representation π we can define a linear functional $\varphi_\pi(a) := \langle \Omega, \pi(a)\Omega\rangle$. If π is a unital $*$-representation, then φ_π is a state. On the other hand, for any state Φ on A, the GNS-representation yields a pointed unital $*$-representation. Given a pointed representation on H with unit vector Ω, we denote by P the projection onto $\mathbb{C}\Omega$ and by id the identity operator on H.

3. Bi-monotone independence

We define the bi-monotone product first for pointed representations of two-faced algebras and afterwards for states on augmented two-faced $*$-algebras.

Definition 3.1. Let A_1, A_2 be two-faced algebras and π_1, π_2 pointed representations of A_1, A_2 on pre-Hilbert spaces H_1, H_2 respectively. Then we define the pointed representation $\pi_1 \bowtie \pi_2$ of $A_1 \sqcup A_2$ on $H_1 \otimes H_2$ with unit vector $\Omega := \Omega_1 \otimes \Omega_2$ by

$$\pi_1 \bowtie \pi_2(a) := \begin{cases} \pi_1(a) \otimes id \; a \in A_1^\ell \\ \pi_1(a) \otimes P \; a \in A_1^r \\ id \otimes \pi_2(a) \; a \in A_2^r \\ P \otimes \pi_2(a) \; a \in A_2^\ell \end{cases}$$

Now suppose that φ_1, φ_2 are states on augmented two-faced $*$-algebras $\widetilde{A}_1, \widetilde{A}_2$ and π_i are pointed $*$-representations of A_i such that $\varphi_i(a) = \langle \Omega_i, \pi_i(a)\Omega_i\rangle$ for all $a \in A_i$. Then we put

$$\varphi_1 \bowtie \varphi_2(a) := \langle \Omega, \pi_1 \bowtie \pi_2(a)\Omega\rangle$$

for all $a \in A_1 \sqcup A_2$ and $\varphi(1) = 1$.

Note that the definition is independent of the choice of the pointed representations, so we can always take $\widetilde{\pi}_i$ to be the GNS-representation of φ_i. By construction, $\varphi_1 \bowtie \varphi_2$ is a state, because $\pi := \widetilde{\pi_1 \bowtie \pi_2}$ is a pointed $*$-representation with $\varphi_1 \bowtie \varphi_2(a) = \langle \Omega, \widetilde{\pi}(a)\Omega\rangle$ for all $a \in \widetilde{A_1 \sqcup A_2}$. But in general $(\pi_1 \bowtie \pi_2)^\sim$ will not be the GNS-representation of $\varphi_1 \bowtie \varphi_2$, because Ω is not necessarily cyclic.

It is easy to check that the bi-monotone product is associative. Indeed

$$\pi_1 \bowtie \cdots \bowtie \pi_n(a) = \begin{cases} P^{\otimes k-1} \otimes \pi_k(a) \otimes id^{\otimes n-k} \; a \in A_k^\ell \\ id^{\otimes k-1} \otimes \pi_k(a) \otimes P^{\otimes n-k} \; a \in A_k^r \end{cases}$$

holds independently of the way one insert parentheses. Associativity on the level of pointed representations clearly implies associativity on the level of states.

Remark 3.1. The bi-monotone product can also be defined for arbitrary linear functionals on algebras. Instead of pointed representations on pre-Hilbert spaces, one can use pointed representations on vector spaces $\widetilde{V}$ with a fixed decomposition $\widetilde{V} = \mathbb{C}\Omega \oplus V$. Then every linear functional admits a pointed representation such that $\Phi(a)\Omega = P\pi(a)\Omega$, where P is the projection onto $\mathbb{C}\Omega$. If $\widetilde{V}_i = \mathbb{C}\Omega_i \oplus V_i$, then $\widetilde{V}_1 \otimes \widetilde{V}_2 = \mathbb{C}(\Omega_1 \otimes \Omega_2) \oplus (\Omega_1 \otimes V_2) \oplus (V_1 \otimes \Omega_2) \oplus (V_1 \otimes V_2)$ yields a direct sum decomposition of $\widetilde{V}_1 \otimes \widetilde{V}_2$. Given linear functionals φ_i on algebras A_i, we can find pointed representations π_1, π_2 as above and define

$$\varphi_1 \breve{\bowtie} \varphi_2(a) := \langle \Omega, \pi_1 \bowtie \pi_2(a)\Omega \rangle$$

for all $a \in A_1 \sqcup A_2$.

Theorem 3.1. *The bi-monotone product of linear functionals is a positive $(1,2)$-u.a.u. product in the sense of* [14], *i.e. for all linear functionals φ_i on two-faced algebras A_i and all two-faced algebra homomorphisms $j_i \colon B_i \to A_i$ it holds that*

- $(\varphi_1 \breve{\bowtie} \varphi_2) \circ \iota_1 = \varphi_1$, $(\varphi_1 \breve{\bowtie} \varphi_2) \circ \iota_2 = \varphi_2$
- $(\varphi_1 \breve{\bowtie} \varphi_2) \bowtie \varphi_3 = \varphi_1 \bowtie (\varphi_2 \bowtie \varphi_3)$
- $(\varphi_1 \breve{\bowtie} \varphi_2) \circ (j_1 \underline{\sqcup} j_2) = (\varphi_1 \circ j_1) \breve{\bowtie} (\varphi_2 \circ j_2)$
- $\widetilde{\varphi_1 \breve{\bowtie} \varphi_2}$ *is a state on $\widetilde{A}_1 \sqcup_1 \widetilde{A}_2$ whenever $\widetilde{\varphi}_1, \widetilde{\varphi}_2$ are states on augmented $*$-algebras $\widetilde{A}_1, \widetilde{A}_2$ respectively.*

where $\iota_i \colon A_i \hookrightarrow A_1 \sqcup A_2$ is the canonical embedding and $j_1 \underline{\sqcup} j_2 := (\iota_1 \circ j_1) \sqcup (\iota_2 \circ j_2)$.

Proof. Straightforward. Note that $\widetilde{\varphi_1 \breve{\bowtie} \varphi_2} = \widetilde{\varphi}_1 \bowtie \widetilde{\varphi}_2$. □

Since we are in the realm of universal products, there are associated notions of independence and Lévy processes for the bi-monotone product, cf. [14] and [7].

Definition 3.2. Let $j_1, \ldots, j_n$ be two-faced random variables. We call $j_1, \ldots, j_n$ *bi-monotonely independent* if

$$\Phi \circ \widetilde{\bigsqcup_i j_i} = (\Phi \circ \widetilde{j}_1) \bowtie \cdots \bowtie (\Phi \circ \widetilde{j}_n). \tag{1}$$

Pairs of $*$-subalgebras are called bi-monotonely independent if their embedding homomorphisms are. Pairs of elements are called bi-monotonely independent if their generated $*$-algebras are.

Lemma 3.1. *Let $j_i\colon B_i \to A$ be two-faced random variables over a non-commutative probability space (A,Φ) with distributions $\varphi_i := \Phi \circ \widetilde{j}_i$. We denote by π_i, Π pointed representations on pre-Hilbert spaces H_i, K with unit vectors Ω_i, Ω respectively such that*

$$\Phi(a) = \langle \Omega, \Pi(a)\Omega \rangle, \quad \varphi_i(b) = \langle \Omega_i, \pi_i(b)\Omega_i \rangle$$

for all $a \in A$, $b \in B_i$, $i \in \{1,\dots,n\}$. Assume that there exists an isometry $i\colon H_1 \otimes \cdots \otimes H_n \to K$ such that

- $i(\Omega_1 \otimes \cdots \otimes \Omega_n) = \Omega$
- $i \circ (\Join_i \pi_i(b)) = \Pi(j_k(b)) \circ i$ *for all* $b \in B_k$, $k \in \{1,\dots,n\}$

Then $j_1,\dots,j_n$ are bi-monotonely independent.

Proof. Since both sides of (1) are states, it is enough to prove equality for elements $a_1 \cdots a_m \in \bigsqcup_i B_i$. Suppose $a_i \in B_{\varepsilon_i}$ for $i \in 1,\dots,n$ with $\varepsilon_1,\dots,\varepsilon_n \in [2]$, $\varepsilon_i \neq \varepsilon_{i+1}$ for $i = 1,\dots,n-1$. In this case we have

$$\begin{aligned}
&\Phi \circ (j_1 \sqcup \cdots \sqcup j_n)(a_1 \cdots a_m)\\
&= \langle \Omega, \Pi\left(\bigsqcup_i j_i(a_1 \cdots a_m)\right)\Omega\rangle\\
&= \langle i\Omega_1 \otimes \cdots \otimes \Omega_n, \Pi(j_{\varepsilon_1}(a_1)) \cdots \Pi(j_{\varepsilon_m}(a_m)) i\Omega_1 \otimes \cdots \otimes \Omega_n\rangle\\
&= \langle \Omega_1 \otimes \cdots \otimes \Omega_n, i^* i\,(\Join_i \pi_i(a_1)) \cdots (\Join_i \pi_i(a_m))\,\Omega_1 \otimes \cdots \otimes \Omega_n\rangle\\
&= \langle \Omega_1 \otimes \cdots \otimes \Omega_n, (\Join_i \pi_i(a_1 \cdots a_m))\,\Omega_1 \otimes \cdots \otimes \Omega_n\rangle\\
&= \varphi_1 \Join \cdots \Join \varphi_n(a_1 \cdots a_m)
\end{aligned}$$

□

4. Bi-monotone Brownian motion

For a set M, put $M^* := \bigcup_{n\in\mathbb{N}_0} M^n$. In particular, $\mathbb{R}^* := \bigcup_{n\in\mathbb{N}_0} \mathbb{R}^n$. We view $\mathbb{R}^*$ as a measure space with the unique measure whose restriction to $\mathbb{R}^n$ is n-dimensional Lebesgue measure. Restriction of this measure to the subset $\Delta := \{(t_1,\dots,t_n) \in \mathbb{R}^* \mid t_1 < \cdots < t_n\}$ makes Δ a measure space. We will also use the subspaces $\Delta_{[s,t]} := \Delta \cap [s,t]^*$ for the interval $[s,t]$, $s < t$.

Definition 4.1. Put $\Gamma := L^2(\Delta)$ and denote by Ω the function with value 1 on the empty tuple $\Lambda := () \in \mathbb{R}^0$ and which vanishes on all $(t_1,\dots,t_n)$

with $n > 0$. We call Γ *monotone Fock space.* The unit vector Ω induces the vector state φ with $\varphi(a) = \langle\Omega, a\Omega\rangle$ on $\mathcal{A} := \mathcal{B}(\Gamma)$ called the *vacuum state.* We also define the monotone Fockspaces over intervals as $\Gamma_{[s,t]} := L^2(\Delta_{[s,t]})$ and view them as subspaces of Γ with respect to the obvious embedding.

We define the following bounded operators on Γ:

Definition 4.2. For every $f \in L^2(\mathbb{R})$, we define the *left creation operator* $\lambda^*(f)$, *left annihilation operator* $\lambda(f)$, *right creation operator* $\rho^*(f)$, and *right annihilation operator* $\rho(f)$ by

$$\begin{aligned}
\lambda^*(f)(g)(t_1,\ldots,t_n) &:= f(t_1)g(t_2,\ldots,t_n), \quad (n>0); \qquad \lambda^*(f)(g)(\Lambda) := 0\\
\lambda(f)(g)(t_1,\ldots,t_n) &:= \textstyle\int_{-\infty}^{t_1} \overline{f(\tau)}g(\tau,t_1,\ldots,t_n)\mathrm{d}\tau\\
\rho^*(f)(g)(t_1,\ldots,t_n) &:= g(t_1,\ldots,t_{n-1})f(t_n), \quad (n>0); \quad \rho^*(f)(g)(\Lambda) := 0\\
\rho(f)(g)(t_1,\ldots,t_n) &:= \textstyle\int_{t_n}^{\infty} \overline{f(\tau)}g(t_1,\ldots,t_n,\tau)\mathrm{d}\tau
\end{aligned}$$

respectively.

All these operators belong to $\mathcal{B}(\Gamma)$. It is easy to check that $\lambda^*(f) = (\lambda(f))^*$ and $\rho^*(f) = (\rho(f))^*$. We abbreviate $\lambda^*(1_{[s,t]})$ as $\lambda^*_{s,t}$ and similarly for the other three families of operators. For every interval $[s,t]$ let $B_{[s,t]}$ denote the two-faced $*$-algebra with $B^{\updownarrow}_{s,t} = * - \mathrm{alg}(\lambda^*_{s,t})$ and $B^{\nabla}_{s,t} = * - \mathrm{alg}(\rho^*_{s,t})$

Theorem 4.1. *The two-faced $*$-subalgebras $B_{t_1,t_2}, B_{t_2,t_3}, \ldots, B_{t_{n-1},t_n}$ are bi-monotonely independent for all choices of $t_1 \leq \ldots \leq t_n$.*

Proof. There is a natural isometry $i\colon \Gamma_{[t_1,t_2]} \otimes \cdots \otimes \Gamma_{[t_{n-1},t_n]} \to \Gamma$ with

$$i(g_1 \otimes \cdots \otimes g_{n-1})(\mathbf{s}) = \begin{cases} g_1(\mathbf{s} \cap [t_1,t_2]) \cdots g_{n-1}(\mathbf{s} \cap [t_{n-1},t_n]) & \mathbf{s} \in \Delta_{[t_1,t_n]} \\ 0 & \textit{else}. \end{cases}$$

Obviously, $i(\Omega_{[t_1,t_2]} \otimes \cdots \otimes \Omega_{[t_{n-1},t_n]}) = \Omega$. Since $\Gamma_{[t_i,t_i+1]}$ is invariant under B_i, we can define a pointed representation $\pi_i\colon B_i \to \mathcal{B}(\Gamma_{[t_i,t_i+1]})$ by restriction. We shortly write $\lambda/\rho_i^{(*)}$ for $\pi_i(\lambda/\rho^{(*)}_{t_i,t_{i+1}})$. So, by Lemma 3.1 and the definition of the bi-monotone product of pointed representations, we are done if we can show

$$\begin{aligned}
\lambda^*_{t_i,t_i+1} \circ i &= i \circ P^{\otimes i-1} \otimes \lambda^*_i \otimes \mathrm{id}^{n-1-i}\\
\lambda_{t_i,t_i+1} \circ i &= i \circ P^{\otimes i-1} \otimes \lambda_i \otimes \mathrm{id}^{n-1-i}\\
\rho^*_{t_i,t_i+1} \circ i &= i \circ \mathrm{id}^{\otimes i-1} \otimes \rho^*_i \otimes P^{n-1-i}\\
\rho_{t_i,t_i+1} \circ i &= i \circ \mathrm{id}^{\otimes i-1} \otimes \rho_i \otimes P^{n-1-i}.
\end{aligned}$$

We check the first equality and leave the remaining ones to the reader, as the calculations are very similar. Suppose that $s_1 < \ldots < s_m$ and that $s_1, \ldots, s_k$ belong to $[t_1, t_i]$, $s_{k+1}, \ldots, s_r$ belong to $[t_i, t_{i+1}]$ and $s_{r+1}, \ldots, s_m$ belong to $[t_{i+1}, t_n]$. For all $g_k \in \Gamma_{[t_k, t_{k+1}]}$, $k \in \{1, \ldots, n-1\}$, we get, using the increasing order of $t_1, \ldots, t_n$ and $s_1, \ldots, s_m$,

$$
\begin{aligned}
&\lambda^*_{t_i,t_i+1}(i(g_1 \otimes \cdots \otimes g_{n-1}))(s_1, \ldots, s_m)\\
&= 1_{[t_i,t_{i+t}]}(s_1) \cdot (g_1 \otimes \cdots \otimes g_{i-1})(s_2, \ldots, s_k) \cdot\\
&\qquad g_i(s_{k+1}, \ldots, s_r) \cdot (g_{i+1} \otimes \cdots \otimes g_{n-1})(s_{r+1}, \ldots, s_m)\\
&= \begin{cases} 0 & k > 0\\ g_1(\Lambda) \cdots g_{i-1}(\Lambda)\, g_i(s_2, \cdots, s_r)\, (g_{i+1} \otimes \cdots g_{n-1})(s_{r+1}, \cdots, s_m) & k = 0, r > 0\\ 0 & k = 0, r = 0 \end{cases}\\
&= \left(P(g_1) \otimes \cdots \otimes P(g_{i-1}) \otimes \lambda^*_i(g_i) \otimes g_{i+1} \otimes \cdots \otimes g_{n-1}\right)(s_1, \ldots, s_m)\\
&= i \circ (P^{\otimes i-1} \otimes \lambda^*_i \otimes \mathrm{id}^{n-1-i})(g_1 \otimes \cdots \otimes g_{n-1})(s_1, \ldots, s_m).
\end{aligned}
$$

□

Put $b^{\updownarrow}_{s,t} := \lambda^*_{s,t} + \lambda_{s,t}$ and $b^{\nabla}_{s,t} := \rho^*_{s,t} + \rho_{s,t}$.

Theorem 4.2. *The* $b_{s,t} = (b^{\updownarrow}_{s,t}, b^{\nabla}_{s,t})$, $s, t \geq 0$ *form an additive Lévy process, i.e.*

- $b_{r,s} + b_{s,t} = b_{r,t}$ *for all* $r \leq s \leq t$ *and* $b_{0,0} = \mathrm{id}$ *(increment property)*
- $\Phi(b^{\delta}_{s,t}) = \Phi(b^{\delta}_{0,t-s})$ *for all* $s \leq t$ *and all* $\delta \in \{\updownarrow, \nabla\}^*$ *(stationarity of increments)*
- $b_{t_1,t_2}, b_{t_2,t_3}, \ldots, b_{t_{n-1},t_n}$ *are bi-monotonely independent for all* $0 \leq t_1 \leq \ldots \leq t_n$ *(independence of increments).*

Proof. The increment property is obvious, independence follows from the previous theorem. Stationarity follows from the fact that the canonical unitary $U_s \colon \Gamma_{s,t} \to \Gamma_{0,t-s}$ with $U_s(f)(t_1, \ldots, t_n) = f(t_1 + s, \ldots, t_n + s)$ fulfills $U_s(\Omega) = \Omega$ and $b_{s,t} = U^*_s b_{0,t-s} U_s$. □

5. Bi-monotone partitions

Definition 5.1. A *partition* of a set X is a set π of subsets of X, called *blocks* of π, such that

- $V \neq \emptyset$ for all blocks $V \in \pi$
- $V_1 \cap V_2 = \emptyset$ for all distinct blocks $V_1, V_2 \in \pi$
- $\bigcup_{V \in \pi} V = X$.

A partition π of X is called a *pair partition* if $\#V = 2$ for all blocks $V \in \pi$.

An *ordered partition* of a set X is a partition π of X with a total order $\leq$ between the blocks (i.e. a total order on the set π).

The special kinds of partitions and ordered partitions we shall define now are important for the study of noncommutative independences, such as freeness (noncrossing partitions), Boolean (interval partitions) and monotone independence (monotona partitions). They appear in moment-cumulant formulas, in the universal coefficients of the corresponding universal product and in the central limit theorems.

Definition 5.2. Let X be a totally ordered set.

A partition π of X is called

- *noncrossing* if $a, c \in V$ and $b, c \in W$ for elements $a < b < c < d$ of X and blocks $V, W \in \pi$ implies $V = W$,
- *interval partition* if $a, c \in V$, $b \in W$ for elements $a < b < c$ of X and blocks $V, W \in \pi$ implies $V = W$.

An ordered partition π of X is called

- *monotone partition* if $a, c \in V$, $b \in W$ for elements $a < b < c$ of X and blocks $V, W \in \pi$ implies $V \geq W$.

We denote the set of all partitions, noncrossing partitions, interval partitions and monotone partitions of an ordered set X by $\mathrm{P}_{\otimes}(X)$, $\mathrm{P}_{\star}(X)$, $\mathrm{P}_{\Diamond}(X)$, and $\mathrm{P}_{\triangleright}(X)$ respectively, according to their associated universal product. An (ordered) partitions is called a *pair partition* if all its blocks have cardinality 2. The set of all pair partitions of type $\odot$ is denoted by $\mathrm{PP}_{\odot}(X)$ for $\odot \in \{\otimes, \star, \Diamond, \triangleright\}$.

Partitions and ordered partitions can be nicely depicted in the following way:

- The elements of X are written on the bottom line of the diagram, in ascending order if X is ordered
- Each block is represented by a horizontal line with vertical "legs" connecting it down to the points of the block
- If π is not ordered, the heights are chosen such that there are no overlaps and as little crossings as possible.
- If π is ordered, the heights are chosen such that greater blocks have higher horizontal lines.

Noncrossing and monotone partitions are simply recognized as the ones that have no crossing lines in the corresponding diagram.

Definition 5.3. Let X be a set. A *two-faced partition* of X is a partition π of X together with a map $\delta\colon X \to \{\updownarrow, \nabla\}$. Similarly, an *ordered two-faced partition* of X is an ordered partition π of X together with a map $\delta\colon X \to \{\updownarrow, \nabla\}$.

The map δ is called *pattern* of the (ordered) two-faced partition. Frequently, we will consider δ as an element of $\{\updownarrow, \nabla\}^X$ and write δ_x instead of $\delta(x)$.

An (ordered) two-faced partition can be drawn like a partition, but the points of X are put on the bottom and on the top line and the vertical leg at $x \in X$ is drawn down to the bottom line if $\delta(x) = \nabla$ and up to the top line if $\delta(x) = \updownarrow$.

Definition 5.4. Let X be an ordered set.

A a two-faced partition of X is called

- *bi-noncrossing partition* if $a, c \in V$ and $b, c \in W$ for elements $a < b < c < d$ with $\delta(b) = \delta(c)$ of X and blocks $V, W \in \pi$ implies $V = W$. (cf. [3])

An ordered two-faced partition π of X is called

- *bi-monotone partition* if $a, c \in V$, $b \in W$ for elements $a < b < c$ of X and blocks $V, W \in \pi$ implies $V \geq W$ if $\delta(b) = \nabla$ and $V \leq W$ if $\delta(b) = \updownarrow$.

Again, the bi-noncrossing and bi-monotone partitions exactly correspond to those diagrams which have no crossing lines. The sets of bi-noncrossing partitions, bi-monotone partitions, bi-noncrossing pair partitions, and bi-monotone pair partitions of X with pattern $\delta\colon X \to \{\updownarrow, \nabla\}$ are denoted by $\mathrm{P}_\star(\delta)$, $\mathrm{P}_{\bowtie}(\delta)$, $\mathrm{PP}_\star(\delta)$, and $\mathrm{PP}_{\bowtie}(\delta)$ respectively.

Example 5.1. The two-faced partition (π, δ) of $\{1, \ldots, 6\}$ with

$$\pi = \{\{1,4\}, \{2,3\}, \{5,6\}\}, \quad \delta = (\nabla, \nabla, \nabla, \updownarrow, \updownarrow, \updownarrow)$$

is bi-noncrossing, as it can be drawn in a noncrossing way, see Figure 1.

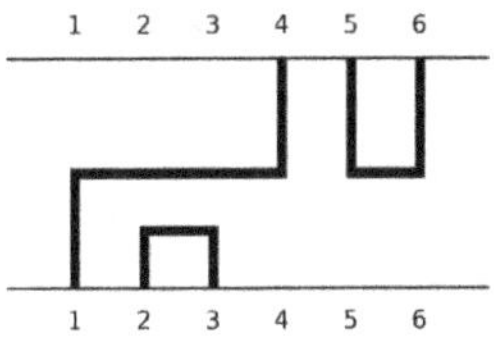

Fig. 1. Bi-noncrossing partition $\{\{1,4\},\{2,3\},\{5,6\}\}$ partition with $\delta = (\tau,\tau,\tau,\ell,\ell,\ell)$.

There are three possible orders between the blocks such that the resulting ordered two-faced partition is bi-monotone. Indeed, the block $\{1,4\}$ has to be smaller than the block $\{2,3\}$, whereas $\{5,6\}$ can be in any position. This is easily seen in the corresponding diagrams in Figure 2.

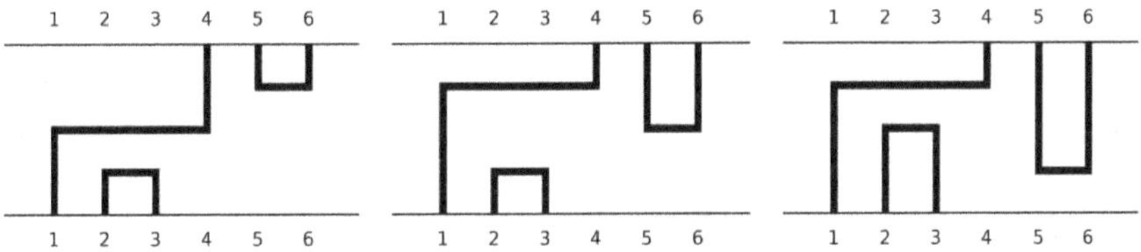

Fig. 2. Bi-monotone orderings of a bi-noncrossing partition.

Remark 5.1. Let X be an ordered $2n$-set. Then one can show that for every pattern $\delta\colon X \to \{\updownarrow, \nabla\}$, there are at most $(2n-1)!!$ bi-monotone pair partitions with pattern δ. A complete proof requires quite a bit of combinatorial background which is not needed for the central limit theorem, so we will present it in a second paper. Here we only sketch the key idea.

A partition π of an ordered set X is called *irreducible* if for all $a \neq \max X$ there is a block $V \in \pi$ with elements $b, c \in V$ such that $b \leq a < c$. For a set M of partitions of X we denote by $\mathrm{I}M$ the set of irreducible partitions in M. For monotone partitions (bi-monotone partitions with constant pattern), we know that $\#\mathrm{PP}(X) = (2n-1)!!$. Also note that the highest block of an irreducible monotone partition has to be $\{\min X, \max X\}$, so deleting the highest block yields a bijection between $\mathrm{IPP}_{\triangleright}(X)$ and $\mathrm{PP}_{\triangleright}('X')$, where $'X' := X \setminus \{\min X, \max X\}$.

For a general pattern $\delta\colon X \to \{\updownarrow, \nabla\}$, it is not difficult to see that

$$\#\mathrm{PP}_{\bowtie}(\delta) = \sum_{\substack{\delta = \delta_1 \smile \cdots \smile \delta_k \\ |\delta_i| \in 2\mathbb{N}}} \#\mathrm{IPP}_{\bowtie}(\delta_1) \cdots \#\mathrm{IPP}_{\bowtie}(\delta_k) \binom{|\delta|/2}{|\delta_1|/2, \ldots, |\delta_k|/2}.$$

On the right-hand side the induction hypothesis can be applied to replace all partitions with shorter patterns by ordinary monotone partitions. Finally, one has to deal with the summand $\mathrm{IPP}_{\bowtie}(\delta)$. The Dyck path description of noncrossing pair partitions (see e.g. [19]) also works for bi-noncrossing pair partitions with fixed pattern, and one can use it to find an injection $\mathrm{IPP}_{\bowtie}(\delta) \hookrightarrow \mathrm{PP}_{\bowtie}('\delta')$, where $'\delta' := \delta \restriction 'X'$. Then the induction hypothesis yields

$$\#\mathrm{IPP}_{\bowtie}(\delta) \leq \#\mathrm{PP}_{\bowtie}('\delta') \leq \#\mathrm{PP}_{\triangleright}('X') = \#\mathrm{IPP}_{\triangleright}(X).$$

6. Bi-monotone central limit theorem

Central limit theorem is based on the central limit theorem for singleton independence as given in [1]. For convenience of the reader, we recall the relevant part of the theorem.

Theorem 6.1. *Let J be a set and (A, Φ) a noncommutative probability space. Assume that $b^{(j)} = (b_n^{(j)})_{n=1}^{\infty}$, $j \in J$, sequences of elements in A such that*

- *each $b_n^{(j)}$ is centered, i.e. $\Phi(b_n^{(j)}) = 0$,*
- *the condition of boundedness of mixed moments is fulfilled, i.e. for each $k \in \mathbb{N}$ there exists a positive constant ν_k such that $\left|\Phi(b_{n_1}^{(j_1)} \cdots b_{n_k}^{(j_k)})\right| \leq \nu_k$ for any choice of $n_1, \ldots, n_k \in \mathbb{N}$ and $j_1, \ldots, j_k \in J$,*
- *the singleton condition is satisfied, i.e. for any choice of $k \in \mathbb{N}$, $j_1, \ldots, j_k \in J$ and $n_1, \ldots, n_k \in \mathbb{N}$*
$$\Phi(b_{n_1}^{(j_1)} \cdots b_{n_k}^{(j_k)}) = 0$$
holds whenever there exists an index n_s which is different from all other ones.

Write $S_N(b^{(j)}) := \sum_{n=1}^{N} b_n^{(j)}$. Then

$$\lim_{N\to\infty} \Phi\left(\frac{S_N(b^{(j_1)})}{N^{1/2}} \cdots \frac{S_N(b^{(j_{2n})})}{N^{1/2}}\right) \tag{2}$$

$$= \lim_{N\to\infty} N^{-n} \sum_{\substack{\pi:\, [2n]\to[n] \\ 2 \text{ to } 1}} \sum_{\substack{\sigma:\, [n]\to[N] \\ \text{order preserving}}} \Phi\left(b_{\sigma\circ\pi(1)}^{(j_1)} \cdots b_{\sigma\circ\pi(2n)}^{(j_{2n})}\right). \tag{3}$$

We will use it in the form of the following Corollary.

Corollary 6.1. *Let $b^{(j)} = (b_n^{(j)})_{n=1}^{\infty}$, $j \in \{\updownarrow, \nabla\}$, be sequences of two-faced random variables such that*

- *each* $b_n^{(j)}$ *is centered*
- *the singleton condition is fulfilled*
- *the distribution is* spreadable, *i.e.*

$$\Phi(b_{n_1}^{j_1}\cdots b_{n_k}^{j_k}) = \Phi(b_{\sigma(n_1)}^{j_1}\cdots b_{\sigma(n_k)}^{j_k})$$

for each order preserving $\sigma\colon [n]\to\mathbb{N}$.

Then

$$\lim_{N\to\infty}\Phi\left(\frac{S_N(b^{(j_1)})}{N^{1/2}}\cdots\frac{S_N(b^{(j_{2n})})}{N^{1/2}}\right) = \frac{1}{n!}\sum_{\substack{\pi\colon [2n]\to[n]\\ 2\ to\ 1}}\Phi\left(b_{\pi(1)}^{(j_1)}\cdots b_{\pi(2n)}^{(j_{2n})}\right).$$

Proof. Spreadability implies boundedness of mixed moments. There are exactly $\binom{N}{n}$ order preserving maps from $[n]$ to $[N]$, and by spreadability the value of $\Phi\left(b_{\sigma\circ\pi(1)}^{(j_1)}\cdots b_{\sigma\circ\pi(2n)}^{(j_{2n})}\right)$ is independent of σ. Finally, note that

$$\lim_{N\to\infty}N^{-n}\binom{N}{n} = \frac{1}{n!}\lim_{N\to\infty}\frac{N(N-1)\cdots(N-n+1)}{N^n} = \frac{1}{n!}.$$

$\square$

Lemma 6.1. *Let* $(B_1^{\updownarrow}, B_1^{\nabla}),\ldots,(B_n^{\updownarrow}, B_n^{\nabla})$ *be bi-monotonely independent pairs of subalgebras. Let* $b_i\in B_{\varepsilon_i}^{\delta_i}$, $i\in\{1,\ldots,m\}$, *with* $\varepsilon_i\in\{1,\ldots,n\}$ *and* $\delta_i\in\{\nabla,\updownarrow\}$. *If the two-faced partition* (π,δ) *with blocks* $V_i := \{k\mid \varepsilon_k = i\}$, $V_1<\ldots<V_n$, *and pattern* δ *is bi-monotone, then*

$$\Phi(b_1\cdots b_m) = \Phi\left(\overrightarrow{\prod_{\varepsilon_k=1}} b_k\right)\cdots\Phi\left(\overrightarrow{\prod_{\varepsilon_k=n}} b_k\right).$$

Proof. Let π_i be pointed representations with $\Phi(b) = \langle\Omega_i, \pi_i(b)\Omega_i\rangle$ for all $b\in B_i$, $i\in[n]$. On the representation space of each π_i, we denote by P the projection onto $\mathbb{C}\Omega_i$ and by id the identity operator. For $b\in B_\varepsilon^\delta$ with $\varepsilon\in[n]$, $\delta\in\{\updownarrow, r\}$, we write $\bowtie_i\pi_i(b)$ as $T_1(b)\otimes\cdots\otimes T_n(b)$ with

$$T_i(b) = \begin{cases} P & \text{if } (i<\varepsilon \text{ and } \delta=\ell) \text{ or } (i>\varepsilon \text{ and } \delta = \tau)\\ id & \text{if } (i<\varepsilon \text{ and } \delta=\tau) \text{ or } (i>\varepsilon \text{ and } \delta = \ell)\\ \pi_i(b) & \text{if } i=\varepsilon\end{cases}$$

With this notation

$$\begin{aligned}\bowtie_i\pi_i(b_1\cdots b_n) &= (\bowtie_i\pi_i(b_1))\cdots(\bowtie_i\pi_i(b_m))\\ &= (T_1(b_1)\cdots T_1(b_n))\otimes\cdots\otimes(T_n(b_1)\cdots T_n(b_n)).\end{aligned}$$

One checks that the fact that (π, δ) is bi-monotone implies that $T_i(b_j) = \mathrm{id}$ whenever there are μ, ν with $\mu < j < \nu$, $\varepsilon_\mu = \varepsilon_\nu = i$ and $\varepsilon_j \neq i$. Then it follows easily that

$$\langle \Omega_i, T_i(b_1) \cdots T_i(b_n)\Omega_i \rangle = \Phi\left(\overrightarrow{\prod_{\varepsilon_k = i}} b_k \right)$$

and therefore

$$\begin{aligned}
\Phi(b_1 \cdots b_m) &= \langle \Omega_1 \otimes \cdots \otimes \Omega_n, \bowtie_i \pi_i(b_1 \cdots b_n)\Omega_1 \otimes \cdots \otimes \Omega_n \rangle \\
&= \langle \Omega_1, T_1(b_1) \cdots T_1(b_n)\Omega_i \rangle \cdots \langle \Omega_n, T_n(b_1) \cdots T_n(b_n)\Omega_i \rangle \\
&= \Phi\left(\overrightarrow{\prod_{\varepsilon_k = 1}} b_k \right) \cdots \Phi\left(\overrightarrow{\prod_{\varepsilon_k = n}} b_k \right)
\end{aligned}$$

as claimed. □

Lemma 6.2. *Let $(b_1^{(\updownarrow)}, b_1^{(\nabla)}), \ldots, (b_n^{(\updownarrow)}, b_n^{(\nabla)}) \in A \times A$ be bi-monotonely independent and centered. For an ordered two-faced pair partition (π, δ) with blocks $V_i := \{k \mid \varepsilon_k = i\}$, $V_1 < \ldots < V_n$, and pattern $\delta = (\delta_1, \ldots, \delta_{2n})$, it holds that*

$$\Phi(b_\pi) = \Phi(b_{\varepsilon_1}^{(\delta_1)} \cdots b_{\varepsilon_{2n}}^{(\delta_{2n})}) = 0$$

whenever (π, δ) is not bi-monotone.

Proof. If (π, δ) is not bi-monotone, there are $\mu < j < \nu$ and $i \in [n]$ with $\varepsilon_\mu = \varepsilon_\nu = i$ and either

(1) $\delta_j = \updownarrow$ and $\varepsilon_j > i$ or
(2) $\delta_j = \nabla$ and $\varepsilon_j < i$.

In either case, using the notation of the previous Lemma,

$$\begin{aligned}
T_i(b_{\varepsilon_\mu}^{(\delta_\mu)}) &= \pi_i(b_{\varepsilon_\mu}^{(\delta_\mu)}) \\
T_i(b_{\varepsilon_j}^{(\delta_j)}) &= P \\
T_i(b_{\varepsilon_\nu}^{(\delta_\nu)}) &= \pi_i(b_{\varepsilon_\nu}^{(\delta_\nu)}),
\end{aligned}$$

so

$$\langle \Omega_i, T_i(b_{\varepsilon_1}^{(\delta_1)}) \cdots T_i(b_{\varepsilon_{2n}}^{(\delta_{2n})})\Omega_i \rangle = \Phi(b_i^{(\delta_\mu)})\Phi(b_i^{(\delta_\nu)}) = 0$$

□

Theorem 6.2. *Let $b^{(j)} = (b_n^{(j)})_{n=1}^\infty$, $j \in \{\updownarrow, \nabla\}$, be sequences of elements in a quantum probability space such that*

- *each $b_n^{(j)}$ is centered*
- *the sequence of pairs $(b_n^{(\updownarrow)}, b_n^{(\nabla)})_{n\in\mathbb{N}}$ is bi-monotonely independent*
- *the pairs $(b_n^{(\updownarrow)}, b_n^{(\nabla)})$ have the same distribution for all $n \in \mathbb{N}$.*

Put $c_{p,q} := \Phi(b_n^{(p)} b_n^{(q)})$ for $p, q \in \{\updownarrow, \nabla\}$. Then for every pattern $\delta = (\delta_1, \ldots, \delta_{2n})$ we have

$$\lim_{N\to\infty} \Phi\left(\frac{S_N(b^{(\delta_1)})}{N^{1/2}} \cdots \frac{S_N(b^{(\delta_{2n})})}{N^{1/2}}\right) = \frac{1}{n!} \sum_{\pi\in \mathrm{PP}_{\bowtie}(\delta)} \prod_{\{k<l\}\in\pi} c_{\delta_k,\delta_l}.$$

Proof. The singleton condition and spreadability of the distribution follow from bi-monotone independence. Therefore we can apply Corollary 6.1 to get

$$\lim_{N\to\infty} \Phi\left(\frac{S_N(b^{(\delta_1)})}{N^{1/2}} \cdots \frac{S_N(b^{(\delta_{2n})})}{N^{1/2}}\right) = \frac{1}{n!} \sum_{\substack{\pi:\ [2n]\to[n] \\ 2\text{ to }1}} \Phi\left(b_{\pi(1)}^{(\delta_1)} \cdots b_{\pi(2n)}^{(\delta_{2n})}\right).$$

With every 2-to-1 map $\pi\colon [2n] \to [n]$ we associate the ordered pair partition $\hat{\pi}$ with blocks $V_i := \pi^{-1}(i)$, $V_1 < \ldots < V_n$ and the two-faced ordered pair partition $(\hat{\pi}, \delta)$. Thus, we can combine Lemma 6.1 and Lemma 6.2 to get

$$\Phi\left(b_{\pi(1)}^{(\delta_1)} \cdots b_{\pi(2n)}^{(\delta_{2n})}\right) = \begin{cases} 0 & \text{if } (\widehat{\pi}, \delta) \text{ is not bi-monotone} \\ \prod\limits_{\{k<\ell\}\in\pi} & \text{if } (\widehat{\pi}, \delta) \text{ is bi-monotone} \end{cases}$$

and the proof is finished. □

7. Distribution of bi-monotone Brownian motion

Next, we will apply the bi-monotone central limit theorem in order to determine the distribution of bi-monotone Brownian motion.

Theorem 7.1. *Put $b_{s,t}^{(\updownarrow)} := \lambda_{s,t}^* + \lambda_{s,t}$ and $b_{s,t}^{(\nabla)} := \rho_{s,t}^* + \rho_{s,t}$. Then*

$$\langle \Omega, b_{0,1}^{(\delta_1)} \cdots b_{0,1}^{(\delta_n)} \Omega\rangle = \frac{1}{n!} \#\mathrm{PP}_{\bowtie}(\delta).$$

Proof. The proof is based on two simple observations:

- For any $N \in \mathbb{N}$, we can write $b_{0,1}^{(j)} = b_{0,\frac{1}{N}}^{(j)} + b_{\frac{1}{N},\frac{2}{N}}^{(j)} + \cdots + b_{\frac{N-1}{N},1}^{(j)}$.
- The two families

$$\left(b_{\frac{i}{n},\frac{i+1}{n}}^{(j)}\right)_{\substack{i\in\mathbb{N} \\ j\in\{\updownarrow,\nabla\}}} \quad \text{and} \quad \left(\frac{b_{i,i+1}^{(j)}}{\sqrt{n}}\right)_{\substack{i\in\mathbb{N} \\ j\in\{\updownarrow,\nabla\}}}$$

have the same distribution, i.e.

$$\Phi\left(b^{(j_1)}_{\frac{i_1}{n},\frac{i_1+1}{n}}\cdots b^{(j_k)}_{\frac{i_k}{n},\frac{i_k+1}{n}}\right) = N^{-\frac{k}{2}}\Phi\left(b^{j_1}_{i_1,i_1+1}\cdots b^{j_k}_{i_k,i_k+1}\right)$$

for all $i_1,\ldots,i_k \in \mathbb{N}$, $j_1,\ldots,j_k \in \{\updownarrow,\nabla\}$. Now put $b_i^{(j)} := b_{i,i+1}^{(j)}$. The sequences $(b_i^{(j)})$ consist of centered bounded operators and are bi-monotonely independent. Therefore, we can apply the bi-monotone central limit theorem. Since $c_{p,q} := \Phi(b_i^{(p)}b_i^{(q)}) = 1$ for all $i \in \mathbb{N}$ and all $p,q \in \{\updownarrow,\nabla\}$, the stated formula follows. □

Corollary 7.1. *Put* $b_{s,t} := \lambda^*_{s,t} + \lambda_{s,t} + \rho^*_{s,t} + \rho_{s,t}$. *Then*

$$\langle\Omega, b_{0,1}^n\Omega\rangle = \frac{1}{n!}\#\mathrm{PP}_{\bowtie}(n),$$

where $\mathrm{PP}_{\bowtie}(n)$ *is the set of all bi-monotone pair partitions of* $[n]$ *with arbitrary pattern.*

Proof. Since $b_{s,t} = b_{s,t}^{(\updownarrow)} + b_{s,t}^{(\nabla)}$, this follows directly from the previous theorem and

$$b_{s,t}^n = (b_{s,t}^{(\updownarrow)} + b_{s,t}^{(\nabla)})^n = \sum_{\delta\in\{\updownarrow,\nabla\}^n} b_{s,t}^{(\delta_1)}\cdots b_{s,t}^{(\delta_n)}.$$

□

This corollary allows us to calculate the number of bi-monotone pair partitions. The number of bi-monotone partitions of a $2n$-set for $n = 0,1,\ldots,10$ are:
1, 4, 48, 928, 24448, 811776, 32460032, 1516774912, 81064953344, 4876115246080, 325959895390976.
We do not know any explicit or recursive formula for these numbers.

Since $b_{0,1}$ is a bounded selfadjoint operator on a Hilbert space, its moments are the moments of a uniquely determined compactly supported probability measure on $\mathbb{R}$. It would be nice to have an explicit formula of this probability measure.

Acknowledgements

The author thanks Prof. Michael Schürmann for several remarks which helped to improve the readability of this paper. He also thanks Volkmar Liebscher for some enlightening discussions. The author acknowledges support by DAAD and MAEDI/MENESR through the PROCOPE programme and by the German Research Foundation (DFG) through the Research Grant "Non-Commutative Stochastic Independence: Algebraic and Analytic Aspects".

References

1. Luigi Accardi, Yukihiro Hashimoto, and Nobuaki Obata, Notions of independence related to the free group, In n. Dimens. Anal. Quantum Probab. Relat. Top. 1 (1998), no. 2, 201–220. MR1628248.
2. Anis Ben Ghorbal and Michael Schurmann, Non-commutative notions of stochastic independence, Math. Proc. Cambridge Philos. Soc. 133 (2002), no. 3, 531–561. MR 1919720 (2003k:46096).
3. Ian Charlesworth, Brent Nelson, and Paul Skoufranis, On two-faced families of non-commutative random variables, Canad. J. Math. 67 (2015), no. 6, 1290–1325. MR 3415654.
4. Uwe Franz, Lévy processes on quantum groups and dual groups, Quantum independent increment processes. II, Lecture Notes in Math., vol. 1866, Springer, Berlin, 2006, pp. 161–257. MR 2213449 (2007g:81063).
5. Yinzheng Gu, Takahiro Hasebe, and Paul Skoufranis, Bi-monotonic independence for pairs of algebras, Preprint, arXiv: 1708.05334, 2017.
6. Malte Gerhold and Stephanie Lachs, Classification and GNS-construction for general universal products, Infin. Dimens. Anal. Quantum Probab. Relat. Top. 18 (2015), no. 1, 1550004, 29. MR3324720 June 30, 2018 12:34 WSPC Proceedings - 9in x 6in BBM page 1818.
7. Malte Gerhold, Stephanie Lachs, and Michael Schurmann, Categorial l evy processes, Preprint, arXiv:1612.05139, 2016.
8. Yinzheng Gu and Paul Skoufranis, Conditionally bi-free independence for pairs of faces, J. Funct. Anal. 273 (2017), no. 5, 1663–1733. MR 3666726.
9. Fumio Hiai and Denes Petz, The semicircle law, free random variables and entropy, Mathematical Surveys and Monographs, vol. 77, American Mathematical Society, Providence, RI, 2000. MR 1746976.
10. Weihua Liu, Free-boolean independence for pairs of algebras, Preprint, arXiv:1710.01374, 2017.
11. __________, Free-free-boolean independence for triples of algebras, Preprint, arXiv:1801.03401, 2018.
12. Y. G. Lu, An interacting free Fock space and the arcsine law, Probab. Math. Statist. 17 (1997), no. 1, Acta Univ. Wratislav. No. 1928, 149-166. MR 1455615.
13. Jean-Louis Loday and Bruno Vallette, Algebraic operads, Grundlehren der Mathematischen Wissenschaften [Fundamental Principles of Mathematical Sciences], vol. 346, Springer, Heidelberg, 2012. MR 2954392.
14. Sarah Manzel and Michael Schurmann, Non-commutative stochastic independence and cumulants, Preprint, arXiv:math.OA/ 1601.06779, 2016, accepted for publication in IDAQP.
15. Naofumi Muraki, A new example of noncommutative \de Moivre-Laplace

theorem, Probability theory and mathematical statistics (Tokyo, 1995), World Sci. Publ., River Edge, NJ, 1996, pp. 353–362. MR 1467953.

16. ________, Noncommutative Brownian motion in monotone Fock space, Comm. Math. Phys. 183 (1997), no. 3, 557–570. MR1462227.
17. ________, The ve independences as natural products, In n. Dimens. Anal. Quantum Probab. Relat. Top. 6 (2003), no. 3, 337–371. MR 2016316 (2005h:46093).
18. ________, A simple proof of the classi cation theorem for positive natural products, Probab. Math. Statist. 33 (2013), no. 2, 315–326. MR3158558.
19. Alexandru Nica and Roland Speicher, Lectures on the combinatorics of free probability, London Mathematical Society Lecture Note Series, vol. 335, Cambridge University Press, Cambridge, June 30, 2018 12:34 WSPC Proceedings - 9in x 6in BBM page 19 19 2006. MR 2266879.
20. Roland Speicher, On universal products, Free probability theory (Waterloo, ON, 1995), Fields Inst. Commun., vol. 12, Amer. Math. Soc., Providence, RI, 1997, pp. 257–266. MR 1426844 (98c:46141).
21. D.-V. Voiculescu, K. J. Dykema, and A. Nica, Free random variables, CRM Monograph Series, vol. 1, American Mathematical Society, Providence, RI, 1992, A noncommutative probability approach to free products with applications to random matrices, operator algebras and harmonic analysis on free groups. MR 1217253.
22. Dan-Virgil Voiculescu, Free probability for pairs of faces I, Comm. Math. Phys. 332 (2014), no. 3, 955–980. MR 3262618.
23. ________, Free probability for pairs of faces II: 2-variables bi-free partial R-transform and systems with rank 1 commutation, Ann. Inst. Henri Poincar e Probab. Stat. 52 (2016), no. 1, 1–15. MR 3449291.
24. ________, Free probability for pairs of faces III: 2-variables bi-free partial S- and T-transforms, J. Funct. Anal. 270 (2016), no. 10, 3623–3638. MR 3478868.

Infinite Dimensional Analysis, Quantum Probability and Related Topics

pp. 77–86

COMPLEX STRUCTURES ON THREE-POINT SPACE

SUVRAJIT BHATTACHARJEE

Stat-Math Unit Indian Statistical Institute 203, B.T. Road Kolkata-700108, India
E-mail: suvra.bh r@isical.ac.in

DEBASHISH GOSWAMI

Stat-Math Unit Indian Statistical Institute 203, B.T. Road Kolkata-700108, India
E-mail: goswamid@isical.ac.in

We discuss notions of almost complex, complex and Kähler structures in the realm of non-commutative geometry and investigate them for a class of finite dimensional spectral triples on the three-point space. We classify all the almost complex structures on this non-commutative manifold, which also turn out to be complex structures, but none of them are Kähler in our sense.

1. Introduction

Almost complex, complex and Kähler structures play an important role in differential and Riemannian geometry. As a generalization of classical Riemannian geometry, non-commutative geometry was formulated by Alain Connes [7,13]. In this context, it is important and natural to formulate analogues of almost complex, complex and Kähler structures. However, it is rather surprising that except for a few attempts [8] in the early days of the subject and more recent works by a few authors [1,3,4,10,11] not too much has been done in this direction. In this article, our humble aim is to propose some formulation of these notions and investigate them in a very simple, finite dimensional non-commutative manifold on the three-point space. We classify all the almost complex structures on this non-commutative manifold, which also turn out to be complex structures, but none of them are Kähler in our sense.

2. Complex Structures

We consider a space made of three points $Y = \{1, 2, 3\}$. The algebra $\mathcal{C}(Y)$ of continuous functions is the direct sum $\mathcal{C}(Y) = \mathbb{C} \oplus \mathbb{C} \oplus \mathbb{C}$ and any element $f \in \mathcal{C}(Y)$ is a triplet of complex numbers (f_1, f_2, f_3), with $f_i = f(i)$ the value of f at the point i. The functions χ_i defined by $\chi_i(j) = \delta_{ij}, \quad i,j = 1,2,3$ form a $\mathbb{C}$-basis for $\mathcal{C}(Y)$.

The following is a simple conceptual tool that will be needed later on.

Definition 2.1. [4] Let A be a $*$-algebra and E be an A bi-module. The conjugate bi-module $\overline{E}$ is defined by declaring:

i) $\overline{E} = E$ as abelian groups;
ii) We write $\overline{e}$ for an element $e \in E$ when we consider it as an element of $\overline{E}$;
iii) The bi-module operations for $\overline{E}$ are $a \cdot \overline{e} = \overline{e \cdot a^*}$ and $\overline{e} \cdot a = \overline{a^* \cdot e}$.

Remark 2.1. If $\theta : E \to F$ is any morphism then we define $\overline{\theta} : \overline{E} \to \overline{F}$ by $\overline{\theta}(\overline{e}) = \overline{\theta(e)}$.

We now introduce the non-commutative analogue of Kähler differentials for commutative algebras.

Definition 2.2. [4] Let A be an arbitrary $\mathbb{C}$-algebra. Let $\Omega_u^1 A$ denote the kernel of the multiplication map $\mu : A \otimes A \to A$. We then define the universal differential graded algebra over A to be the tensor algebra

$$\Omega_u A = T_A(\Omega_u^1 A)$$

endowed with the unique degree one derivation such that

$$d(a) := 1 \otimes a - a \otimes 1$$

for $a \in A$.

Definition 2.3. [4] A differential calculus or differential structure on A is a differential graded algebra $(\Omega A, d)$ that is a quotient of $(\Omega_u A, d)$ by a differential graded ideal whose degree-zero component is zero.

It follows that ΩA is generated by A and $\Omega^1 A$, and $\Omega^n A$ is the $\mathbb{C}$-span of

$$\{\, a_0 da_1 \wedge \cdots \wedge da_n \mid a_0 \ldots a_n \in A\}.$$

The notion of a differential $*$-calculus first appeared in [19].

Definition 2.4. [4] A differential calculus $(\Omega A, d)$ on a $*$-algebra A is compatible with the star operation on A if the star operation on $\Omega^0 A = A$

extends to an involution $\xi \mapsto \xi^*$ on ΩA that preserves the grading and has the property that $d(\xi)^* = d(\xi^*)$ and $(\xi \wedge \eta)^* = (-1)^{|\xi||\eta|}\eta^* \wedge \xi^*$ for all homogeneous $\eta, \xi \in \Omega A$. When the conditions hold we call $(\Omega A, d, *)$ a differential $*$-calculus on A.

Remark 2.2. The definition implies $(a\xi b)^* = b^*\xi^*a^*$ for all $a, b \in A$ and $\xi \in \Omega A$.

Proposition 2.1. [4] *The map* $\star : \Omega A \to \overline{\Omega A}$ *defined by* $\star(\xi) = \overline{\xi^*}$ *is an* A*-bimodule homomorphism.*

We will now produce a differential $*$-calculus on the $*$-algebra $\mathcal{C}(Y)$ of continuous functions on the three-point space.

The notion of a spectral triple is the non-commutative analogue of a manifold. It is a basic ingredient in non-commutative geometry as developed in [6].

Definition 2.5. [7,13] A spectral triple (A, H, D) is given by a $*$-algebra with a faithful representation $\pi : A \to B(H)$ on the Hilbert space H together with a self-adjoint operator $D = D^*$ on H with the following properties:

Definition 2.6. The resolvent $(D - \lambda)^{-1}$, $\lambda \notin \mathbb{R}$, is a compact operator on H;

$[D, a] := D\pi(a) - \pi(a)D \in B(H)$, for any $a \in A$.

Given a spectral triple (A, H, D) one constructs [13] a compatible differential calculus on A by means of a suitable representation of the universal algebra $\Omega_u A$ in the algebra of bounded operators on H. The map

$$\pi_u : \Omega_u A \to B(H)$$

given by

$$\pi_u(a_0 da_1 \cdots da_p) := a_0[D, a_1] \cdots [D, a_p], \qquad a_j \in A$$

is a $*$-homomorphism of algebras.

Proposition 2.2. [13] *Let* $J_0 := \oplus_p J_0^p$ *be the graded two sided ideal of* $\Omega_u A$ *given by*

$$J_0^p := \{\omega \in \Omega_u^p A \mid \pi_u(\omega) = 0\}.$$

Then $J := J_0 + dJ_0$ *is graded differential ideal of* $\Omega_u A$.

Remark 2.3. The elements of J are called junk forms.

Definition 2.7. The differential graded algebra of Connes' forms over the algebra A is defined by

$$\Omega_D A := \Omega_u A/J \cong \pi_u(\Omega_u A)/\pi_u(J).$$

Remark 2.4. This is a differential $*$-calculus on the $*$-algebra A.

Let us now describe a spectral triple on the three-point space $Y = \{1, 2, 3\}$. This is a special case of a class of spectral triples considered in [5,18] on compact metric spaces.

Proposition 2.3. *Put $A = \mathcal{C}(Y)$ and $H = \mathbb{C}^2_{12} \oplus \mathbb{C}^2_{23} \oplus \mathbb{C}^2_{13}$ (the subscript ij says that the Hilbert space is along the "edge" connecting the point i with j). Define $\pi : A \to B(H)$ by*

$$\pi(f) = \begin{bmatrix} f(1) & 0 \\ 0 & f(2) \end{bmatrix}_{12} \oplus \begin{bmatrix} f(2) & 0 \\ 0 & f(3) \end{bmatrix}_{23} \oplus \begin{bmatrix} f(1) & 0 \\ 0 & f(3) \end{bmatrix}_{13},$$

for $f \in A$. And finally, define the operator D as

$$D = \begin{bmatrix} 0 & 1 \\ 1 & 0 \end{bmatrix}_{12} \oplus \begin{bmatrix} 0 & 1 \\ 1 & 0 \end{bmatrix}_{23} \oplus \begin{bmatrix} 0 & 1 \\ 1 & 0 \end{bmatrix}_{13}.$$

Then (A, H, D), as constructed above, is a spectral triple on the three-point space.

Proof. The conditions of Theorem 2.5 are satisfied since H is a finite dimensional Hilbert space (D is manifestly self-adjoint). □

Remark 2.5. Let us denote the differential graded algebra of Connes' forms over the algebra $A = \mathcal{C}(Y)$ simply by Ω.

Theorem 2.1. *Let (A, H, D) be the spectral triple on the three-point space described in Proposition 2.3. Then $e_i = [D, \chi_i], i = 1, 2$ is a free right basis for Ω^1. The bi-module structures are given by*

$$\chi_i e_i = e_i(1 - \chi_i)$$
$$\chi_i e_j = -e_i \chi_j, \qquad i \neq j$$

and

$$e_i \chi_i = (1 - \chi_i) e_i$$
$$e_i \chi_j = -\chi_i e_j \qquad i \neq j.$$

Proof. We note that, by definition, the space of 1-forms consists of bounded operators on H of the form $\sum_j a_0^j[D, a_1^j]$, where $a_i^j \in A$. We recall that $\chi_i, \quad i = 1, 2, 3$ form a $\mathbb{C}$ basis of A and satisfies $\chi_1+\chi_2+\chi_3 = 1$. Since $[D, 1] = 0$, we get the first conclusion. The bi-module structure follows from Leibniz rule and the observations $\chi_i^2 = \chi_i$, $\chi_i\chi_j = 0$. □

Remark 2.6. The basis described above is also a left basis for Ω^1.

Remark 2.7. It can be shown by computation that there are no junk forms. Also the higher spaces of forms are finite dimensional vector spaces.

The following definition is the beginning of non-commutative complex geometry and one can go very far mimicking the classical theory along the lines [9].

Definition 2.8. [4,2] Let $(\Omega A, d, *)$ be a $*$-differential calculus on A. An almost complex structure on $(\Omega A, d, *)$ is a degree zero derivation $J : \Omega A \to \Omega A$ such that

J is identically 0 on A and hence an A-bimodule endomorphism of ΩA;
$J^2 = -1$ on $\Omega^1 A$; and
$J(\xi^*) = J(\xi)^*$ for $\xi \in \Omega^1 A$ (J preserves $*$, i.e., $\overline{J}\star = \star J$).

Because $J^2 = -1$ on $\Omega^1 A$, there is an A-bimodule decomposition

$$\Omega^1 A = \Omega^{1,0} A \oplus \Omega^{0,1} A$$

where

$$\Omega^{1,0} A = \{\ \omega \in \Omega^1 A \mid J\omega = \iota\omega\}$$

and

$$\Omega^{0,1} A = \{\ \omega \in \Omega^1 A \mid J\omega = -\iota\omega\}.$$

Condition iii) implies $(\Omega^{0,1} A)^* = \Omega^{1,0} A$.[a]

For all $p, q \geq 0$ we define

$$\Omega^{p,q} A := \{\ \xi \in \Omega^{p+q} A \mid J\xi = (p-q)\iota\xi\}.$$

Elements in $\Omega^{p,q} A$ are called (p, q) forms. It is a theorem that

$$\Omega^n A = \bigoplus_{p+q=n} \Omega^{p,q} A.$$

[a]The map $\star : \Omega^{1,0} A \to \overline{\Omega^{0,1} A}$ is an isomorphism of A bi-modules.

For all pairs of non-negative integers (p,q), let

$$\pi^{p,q} : \Omega^{p+q}A \to \Omega^{p,q}A$$

be the projections associated to the direct sum decomposition $\Omega^n A = \bigoplus_{p+q=n} \Omega^{p,q}A$.

Remark 2.8. We call an endomorphism $J : \Omega^1 A \to \Omega^1 A$ satisfying the above conditions a first order almost complex structure. In most of the examples studied so far, one defines the endomorphism on $\Omega^1 A$ and then extends it to whole of ΩA using the derivation property and some "basis" of higher forms. In our case we use the free basis for Ω^1 and vector space basis for higher forms.

Definition 2.9. [4,2] Let

$$\partial : \Omega^{p,q}A \to \Omega^{p+1,q}A$$

be the composition $\pi^{p+1,q}d : \Omega^{p,q}A \to \Omega^{p+1,q}A$. Let

$$\bar{\partial} : \Omega^{p,q}A \to \Omega^{p,q+1}A$$

be the composition $\pi^{p,q+1}d : \Omega^{p,q}A \to \Omega^{p,q+1}A$.

The following is the analogue of Newlander-Nirenberg theorem [9].

Definition 2.10. [4,2] An almost complex structure J on $(\Omega A, d, *)$ is integrable if $d\Omega^{1,0}A \subset \Omega^{2,0}A \oplus \Omega 1, 1A$.

Definition 2.11. [4,2] A complex structure on $(\Omega A, d, *)$ is an almost complex structure J which is integrable.

Now we explicitly determine all the complex structures on the three-point space.

Theorem 2.2. *Let (A, H, D) be the spectral triple on the three-point space as described in Proposition 2.3. Then there are 8 complex structures for the calculus obtained in Theorem 2.1, as enumerated below:*

$$i\begin{bmatrix} 1-2\chi_3 & 0 \\ 2\chi_1 & 2\chi_2 - 1 \end{bmatrix}, \qquad -i\begin{bmatrix} 1-2\chi_3 & 0 \\ 2\chi_1 & 2\chi_2 - 1 \end{bmatrix};$$

$$i\begin{bmatrix} 1-2\chi_3 & -2\chi_2 \\ 2\chi_1 & 2\chi_3 - 1 \end{bmatrix}, \qquad -i\begin{bmatrix} 1-2\chi_3 & -2\chi_2 \\ 2\chi_1 & 2\chi_3 - 1 \end{bmatrix};$$

$$i\begin{bmatrix} 2\chi_1 - 1 & 2\chi_2 \\ 0 & 1-2\chi_3 \end{bmatrix}, \qquad -i\begin{bmatrix} 2\chi_1 - 1 & 2\chi_2 \\ 0 & 1-2\chi_3 \end{bmatrix};$$

$$i\begin{bmatrix} 2\chi_1 & 0 \\ 0 & 1-2\chi_2 \end{bmatrix}, \qquad -i\begin{bmatrix} 2\chi_1 & 0 \\ 0 & 1-2\chi_2 \end{bmatrix}.$$

Proof. By Remark 2.8, we let J be a first order almost complex structure. Let $Je_i = e_1 J_{1i} + e_2 J_{2i}, \quad i = 1, 2$ and extend right linearly. Then J is a left module morphism reads as $J(\chi_i e_j) = \chi_i J(e_j)$ (we use the bi-module rules to take the χ_i to the other side). In coordinates, $J^2 = -1$ is $\sum_{j,k} e_k J_{kj} J_{ji} = e_i$. Finally, J preserves $*$ reads as $\sum_j e_j J_{ji} = \sum_j \overline{J_{ji}} e_j$.

Now comparing coefficients and solving for J_{ij} from the above equations, we get the first order almost complex structures. Surprisingly, there are no more first order almost complex structures than the listed ones. Next we extend these according to Remark 2.8 and note that the restriction of any almost complex structure to the space of one forms has to be one of these and since there is only one way to (derivation property) extend the first order ones, we get all the almost complex structures.

For the integrability condition, we check it individually case by case. For example, for $J = i \begin{bmatrix} 2\chi_1 & 0 \\ 0 & 1 - 2\chi_2 \end{bmatrix}$, $\Omega^{1,0}$ consists of elements of the form $\alpha(e_1\chi_1) + \beta(e_2\chi_1) + \gamma(e_2\chi_3), \quad \alpha, \beta, \gamma \in \mathbb{C}$.

We apply D to an element of that form, followed by J, and find that the result is 0, i.e., the element lies in $\Omega^{1,1}$. So this J is integrable. And similarly for the other J's, which concludes the proof. □

3. Kähler Structures

Let us now go over to a possible non-commutative version of Kähler geometry. We begin with a basic ingredient in the classical theory, namely, that of a metric.

Definition 3.1. Let $(\Omega A, d, *)$ be a $*$-differential calculus on A. A metric g on A is a non-degenerate[b] bi-module morphism

$$g : \Omega^1 A \otimes_A \Omega A^1 \to A$$

and a hermitian metric is a non-degenerate bi-module morphism

$$h : \Omega^1 A \otimes_A \overline{\Omega^1 A} \to A.$$

[b]Non-degeneracy means $g(\xi, \_)$ induces a right A module isomorphism $\Omega^1 A \to (\Omega^1 A)'$, the $()'$ denoting the right A dual.

Remark 3.1. Given a hermitian metric $h : \Omega^1 A \otimes_A \overline{\Omega^1 A} \to A$, we get a metric $g : \Omega^1 A \otimes_A \Omega A^1 \to A$ by defining $g = h \circ (id \otimes \star)$. We say h is induced by g.

Remark 3.2. Assume that $\Omega^1 A$ is projective as a left module. Then using a dual basis, we get a bi-module morphism

$$\mathfrak{g} : A \to \Omega^1 A \otimes_A \Omega A^1.$$

We identify the morphism with the element $\mathfrak{g}(1) \in \Omega^1 A \otimes_A \Omega A^1$. We call this element the inverse pf g

Along the classical lines we have,

Definition 3.2. Let $(\Omega A, d, *)$ be a $*$-differential calculus with almost complex structure J. We say that a metric $g : \Omega^1 A \otimes_A \Omega A^1 \to A$ (respectively, a hermitian metric $h : \Omega^1 A \otimes_A \overline{\Omega^1 A} \to A$) is compatible with J if $g \circ (J \otimes J) = g$ (respectively, $h \circ (J \otimes \overline{J}) = h$).

Remark 3.3. If g is induced by h and h is compatible with J, then g is automatically compatible with J.

Remark 3.4. If a metric g is compatible with J then g is identically 0 on $\Omega^{1,0} A \otimes_A \Omega^{1,0}$ and $\Omega^{0,1} A \otimes_A \Omega^{0,1} A$.

The very algebraic definition given in [9] has the following counterpart.

Definition 3.3. Let $(\Omega A, d, *)$ be a $*$-differential calculus with a complex structure J and g be a compatible metric on A. Assume $\Omega^1 A$ is projective as a left module. Let $\mathfrak{g} \in \Omega^1 A \otimes_A \Omega A^1$ be the corresponding inverse of g. Then the fundamental form ω is defined to be the form $\wedge (J \otimes id)(\pi^{1,0} \otimes \pi^{0,1})(\mathfrak{g}) \in \Omega^{1,1} A$. The metric is said to be Kähler if $d\omega = 0$.

Unfortunately, for the three-point space we have the following

Theorem 3.1. *For the spectral triple (A, H, D) on the three-point space as above in Proposition 2.3 and the calculus obtained in Theorem 2.1, there are no compatible Kähler metric for the complex structures enumerated in Theorem 2.3.*

Proof. This is also proved by case by case analysis. We outline the overall scheme.

We first find J compatible metric g. Let $g(e_i, e_j) = g_{ij}$. Since $e_i, i = 1, 2$ is both a right and left basis for Ω^1 and g is a bi-module morphism, g is determined by g_{ij}'s. The non-degeneracy condition turns into the invertibility of the matrix $\begin{bmatrix} g_{11} & g_{12} \\ g_{21} & g_{22} \end{bmatrix}$ (since A is commutative, this is equivalent to the determinant $g_{11}g_{22} - g_{12}g_{21}$ being an unit).

Then the compatibility condition reads as $g_{ij} = g(\sum_k e_k J_{ki}, \sum_l e_l J_{lj})$. We take the J_{ki}'s to the right entry of g and then use the bi-module rule to get those out of g. Solving, we get J compatible metrics.

The inverse $\mathfrak{g}$ of g takes the form $\sum_i e_i \otimes (\sum_j e_j g^{ji})$, where $\begin{bmatrix} g^{11} & g^{12} \\ g^{21} & g^{22} \end{bmatrix}$ is the inverse of $\begin{bmatrix} g_{11} & g_{12} \\ g_{21} & g_{22} \end{bmatrix}$.

According to Definition 3.3, we apply the projections and then J on the first tensorand (which is just multiplication by i) and multiply them, thus getting the fundamental form. The condition $d\omega = 0$ then contradicts the non-degeneracy condition of g as can be seen by direct computation. □

Remark 3.5. In the paper [1] a similar definition of the fundamental form was proposed. But in [3], it is defined more conceptually and many of the standard classical theorems are proved in that paper. It is of interest to know whether the three point space admits a Kähler structure in the set-up of the paper [3].

Remark 3.6. It remains to study several other examples. The quantum homogeneous spaces arising from quantum groups are studied in [2,3] and are still being studied by several people. The Podles' sphere [14] and higher projective spaces are studied in [11,10]. We wish to study the non-commutative complex geometry on the very basic example of a non-commutative manifold, namely, the non-commutative torus [17]. There has not been done much for this particular example except for the papers [16,15,12] that we know of.

References

1. Rèamonn Ó Buachalla, Noncommutative Kahler Geometry of the Standard Podles Sphere, arXiv preprint arXiv:1401.1459 (2014).
2. ________, Noncommutative complex structures on quantum homogeneous spaces, J. Geom. Phys. 99 (2016), 154–173, DOI 10.1016/j.geomphys.2015.10.003. MR3428362.
3. ________, Noncommutative Kahler structures on quantum homogeneous

spaces, Adv. Math. 322 (2017), 892{939, DOI 10.1016/j.aim.2017.09.031. MR3720811.

4. Edwin Beggs and S. Paul Smith, Non-commutative complex di erential geometry, J. Geom. Phys. 72 (2013), 7–33, DOI 10.1016/j.geomphys.2013.03.018. MR3073899.
5. Erik Christensen and Cristina Ivan, Sums of two-dimensional spectral triples, Math. Scand. 100 (2007), no. 1, 35–60, DOI 10.7146/math.scand.a-15015. MR2327719.
6. Alain Connes, Noncommutative di erential geometry, Inst. Hautes Etudes Sci. Publ. Math. 62 (1985), 257–360. MR823176.
7. _________, Noncommutative geometry, Academic Press, Inc., San Diego, CA, 1994. MR1303779.
8. J. Frohlich, O. Grandjean, and A. Recknagel, Supersymmetric quantum theory and non-commutative geometry, Comm. Math. Phys. 203 (1999), no. 1, 119–184, DOI 10.1007/s002200050608. MR1695097.
9. Daniel Huybrechts, Complex geometry, Universitext, Springer-Verlag, Berlin, 2005. An introduction. MR2093043.
10. Masoud Khalkhali, Giovanni Landi, and Walter Daniel van Suijlekom, Holomorphic structures on the quantum projective line, Int. Math. Res. Not. IMRN 4 (2011), 851–884, DOI 10.1093/imrn/rnq097. MR2773332.
11. Masoud Khalkhali and Ali Moatadelro, Noncommutative complex geometry of the quantum projective space, J. Geom. Phys. 61 (2011), no. 12, 2436–2452, DOI 10.1016/j.geomphys.2011.08.004. MR2838520.
12. _________, A Riemann-Roch theorem for the noncommutative two torus, J. Geom. Phys. 86 (2014), 19–30, DOI 10.1016/j.geomphys.2014.06.005. MR3282309.
13. Giovanni Landi, An introduction to noncommutative spaces and their geometries, Lecture Notes in Physics. New Series m: Monographs, vol. 51, Springer-Verlag, Berlin, 1997. MR1482228.
14. P. Podleś, Quantum spheres, Lett. Math. Phys. 14 (1987), no. 3, 193–202, DOI 10.1007/BF00416848. MR919322.
15. A. Polishchuk, Classi cation of holomorphic vector bundles on noncommutative twotori, Doc. Math. 9 (2004), 163–181. MR2054986.
16. A. Polishchuk and A. Schwarz, Categories of holomorphic vector bundles on noncommutative two-tori, Comm. Math. Phys. 236 (2003), no. 1, 135–159, DOI 10.1007/s00220-003-0813-9. MR1977884.
17. Marc A. Rieffel, C-algebras associated with irrational rotations, Pacific J. Math. 93 (1981), no. 2, 415–429. MR623572.
18. _________, Metrics on state spaces, Doc. Math. 4 (1999), 559–600. MR1727499.
19. S. L. Woronowicz, Differential calculus on compact matrix pseudogroups (quantum groups), Comm. Math. Phys. 122 (1989), no. 1, 125–170. MR994499.

Infinite Dimensional Analysis, Quantum Probability and Related Topics

pp. 87–101

UNCERTAINTY RELATIONS AND INFORMATION LOSS FOR SPIN-1/2 MEASUREMENTS

ALBERTO BARCHIELLI

Politecnico di Milano, Dipartimento di Matematica,
Piazza Leonardo da Vinci 32, I-20133 Milano, Italy;
Istituto Nazionale di Alta Matematica (INDAM-GNAMPA), Italy;
Istituto Nazionale di Fisica Nucleare (INFN), Sezione di Milano, Italy

MATTEO GREGORATTI

Politecnico di Milano, Dipartimento di Matematica,
Piazza Leonardo da Vinci 32, I-20133 Milano, Italy;
Istituto Nazionale di Alta Matematica (INDAM-GNAMPA), Italy;
E-mail: matteo.gregoratti@polimi.it

We formulate entropic measurements uncertainty relations (MURs) for a spin-1/2 system. When incompatible observables are approximatively jointly measured, we use relative entropy to quantify the information lost in approximation and we prove positive lower bounds for such a loss: there is an unavoidable information loss. Firstly we allow only for covariant approximate joint measurements and we find state-dependent MURs for two or three orthogonal spin-1/2 components. Secondly we consider any possible approximate joint measurement and we find state-independent MURs for two or three spin-1/2 components. In particular we study how MURs depend on the angle between two spin directions. Finally, we extend our approach to infinitely many incompatible observables, namely to the spin components in all possible directions. In every scenario, we always consider also the characterization of the optimal approximate joint measurements.

Keywords: Measurement uncertainty relations, spin, relative entropy, information loss.

1. Introduction

Measurement uncertainty relations (MURs) quantify to which extent one can approximate a set of measurements of incompatible observables by a single joint measurement; for a recent presentation of the modern theory of quantum measurements see, e.g., [1]. Various approaches have been proposed to quantify this approximation, such as distances for probability measures [2,3,4,5,6,7,8] or conditional entropies [9,10,11]. Our approach is to see the joint measurement approximation of incompatible observables as a loss of information and to quantify it by the use of the relative entropy [12,13].

Here we describe the spin-1/2 variant of Ref. [13] with some new results (proved in Section 6). We formulate both state-dependent and state-independent entropic MURs for spin components, which are always incompatible when different directions are measured. The state-dependent MURs are new and they arise, as done in [12] for position and momentum, from a selection of the admissible approximate joint measurements based on the covariance properties of the target observables.

We start with two or three incompatible spin components, and then we extend our approach to an approximate joint measurement of infinite observables, the components of the spin along all the possible directions.

Let us remark that the entropic approach naturally selects covariant approximate joint measurements and, in several remarkable cases, gives uniqueness of the optimal one and completely characterises it.

2. Information lost by approximate measurements

Given a spin-1/2 quantum system, we want to quantify the information lost when two *incompatible* sharp observables A and B, the spin components along independent unit vectors $\vec{a}$ and $\vec{b}$, are approximated by two *compatible* POVM observables $\mathsf{M}_{[1]}$ and $\mathsf{M}_{[2]}$. For general $\vec{a}$ and $\vec{b}$, the corresponding sharp observables are identified by the pvm's

$$\mathsf{A}(x) = \frac{1}{2}\left(\mathbf{1} + x\,\vec{a}\cdot\vec{\sigma}\right), \qquad \mathsf{B}(y) = \frac{1}{2}\left(\mathbf{1} + y\,\vec{b}\cdot\vec{\sigma}\right), \qquad x, y = \pm 1, \quad (1)$$

while $\mathsf{M}_{[1]}$ and $\mathsf{M}_{[2]}$ are the marginals of some bi-observable M

$$\mathsf{M}(x,y) \geq 0, \quad x, y = \pm 1, \qquad \sum_{x,y} \mathsf{M}(x,y) = \mathbf{1}.$$

Given the system state

$$\rho = \frac{1}{2}\left(\mathbf{1} + \vec{r}\cdot\vec{\sigma}\right), \qquad r \equiv |\vec{r}| \leq 1, \quad (2)$$

if the target observable A is approximated by $\mathsf{M}_{[1]}$, then the target probability distribution $\mathsf{A}^\rho(x) = \operatorname{Tr}\{\mathsf{A}(x)\rho\}$ is approximated by the probability distribution $\mathsf{M}^\rho_{[1]}(x) = \operatorname{Tr}\{\mathsf{M}_{[1]}(x)\rho\}$. This causes a loss of information that is quantified by the relative entropy $S\big(\mathsf{A}^\rho \| \mathsf{M}^\rho_{[1]}\big)$ [14,15].

Given two probabilities p and q on a same discrete space, the relative entropy of p with respect to q is

$$S(p\|q) = \sum_x p(x) \log \frac{p(x)}{q(x)}, \tag{3}$$

which is always non-negative, it vanishes if and only if $p = q$, while it gives $+\infty$ whenever $\operatorname{supp} p \subsetneq \operatorname{supp} q$; the usual convention $0 \log(0/0) = 0$ is understood. We use base 2 logarithms, so that the entropy will be measured in bits. The interpretation in terms of information loss can be understood in the framework of data compression theory [15]: $S(p\|q)$ is the increase in the expected description length of a Shannon code when the latter is optimized assuming that a random signal has distribution q, but actually the true distribution is p.

Within this informational approach, a Heisenberg-type uncertainty relation for A and B is a trade-off relation between $S(\mathsf{A}^\rho\|\mathsf{M}^\rho_{[1]})$ and $S(\mathsf{B}^\rho\|\mathsf{M}^\rho_{[2]})$ such that no M can make them both too small. Any trade-off of this kind is a consequence of the incompatibility of A and B, and so it reveals their quantum nature.

Depending on the framework and on the aim, one can look for relations in a specific state ρ (state dependent entropic MURs), or one can first process these two relative entropies into a state-independent quantification of the quality of the approximating device M and then look for nontrivial constraints (state independent entropic MURs).

In both cases, we sum the two information losses and then look for some positive lower bound for such a sum. Thus, we introduce the state-dependent *error function*

$$S[\mathsf{A},\mathsf{B}\|\mathsf{M}](\rho) = S(\mathsf{A}^\rho\|\mathsf{M}^\rho_{[1]}) + S(\mathsf{B}^\rho\|\mathsf{M}^\rho_{[2]}) = S(\mathsf{A}^\rho \otimes \mathsf{B}^\rho\|\mathsf{M}^\rho_{[1]} \otimes \mathsf{M}^\rho_{[2]}). \tag{4}$$

The last equality gives the informational interpretation of the error function: it is the increase in expected length of a Shannon code describing the random signal (A,B), if the signal has distribution $\mathsf{A}^\rho \otimes \mathsf{B}^\rho$, but it is coded using the wrong distribution $\mathsf{M}^\rho_{[1]} \otimes \mathsf{M}^\rho_{[2]}$. Thus, it is the total loss of information due to the approximations $\mathsf{A}^\rho \simeq \mathsf{M}^\rho_{[1]}$ and $\mathsf{B}^\rho \simeq \mathsf{M}^\rho_{[2]}$. Note that $\mathsf{A}^\rho \otimes \mathsf{B}^\rho$ is the distribution of a measurement of A and B (possibly incompatible observables) in two independent preparations of the same state ρ; coherently, when M^ρ is available in place of $\mathsf{A}^\rho \otimes \mathsf{B}^\rho$, we take into account this independence and, so, first we compute the marginals $\mathsf{M}^\rho_{[1]}$ and $\mathsf{M}^\rho_{[2]}$, and then we optimize the Shannon code according to $\mathsf{M}^\rho_{[1]} \otimes \mathsf{M}^\rho_{[2]}$ and not to M^ρ. Indeed, if A and B are incompatible, correlations between their values in a single experiment simply do not exist, and so we do not care about correlations emerging from joint measurements of $\mathsf{M}_{[1]}$ and $\mathsf{M}_{[2]}$. In other words, different approximate joint measurements M and M' with the same marginals $\mathsf{M}_{[1]} = \mathsf{M}'_{[1]}$ and $\mathsf{M}_{[2]} = \mathsf{M}'_{[2]}$ give the same error function.

Given the target observables A and B, our aim is to minimize $S[\mathsf{A},\mathsf{B}\|\mathsf{M}](\rho)$: we want to find out the minimal amount of information that is unavoidably lost in an approximate joint measurement (entropic MUR), and, possibly, we want to characterize the optimal devices M minimizing such information loss.

Similarly, if we approximate three sharp spin components $\mathsf{A},\mathsf{B},\mathsf{C}$ with a joint measurement M of marginals $\mathsf{M}_{[1]},\mathsf{M}_{[2]},\mathsf{M}_{[3]}$, the error function is

$$S[\mathsf{A},\mathsf{B},\mathsf{C}\|\mathsf{M}](\rho) = S\big(\mathsf{A}^\rho\|\mathsf{M}^\rho_{[1]}\big) + S\big(\mathsf{B}^\rho\|\mathsf{M}^\rho_{[2]}\big) + S\big(\mathsf{C}^\rho\|\mathsf{M}^\rho_{[3]}\big). \tag{5}$$

3. State dependent entropic MURs

Any study of state dependent MURs between A and B must be preceded by a consideration: given a state ρ_*, we can always find a bi-observable M_* such that $S[\mathsf{A},\mathsf{B}\|\mathsf{M}_*](\rho_*) = 0$. It suffices to take $\mathsf{M}_* = (\mathsf{A}^{\rho_*}\otimes\mathsf{B}^{\rho_*})\,\mathbf{1}$, the trivial bi-observable giving an output distributed as $\mathsf{A}^{\rho_*}\otimes\mathsf{B}^{\rho_*}$, independently of the system state ρ. For this reason it make sense to look for state-dependent MURs only if some criterion reduces the class of the approximate joint measurements M. A typical criterion is to allow only for POVMs M sharing the same covariance properties of the target observables A and B.

For a spin-1/2 quantum system, let us start with maximally incompatible A and B, i.e. the case of orthogonal $\vec{a}$ and $\vec{b}$. To fix the notations, let us set

$$\mathsf{A}(x) = \mathsf{X}(x) = \frac{1}{2}\,(\mathbf{1} + x\sigma_1), \qquad \mathsf{B}(y) = \mathsf{Y}(y) = \frac{1}{2}\,(\mathbf{1} + y\sigma_2). \tag{6}$$

Then, their symmetry properties [13] with respect to the order 8 dihedral group D_4, generated by the $180°$ rotations around the x-axis and around the bisector of the first and third quadrant, lead us to consider D_4-covariant bi-observable on $\{-1,+1\}^2$, that is [13]

$$\mathsf{M}(x,y) = \frac{1}{4}\Big[\mathbf{1} + c\,(x\sigma_1 + y\sigma_2)\Big], \qquad |c| \leq 1/\sqrt{2}, \tag{7}$$

Then we have:

- for every state ρ there is a D_4-covariant bi-observable that minimizes $S[\mathsf{X},\mathsf{Y}\|\mathsf{M}](\rho)$,

$$\mathsf{M}_0(x,y) = \frac{1}{4}\left(\mathbf{1} + \frac{x}{\sqrt{2}}\,\sigma_1 + \frac{y}{\sqrt{2}}\,\sigma_2\right), \tag{8}$$

- the optimal approximate joint measurement M_0 is independent of ρ,
- the marginals of M_0 are equally noisy versions of the target observables,

$$\mathsf{M}_{0\,[1]} = \frac{1}{\sqrt{2}}\mathsf{X} + \left(1 - \frac{1}{\sqrt{2}}\right)\frac{\mathbf{1}}{2}, \qquad \mathsf{M}_{0\,[2]} = \frac{1}{\sqrt{2}}\mathsf{Y} + \left(1 - \frac{1}{\sqrt{2}}\right)\frac{\mathbf{1}}{2},$$

- M_0 provides a lower bound for the error function:

$$S[\mathsf{X},\mathsf{Y}\|\mathsf{M}](\rho)[\mathsf{X},\mathsf{Y}\|\mathsf{M}_0](\rho) = \sum_{i=1}^{2}\left(\frac{1+r_i}{2}\log\frac{1+r_i}{1+r_i/\sqrt{2}} + \frac{1-r_i}{2}\log\frac{1-r_i}{1-r_i/\sqrt{2}}\right) \tag{9}$$

  for every state ρ and every D_4-covariant approximate joint measurement M,
- if the state vector $\vec{r}$ is not perpendicular to the plane xy, then the lower bound (9) is strictly positive and M_0 is the unique optimal D_4-covariant approximate joint measurement,
- if $\vec{r}$ is perpendicular to the plane xy, then the lower bound (9) vanishes and every D_4-covariant bi-observable M is optimal.

The inequality (9) is our state-dependent entropic MUR for two orthogonal spin components. Depending on the state ρ it can be significant or trivial. Changing the point of view, when ρ is given, the MUR (9) indicates an unavoidable information loss if and only if we perform a covariant approximate joint measurement of two spin components in a plane non perpendicular to $\vec{r}$. Otherwise A^ρ, $\mathsf{M}^\rho_{[1]}$, B^ρ and $\mathsf{M}^\rho_{[2]}$ are all equal to the uniform distribution on $\{-1,+1\}$ and no criterion based on the output distributions can detect the approximation.

A similar result holds for approximate joint measurements of three maximally incompatible spin-1/2 components, say

$$\mathsf{X}(x) = \frac{1}{2}\,(\mathbf{1} + x\sigma_1), \qquad \mathsf{Y}(y) = \frac{1}{2}\,(\mathbf{1} + y\sigma_2), \qquad \mathsf{Z}(z) = \frac{1}{2}\,(\mathbf{1} + z\sigma_3). \tag{10}$$

Then, their symmetry properties [13] with respect to the order 24 octahedron group O, generated by the 90° rotations around the three coordinate axes, lead us to consider O-covariant tri-observable on $\{-1,+1\}^3$, that is [13]

$$\mathsf{M}(x,y,z) = \frac{1}{8}\Big[\mathbf{1} + c\,(x\sigma_1 + y\sigma_2 + z\sigma_3)\Big], \qquad |c| \leq 1/\sqrt{3}. \tag{11}$$

Then we have:

- for every ρ there is a O-covariant tri-observable that minimizes $S[\mathsf{X},\mathsf{Y},\mathsf{Z}\|\mathsf{M}](\rho)$,

$$\mathsf{M}_0(x,y,z) = \frac{1}{8}\left(\mathbb{1} + \frac{x}{\sqrt{3}}\sigma_1 + \frac{y}{\sqrt{3}}\sigma_2 + \frac{z}{\sqrt{3}}\sigma_3\right), \qquad (12)$$

- the optimal approximate joint measurement M_0 is independent of the state ρ,
- the marginals of M_0 are equally noisy versions of the target observables,

$$\mathsf{M}_{0\,[1]} = \frac{1}{\sqrt{3}}\mathsf{X} + \left(1 - \frac{1}{\sqrt{3}}\right)\frac{\mathbb{1}}{2}, \qquad \mathsf{M}_{0\,[2]} = \frac{1}{\sqrt{3}}\mathsf{Y} + \left(1 - \frac{1}{\sqrt{3}}\right)\frac{\mathbb{1}}{2},$$
$$\mathsf{M}_{0\,[3]} = \frac{1}{\sqrt{3}}\mathsf{Z} + \left(1 - \frac{1}{\sqrt{3}}\right)\frac{\mathbb{1}}{2},$$

- M_0 provides a lower bound for the error function:

$$S[\mathsf{X},\mathsf{Y},\mathsf{Z}\|\mathsf{M}](\rho)[\mathsf{X},\mathsf{Y},\mathsf{Z}\|\mathsf{M}_0](\rho)$$
$$= \sum_{k=1}^{3}\left\{\frac{1+r_k}{2}\log\frac{1+r_k}{1+r_k/\sqrt{2}} + \frac{1-r_k}{2}\log\frac{1-r_k}{1-r_k/\sqrt{2}}\right\} \qquad (13)$$

  for every state ρ and every O-covariant approximate joint measurement M,
- for every $\rho \neq \frac{1}{2}$ the lower bound (13) is strictly positive and M_0 is the unique optimal O-covariant approximate joint measurement,
- for $\rho = \frac{1}{2}$ the lower bound (13) vanishes and every O-covariant tri-observable M is optimal.

The inequality (13) is our state-dependent entropic MUR for three orthogonal spin components. Differently from the two components case, the lower bound now vanishes only for the maximally chaotic state.

4. State independent entropic MURs

In this section, every POVM on $\{-1,+1\}^2$ will be considered as a possible approximate joint measurement of the target observables A and B.

If one is interested in the quality of a joint measurement device M, whatever the state ρ of the observed system, then a state-independent quantification of the approximation error is needed.

Note that the minimum error is not significative, as it can be null for an approximate joint measurement M of incompatible A and B (as we have seen in the previous section). That is, it can be null for some ρ even if $\mathsf{M}_{[1]} \neq \mathsf{A}$ or $\mathsf{M}_{[2]} \neq \mathsf{B}$.

Then typical approaches consist in computing some mean error or the worst error. Following [5], we maximize our error function. The *divergence* of M from (A, B) is the worst-case information loss

$$D(\mathsf{A}, \mathsf{B} \| \mathsf{M}) = \sup_{\rho} S[\mathsf{A}, \mathsf{B} \| \mathsf{M}](\rho). \tag{14}$$

Note that $D(\mathsf{A}, \mathsf{B} \| \mathsf{M})$ takes into account any possible balancing and compensation between the information losses in the two approximations $\mathsf{A}^\rho \simeq \mathsf{M}^\rho_{[1]}$ and $\mathsf{B}^\rho \simeq \mathsf{M}^\rho_{[2]}$. It is always non-negative and it vanishes exactly when A and B are compatible and M is one of their joint measurements. Given A and B, our aim is to make the divergence $D(\mathsf{A}, \mathsf{B} \| \mathsf{M})$ as small as possible.

Again, explicit computations can be done in the case of maximally incompatible A and B, i.e. the observables X and Y (6). Then [13]

- there is a unique optimal approximate joint measurement of X and Y, that is the covariant bi-observable M_0 (8),
- if ρ_e is the projection on any eigenvector of σ_1 or σ_2, then

$$D(\mathsf{X}, \mathsf{Y} \| \mathsf{M}_0) = S[\mathsf{X}, \mathsf{Y} \| \mathsf{M}_0](\rho_e) = \log \left[2 \left(2 - \sqrt{2} \right) \right], \tag{15}$$

- M_0 provides a lower bound for the divergence:

$$D(\mathsf{X}, \mathsf{Y} \| \mathsf{M}) \geq \log \left[2 \left(2 - \sqrt{2} \right) \right] \tag{16}$$

  for every approximate joint measurement M.

The inequality (16) is our state-independent entropic MUR, which can be stated also as a statement about the total loss of information that occurs in a single preparation of the system: for every approximate joint measurement M of X and Y, there exists a state ρ such that

$$S(\mathsf{X}^\rho \| \mathsf{M}^\rho_{[1]}) + S(\mathsf{Y}^\rho \| \mathsf{M}^\rho_{[2]}) \geq \log \left[2 \left(2 - \sqrt{2} \right) \right]. \tag{17}$$

So, in an approximate joint measurement of X and Y, the total loss of information can not be arbitrarily reduced by the choice of the device M: it depends on the state ρ, but potentially it can always be as large as $\log \left[2 \left(2 - \sqrt{2} \right) \right]$.

A similar result holds for approximate joint measurements of three maximally incompatible spin-1/2 components, i.e. the observables X, Y, Z (10). The definition of $D(\mathsf{X}, \mathsf{Y}, \mathsf{Z} \| \mathsf{M})$ is obvious. Then [13]

- there is a unique optimal O-covariant approximate joint measurement of X, Y, Z, that is the tri-observable M_0 (12),
- if ρ_e is the projection on any eigenvector of σ_1 or σ_2 or σ_3, then

$$D(\mathsf{X},\mathsf{Y},\mathsf{Z}\|\mathsf{M}_0) = S[\mathsf{X},\mathsf{Y},\mathsf{Z}\|\mathsf{M}_0](\rho_e) = \log\left(3-\sqrt{3}\right), \tag{18}$$

- M_0 provides a lower bound for the divergence:

$$D(\mathsf{X},\mathsf{Y},\mathsf{Z}\|\mathsf{M}) \geq \log\left(3-\sqrt{3}\right) \tag{19}$$

  for every approximate joint measurement M.

Note that, differently from the two components case, the uniqueness of the optimal approximate joint measurement is stated only in the class of O-covariant bi-observables; indeed, a non-covariant tri-observable with the same marginals as M_0 can be easily exhibited [16].

For two general components A and B, let $0 \leq \alpha \leq \pi$ be the angle between $\vec{a}$ and $\vec{b}$. We introduce the *entropic incompatibility degree* of the observables A and B,

$$c_{\text{inc}}(\alpha) = \inf_{\mathsf{M}} D(\mathsf{A},\mathsf{B}\|\mathsf{M}) = \inf_{\mathsf{M}} \sup_{\rho} S[\mathsf{A},\mathsf{B}\|\mathsf{M}](\rho). \tag{20}$$

Indeed, the infimum (20) depends only on the angle α. Since $c_{\text{inc}}(\alpha)$ vanishes if and only if $\alpha = 0, \pi$, that is for compatible A and B, we have the state independent entropic MUR:

$$D(\mathsf{A},\mathsf{B}\|\mathsf{M}) \geq c_{\text{inc}}(\alpha) \tag{21}$$

for every approximate joint measurement M. By construction, the lower bound $c_{\text{inc}}(\alpha)$ is tight. So its name is justified.

In this general case, we choose the coordinate axes in such a way that the bisector of α coincides with the bisector of the first quadrant. Then the symmetry group of A and B is the order 4 dihedral group D_2 generated by the 180° rotations around the two bisectors of the quadrants.

We do not know if the optimal approximate joint measurement M of A and B is unique and if it needs to be D_2-covariant. Anyway, we know that there is at least an optimal one in the family, consisting of D_2-covariant bi-observables: for $|\gamma| \leq 1$,

$$\mathsf{M}_\gamma(x,y) = \frac{1}{4}\left[(1+\gamma xy)\,\mathbb{1} + \frac{1}{\sqrt{2}}\,(x\sigma_1 + y\sigma_2) + \frac{\gamma}{\sqrt{2}}\,(y\sigma_1 + x\sigma_2)\right]. \tag{22}$$

Then numerical computations give the following graph of $c_{\text{inc}}(\alpha)$ as a function of the angle α. Just as one would expect, the entropic incompatibility

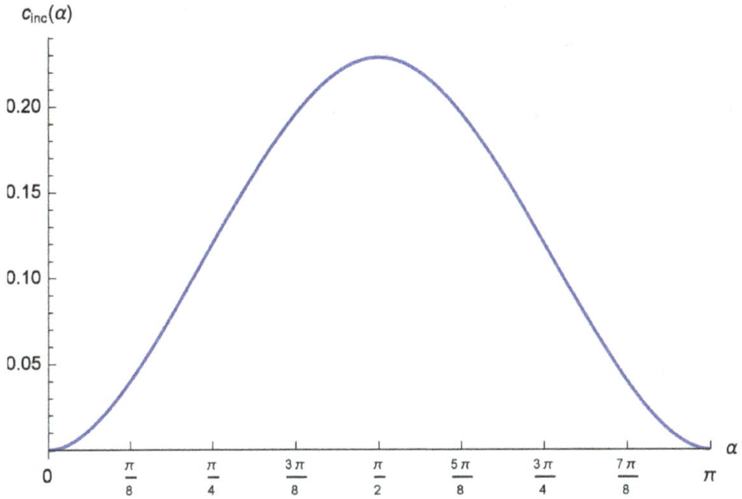

Fig. 1. Numerical evaluations of $c_{\text{inc}}(\alpha)$ as a function of α, from (20), (22): $c_{\text{inc}}(\alpha) = \inf_\gamma \sup_\rho S[\mathsf{A}, \mathsf{B} \| \mathsf{M}_\gamma](\rho)$.

degree $c_{\text{inc}}(\alpha)$ is null for compatible observables, $\alpha = 0$, and then continuously increases to its maximum value given by maximally incompatible observables, $\alpha = \pi/2$. Then the graph is symmetric, as it should be, with respect to $\alpha = \pi/2$.

Numerical computations also provide an optimal γ_0 for every chosen α and it turns out that, differently from the orthogonal cases, for a general α the marginals of the optimal approximate joint measurement M_{γ_0} are not noisy versions of the target observables A and B [13]; similar results hold also for different quantifications of the uncertainty [18,17,8]. This shows that it is not always possible to reduce the study of approximate joint measurements of sharp observables to the study of their noisy versions.

5. Entropic MURs for infinite components

Now we consider an approximate joint measurement of all the spin components A, $|\vec{a}| = 1$. Since $\vec{a}$ is not fixed anymore, we change notation and, for every $\vec{a}$, we denote by $\mathsf{A}_{\vec{a}}$ the corresponding observable.

The following is inspired by [6], where the same problem is considered for a generic spin s, but the approximation error is quantified by Wasserstein distances between target and approximating distributions, and uncertainty regions are introduced.

First we consider a POVM observable F with output $\vec{\xi} \in \mathbb{R}^3$, $|\vec{\xi}| = 1$. From this, by post-processing we derive our approximate joint measurement M of all spin components $\mathsf{A}_{\vec{a}}$: if $\vec{\xi}$ is the output of F, then the output $x = \pm 1$ of the approximate measurement $\mathsf{M}_{[\vec{a}]}$ along $\vec{a}$ is the signum the $\vec{a}$-component $\vec{\xi}\cdot\vec{a}$. Thus the marginals $\mathsf{M}_{[\vec{a}]}$ are all compatible by construction and, of course, their outputs are all "classically coherent".

Just as in [6], we want F to represent an angular momentum and, so, we consider only $SO(3)$-covariant F, that is (Eq. (108) of [6])

$$\mathsf{F}_\epsilon(\mathrm{d}\theta, \mathrm{d}\phi) = (1-\epsilon)\mathsf{F}_+(\mathrm{d}\theta, \mathrm{d}\phi) + \epsilon \mathsf{F}_-(\mathrm{d}\theta, \mathrm{d}\phi), \qquad \epsilon \in [0,1], \tag{23}$$

$$\mathsf{F}_\pm(\mathrm{d}\theta, \mathrm{d}\phi) := \frac{\sin\theta \mathrm{d}\theta \mathrm{d}\phi}{4\pi} \left[\mathbf{1} \pm \sin\theta\left(\cos\phi\, \sigma_1 + \sin\phi\, \sigma_2\right) \pm \cos\theta\, \sigma_3\right]. \tag{24}$$

We are using spherical coordinates with the usual conventions $\vec{\xi} = (\sin\theta\cos\phi, \sin\theta\sin\phi, \cos\theta)$, $0 \leq \theta \leq \pi$, $0 \leq \phi < 2\pi$; the choice of the z-axis is arbitrary. Note that $\frac{\sin\theta}{4\pi}\, \mathrm{d}\theta\mathrm{d}\phi$ is the uniform distribution on the unit sphere. Then, the marginals of the corresponding approximate joint measurement M_ϵ turn out to be

$$\mathsf{M}_{\epsilon\,[\vec{a}]}(x) = \frac{1}{2}\left(\mathbf{1} + \frac{1-2\epsilon}{2}\, x\vec{a}\cdot\vec{\sigma}\right), \qquad x = \pm 1. \tag{25}$$

In this case we can not sum infinite relative entropies, but we can slightly modify our approach and, instead of computing the total information loss, we consider the mean information loss. Thus, our *error function* now is

$$\begin{aligned} \overline{S}[\mathcal{A}\|\mathsf{M}_\epsilon](\rho) &= \int_0^\pi \mathrm{d}\theta\, \frac{\sin\theta}{4\pi} \int_0^{2\pi} \mathrm{d}\phi\, S\big(\mathsf{A}^\rho_{\vec{a}} \| \mathsf{M}^\rho_{\epsilon\,[\vec{a}]}\big), \\ \vec{a} &= \begin{pmatrix} \sin\theta\cos\phi \\ \sin\theta\sin\phi \\ \cos\theta \end{pmatrix}, \end{aligned} \tag{26}$$

where $\mathcal{A} = \{\mathsf{A}_{\vec{a}}\}_{\vec{a}}$ denotes the collection of all the possible spin components. Then

- for every state ρ there is a $SO(3)$-covariant observable F_0 such that the corresponding M_0 minimizes $\overline{S}[\mathcal{A}\|\mathsf{M}_\epsilon](\rho)$,

$$\begin{aligned} \mathsf{F}_0(\mathrm{d}\theta,\, \mathrm{d}\phi) &= \mathsf{F}_+(\mathrm{d}\theta,\, \mathrm{d}\phi) \\ &= \Big\{\mathbf{1} + \sin\theta\left(\cos\phi\, \sigma_1 + \sin\phi\, \sigma_2\right) + \cos\theta\, \sigma_3\Big\} \frac{\sin\theta}{4\pi}\, \mathrm{d}\theta\mathrm{d}\phi, \end{aligned} \tag{27}$$

- the optimal observable F_0 is independent of ρ,

- the marginals of M_0 are equally noisy versions of the target observables,

$$\mathsf{M}_{0\,[\vec{a}]} = \frac{1}{2}\mathsf{A}_{\vec{a}} + \frac{1}{2}\frac{\mathbb{1}}{2},$$

- F_0 provides a lower bound for the error function:

$$\overline{S}[\mathcal{A}\|\mathsf{M}_\epsilon](\rho) \geq \overline{S}[\mathcal{A}\|\mathsf{M}_0](\rho) = \frac{(1+r)^2}{4r}\log\frac{2(1+r)}{2+r} + \frac{(1-r)^2}{4r}\log\frac{2-r}{2(1-r)} + \frac{1}{4r}\log\frac{2+r}{2-r} - \frac{\log e}{2} \quad (28)$$

  for every ρ and every approximate joint measurement M_ϵ coming from a $SO(3)$-covariant F_ϵ,
- for every $\rho \neq \frac{1}{2}$ the lower bound (28) is strictly positive and F_0 is the unique optimal $SO(3)$-covariant observable,
- for $\rho = \frac{1}{2}$ the lower bound (28) vanishes and every $SO(3)$-covariant observable F_ϵ is optimal.

The inequality (28) is our state-dependent entropic MUR. The lower bound is always non negative and it vanishes only for the maximally chaotic state.

Let us note that [6] characterizes the error of the approximating device with respect to the target reference by the worst direction $\vec{a}$. Similarly, we could find an alternative entropic MUR by studying $\max_{\vec{a}} S\big(\mathsf{A}^\rho_{\vec{a}}\|\mathsf{M}^\rho_{\epsilon\,[\vec{a}]}\big)$, but we prefer our error function because it allows a direct comparison with the previous sections.

In this framework, we introduce the *divergence* of M_ϵ from the collection $\mathcal{A}$ of all the spin components as the worst-case mean information loss,

$$\overline{D}(\mathcal{A}\|\mathsf{M}_\epsilon) = \sup_\rho \overline{S}[\mathcal{A}\|\mathsf{M}_\epsilon](\rho). \quad (29)$$

Then

- $\overline{D}(\mathcal{A}\|\mathsf{M}_\epsilon) = \overline{S}[\mathcal{A}\|\mathsf{M}_\epsilon](\rho_*)$ for every *pure* state ρ_*,
- there is a unique $SO(3)$-covariant observable F_ϵ giving an optimal approximate joint measurement M of all spin components $\mathsf{A}_{\vec{a}}$, that is F_0 (27),
- the corresponding measurement M_0 provides a lower bound for the divergence: for every approximate joint measurement M_ϵ coming from a $SO(3)$-covariant observable F_ϵ we have

$$\overline{D}(\mathcal{A}\|\mathsf{M}_\epsilon) \geq \overline{D}(\mathcal{A}\|\mathsf{M}_0) = \frac{3}{4}\log\frac{4}{3} - \frac{\log e - 1}{2}. \quad (30)$$

The inequality (30) is our state independent entropic MUR, which now can be formulated also as a statement about the mean loss of information that occurs in every *pure* preparation of the system: for every approximate joint measurement M_ϵ of all the spin components,

$$\overline{S}[\mathcal{A}\|\mathsf{M}_\epsilon](\rho) \geq \frac{3}{4}\log\frac{4}{3} - \frac{\log \mathrm{e} - 1}{2}, \tag{31}$$

for all the pure states ρ.

In order to compare the lower bound in the previous MURs (16) and (19) with (30), we divide each member in (16) and (19) by the number of the target observables, so that every time the mean information loss is considered. The lower bounds in the three resulting MURs are:

$$\overline{c}_{\mathrm{inc}}(\mathsf{X},\mathsf{Y}) = \frac{1}{2}\log\left[2\left(2-\sqrt{2}\right)\right] \simeq 0.114223$$

$$\overline{c}_{\mathrm{inc}}(\mathsf{X},\mathsf{Y},\mathsf{Z}) = \frac{1}{3}\log\left(3-\sqrt{3}\right) \simeq 0.114166$$

$$\overline{c}_{\mathrm{inc}}(\mathcal{A}) = \frac{3}{4}\log\frac{4}{3} - \frac{\log \mathrm{e} - 1}{2} = 2 - \frac{3}{4}\log 3 - \frac{1}{2}\log \mathrm{e} \simeq 0.0899306040$$

so that

$$\overline{c}_{\mathrm{inc}}(\mathcal{A}) < \overline{c}_{\mathrm{inc}}(\mathsf{X},\mathsf{Y},\mathsf{Z}) < \overline{c}_{\mathrm{inc}}(\mathsf{X},\mathsf{Y}),$$

which shows that the higher mean loss of information is in the first step: to make compatible two orthogonal components.

6. Proofs

The results of Sections 3 and 5 are new and they are proved in the following.

Proofs of Section 3. The results for the two orthogonal spin components X and Y come from the direct computation of the error function for D_4-covariant bi-observables M (7):

$$S[\mathsf{X},\mathsf{Y}\|\mathsf{M}](\rho) = \sum_{k=1}^{2}\left\{\frac{1+r_k}{2}\log\frac{1+r_k}{1+cr_k} + \frac{1-r_k}{2}\log\frac{1-r_k}{1-cr_k}\right\} = \sum_{k=1}^{2} s_k(c).$$

If $\vec{r}$ is perpendicular to the plane xy, then $r_1 = r_2 = 0$ and the error function $S[\mathsf{X},\mathsf{Y}\|\mathsf{M}](\rho)$ is null for every c, so that (8) trivially holds. On the other hand, for every $r_k \neq 0$, the function $s_k(c)$ is decreasing in c because of general properties of relative entropy: the function is convex in $c \in [-1,1]$, vanishes in $c = 1$ and is positive elsewhere. As we can allow only for $|c| \leq 1/\sqrt{2}$, the optimal observable M_0 is given by $c = 1/\sqrt{2}$, that is (8). In both cases, (8) implies all the other results for X and Y.

Similarly, the results for the three orthogonal spin components X, Y and Z come from the direct computation of the error function for O-covariant bi-observables M (11):

$$S[\mathsf{X},\mathsf{Y},\mathsf{Z}\|\mathsf{M}](\rho) = \sum_{k=1}^{3}\left\{\frac{1+r_k}{2}\log\frac{1+r_k}{1+cr_k} + \frac{1-r_k}{2}\log\frac{1-r_k}{1-cr_k}\right\}.$$

Proofs of Section 5. Given the $SO(3)$-covariant observable F_ϵ (23), the marginals (25) come from direct computation. Then the error function $\overline{S}[\mathcal{A}\|\mathsf{M}_\epsilon](\rho)$ (26) can be explicitly computed and its monotone properties with respect to r and ϵ can be studied.

By rotation invariance we can take the z-axis parallel to $\vec{r}$; then $\vec{a}\cdot\vec{r} = r\cos\theta$, and the relative entropy of each approximation is

$$S(\mathsf{A}^{\rho}_{\vec{a}}\|\mathsf{M}^{\rho}_{\epsilon\,[\vec{a}]}) = \sum_{x=\pm 1}\frac{1+xr\cos\theta}{2}\log\frac{1+xr\cos\theta}{1+\lambda xr\cos\theta}\,, \lambda = \frac{1-2\epsilon}{2} \in [-1/2, 1/2].$$

This quantity is decreasing in $\lambda \in [-1/2, 1/2]$ if $r\cos\theta \neq 0$, as it is convex in $\lambda \in [-1, 1]$ and it vanishes if and only if $\lambda = 1$, and it is increasing in $r \in [0, 1]$, as it is convex in $r \in [0, 1]$ and it vanishes if and only if $r = 0$. Then the error function (26) becomes

$$\overline{S}[\mathcal{A}\|\mathsf{M}_\epsilon](\rho) = \frac{1}{4}\sum_{x=\pm 1}\int_0^{\pi}\sin\theta\,(1+xr\cos\theta)\log\frac{1+xr\cos\theta}{1+\lambda xr\cos\theta}\,\mathrm{d}\theta\,,$$

which is increasing in $r \in [0, 1]$ if $\lambda \neq 0$, and, if $r \neq 0$, it is decreasing in $\lambda \in [-1/2, 1/2]$ and increasing in $\epsilon \in [0, 1]$. By the changes of variable $z = xr\cos\theta$ we get

$$\overline{S}[\mathcal{A}\|\mathsf{M}_\epsilon](\rho) = \frac{\log \mathrm{e}}{2r}\int_{-r}^{r}(1+z)\ln\frac{1+z}{1+\lambda z}\,\mathrm{d}z,$$

and using

$$\frac{1}{2}\frac{\mathrm{d}}{\mathrm{d}z}\left\{(1+z)^2\ln\frac{1+z}{1+\lambda z} + \frac{1-\lambda}{\lambda}\left[\frac{1-\lambda}{\lambda}\ln(1+\lambda z) - z\right]\right\} = (1+z)\ln\frac{1+z}{1+\lambda z}\,,$$

finally we get

$$\overline{S}[\mathcal{A}\|\mathsf{M}_\epsilon](\rho) = \frac{(1+r)^2}{4r}\log\frac{2(1+r)}{2+(1-2\epsilon)r} - \frac{(1-r)^2}{4r}\log\frac{2(1-r)}{2-(1-2\epsilon)r} + \frac{(1+2\epsilon)^2}{4(1-2\epsilon)^2 r}\log\frac{2+(1-2\epsilon)r}{2-(1-2\epsilon)r} - \frac{1+2\epsilon}{2(1-2\epsilon)}\log \mathrm{e}. \quad (32)$$

Equality (32) and the monotonicity properties with respect to r and ϵ imply all the other results.

References

1. P. Busch, P. Lahti, J.-P. Pellonpää, K. Ylinen, *Quantum Measurement*, Theoretical and Mathematical Physics (Springer International Publishing, Switzerland, 2016).
2. R.F. Werner, *The uncertainty relation for joint measurement of position and momentum*, Quantum Inf. Comput. **4** (2004) 546–562.
3. P. Busch, P. Lahti, R.F. Werner, *Measurement uncertainty relations*, J. Math. Phys. **55** (2014) 042111.
4. P. Busch, P. Lahti, R.F. Werner, *Colloquium: Quantum root-mean-square error and measurement uncertainty relations*, Rev. Mod. Phys. **86** (2014) 1261–1281.
5. P. Busch, P. Lahti, R.F. Werner, *Heisenberg uncertainty for qubit measurements*, Phys. Rev. A **89** (2014) 012129.
6. L. Dammeier, R. Schwonnek, R.F. Werner, *Uncertainty relations for angular momentum*, New J. Phys. **17** (2015) 093046.
7. R.F. Werner, *Uncertainty relations for general phase spaces*, Front. Phys. **11** (2016) 110305.
8. T. Bullock, P. Busch, *Measurement uncertainty relations: characterising optimal error bounds for qubits*, (2018) arXiv:1512.00104v3.
9. F. Buscemi, M.J.W. Hall, M. Ozawa, M.M. Wilde, *Noise and disturbance in quantum measurements: an information-theoretic approach*, Phys. Rev. Lett. **112** (2014) 050401.
10. P.J. Coles, F. Furrer, *State-dependent approach to entropic measurement disturbance relations*, Phys. Lett. A **379** (2015) 105–112.
11. A.A. Abbot, C. Branciard, *Noise and disturbance of Qubit measurements: An information-theoretic characterisation*, Phys. Rev. A **94** (2016) 062110.
12. A. Barchielli, M. Gregoratti, A. Toigo, *Measurement uncertainty relations for position and momentum: Relative entropy formulation*, Entropy **19** (2017) 301.
13. A. Barchielli, M. Gregoratti, A. Toigo, *Measurement uncertainty relations for discrete observables: Relative entropy formulation*, Commun. Math. Phys. **357** (2018) 1253–1304.
14. K.P. Burnham, D.R. Anderson, *Model Selection and Multi-Model Inference*, 2nd edition (Springer, New York, 2002).

15. T.M. Cover, J.A. Thomas, *Elements of Information Theory*, 2nd edition (Wiley, Hoboken, New Jersey, 2006).
16. T. Heinosaari, M.A. Jivulescu, D. Reitzner, M. Ziman, *Approximating incompatible von Neumann measurements simultaneously*, Phys. Rev. A **82** (2010) 032328.
17. T. Heinosaari, P. Stano, D. Reitzner, *Approximate joint measurability of spin along two directions*, Int. J. Quantum Inf. **6** (2008) 975–980.
18. P. Busch, T. Heinosaari, *Approximate joint measurements of qubit observables*, Quantum Inf. Comp. **8** (2008) 797–818.

Infinite Dimensional Analysis, Quantum Probability and Related Topics

pp. 103–112

A COMBINED QUANTUM ALGORITHM AND ITS COMPUTATIONAL COMPLEXITY

SATOSHI IRIYAMA

Department of Information Sciences, Tokyo University of Science, Japan
E-mail: iriyama@is.noda.tus.ac.jp

(dedicated to L. Accardi, I. V. Volovich and T. Hida)

A quantum algorithm for the general search problem is proposed by Ohya *et al.* in 2015. An NP-hard problem can be solved in polynomial steps with high probability by the quantum algorithm which implicates multiple quantum computation. Each quantum algorithm amplifies the success probability using the special dynamics, so called the Chaos amplifier, which is defined by the logistic map. In 2017, we showed that the Chaos amplifier and more general amplification process can be represented by the GKSL master equation in a two-qubit system. In this study, we explain our notion of combined quantum algorithm, and discuss on its computational complexity for the difficult problem.

1. Quantum Algorithm

In 2000, Ohya and Volovich proposed a quantum algorithm with the amplification process based on the logistic map, so called, Chaos amplifier. In the case that success probability is very low, it requires many trials to obtain correct result for the computation. The Chaos amplifier is effective to increase the success probability within the polynomial trials defined as the map on the Abelian state space, not on the quantum state space[1,2]. However it was not clear to implement using unitary gates. Accardi *et al.* showed that there exists unitary process to describe such amplification using stochastic limit[3].

The quantum algorithm with effective amplification process has a large computational ability to solve NP complete problems, and we showed that cooperative quantum algorithm can solve the NP-hard problems, which is search problem, in polynomial time[4]. This quantum algorithm is constructed by quantum algorithms with different initial vectors. Each result is used to construct the next initial vectors of the quantum algorithms. This method is not described by a single quantum algorithm with one measurement. We have to measure the each final states of the quantum algorithm. However the number of total steps is polynomial of the input size[4]. We call this quantum algorithm a combined quantum algorithm.

In this paper, we explain our approach of the quantum algorithm, and show that there exists a two-qubit system to achieve the Chaos amplifier according to [5].

A quantum algorithm is constructed by the following steps:

(1) Prepare a Hilbert space $\mathcal{H}$, for example, $\mathcal{H} = \left(\mathbb{C}^2\right)^{\otimes n}$
(2) Construct an initial vector $|\psi_{in}\rangle \in \mathcal{H}$
(3) Construct unitary operators U to solve the problem
(4) Apply them for the initial vector and obtain a result state $|\psi_{out}\rangle = U|\psi_{in}\rangle$
(5) If necessary, amplify the probability of correct result
(6) Measure an observable with the result state.

Quantum algorithm is described by the dynamical semi-group including one parameter sub-group[6]. Let $\mathcal{H} = \mathbb{C}^2$ be a Hilbert space spanned by $|0\rangle = \binom{1}{0}$ and $|1\rangle = \binom{0}{1}$. A normalized vector $|\psi\rangle = \alpha|0\rangle + \beta|1\rangle \in \mathcal{H}$ is called a qubit. Hadamard transformation given by

$$U_H = \frac{1}{\sqrt{2}}\begin{pmatrix} 1 & 1 \\ 1 & -1 \end{pmatrix}$$

is useful to construct a superposition of all numbers from 0 to $2^n - 1$ on n qubits as follows

$$(U_H)^{\otimes n}|0\rangle \otimes \cdots \otimes |0\rangle = \frac{1}{\sqrt{2^n}}\sum_{i=0}^{2^n-1}|e_i\rangle,$$

where

$$\begin{aligned} |e_0\rangle &= |0\rangle \otimes \cdots \otimes |0\rangle \\ |e_1\rangle &= |1\rangle \otimes |0\rangle \otimes \cdots \otimes |0\rangle \\ &\cdots \\ |e_{2^n-1}\rangle &= |1\rangle \otimes \cdots \otimes |1\rangle \end{aligned}$$

NOT gate U_{NOT}, Controlled gate U_{CN}, Controlled-Controlled-NOT gate U_{CCN}, which are called as fundamental gates, are given by

$$U_{NOT} = |1\rangle\langle 0| + |0\rangle\langle 1|$$

$$U_{CN} = |0\rangle\langle 0| \otimes I + |1\rangle\langle 1| \otimes U_{NOT}$$

$$U_{CCN} \underbrace{|x\rangle \otimes |y\rangle}_{\text{input}} \otimes |0\rangle = \underbrace{|x\rangle \otimes |y\rangle}_{\text{dust}} \otimes \underbrace{|x \wedge y\rangle}_{\text{result}}$$

Logical NOT, AND and OR operations are achieved by products of these gates. Note that U_{CCN} works as an AND operation with dust qubits.

1.1. *Computational complexity*

Definition 1.1. We say the computational complexity is n if the unitary operator U is constructed by a product of n fundamental gates.

Following this definition, decomposition of the unitary operator in quantum algorithm into fundamental gates is interesting problem. There are several language classes defined by a classical Turing machine.

- P: solved in polynomial time by a deterministic Turing machine (DTM).
- NP: verifiable in polynomial time by a DTM (prime factorization).
- NP-complete: belonging to NP and reduced to in polynomial time to NP. (SAT problem)
- NP-hard: if and only if there exists the NP-complete problem which is reduced to. (search problem)

It has the following relation.

$$P \subseteq NP \leq_p NP\text{-complete} \leq_p NP\text{-hard}$$

One of essential problems in NP-hard is searching problem. Let n be a positive integer, and $f : \{0, 1, \cdots, 2^n - 1\} \to \{0, 1\}$ the map holding one of the following conditions:

(A) there exists one $x \in \{0, 1, \cdots, 2^n - 1\}$ such that $f(x) = 1$.
(B) for all x, $f(x) = 0$.

Problem 1.1. (searching problem) Determine whether f satisfies (A) or (B).

In a classical algorithm, one can adopt the Brute-Force algorithm which is a full search algorithm for all x. Then the computational complexity is 2^n. Here we suppose that f is also given as a unitary operator U_f

$$U_f \left|x\right\rangle \otimes \left|0\right\rangle = \left|x\right\rangle \otimes \left|f(x)\right\rangle .$$

We apply it to a superposition state vector, and obtain

$$\begin{aligned} U_f \left|\psi_{in}\right\rangle \otimes \left|0\right\rangle &= U_f \sum_{i=0}^{2^n-1} \left|e_i\right\rangle \otimes \left|0\right\rangle \\ &= \sum_{i=0}^{2^n-1} \left|e_i\right\rangle \otimes \left|f(e_i)\right\rangle = \left|\psi_f\right\rangle . \end{aligned}$$

We put $\rho = \left|\psi_f\right\rangle \left\langle\psi_f\right|$. Taking a partial trace from first to n-th qubits, we have

$$tr_{1,\cdots,n}\rho = (1-p)\left|0\right\rangle\left\langle 0\right| + p\left|1\right\rangle\left\langle 1\right| .$$

If f satisfies (A), then

$$p = \frac{1}{2^n}$$

(B), then

$$p = 0.$$

It implies that the number of trials to solve the problem using quantum algorithm is same as a classical algorithm, 2^n. Then we have to amplify p until enough large to measure, for example, $p \geq \frac{1}{2}$.

2. Amplification Process

Ohya and Volovich [1,2] proposed the channel Λ_{CA} for a positive integer k and a classical state, namely a final state of quantum algorithm, $\rho = (1-p)\left|0\right\rangle\left\langle 0\right| + p\left|1\right\rangle\left\langle 1\right|$

$$\Lambda_{CA}^k(\rho) = \frac{\left(I + g_a^k(p)\,\sigma_3\right)}{2}$$

where σ_3 is the z-component of Pauli matrices, and

$$g_a(x) = ax(1-x), a \in [0,4]$$

$$\mathrm{tr}\Lambda_{CA}^k(\rho)\,\sigma_3 \geq \frac{1}{2}.$$

For $\rho = |0\rangle\langle 0|$ or $|1\rangle\langle 1|$

$$\Lambda_{CA}(\rho) = \rho.$$

Note that the dynamics depends on the initial state essentially, it is an example of state adaptive. The amplification process increases the von Neumann entropy of the system. Let

$$\rho = (1-p)|0\rangle\langle 0| + p|1\rangle\langle 1|$$

be a state at time $t = 0$ with very small $p \simeq \frac{1}{2^n}$ where the von Neumann entropy is also $\simeq 0$. The amplification process increases p and von Neumann entropy. Then this process is an entropy increasing process. After suitable time, p becomes larger than $\frac{1}{2}$, and it decreases the von Neumann entropy. Then this process is changed to an entropy decreasing process. Here we calculate the condition of entropy increasing or decreasing of dynamical process.

2.1. *GKSL master equation*

The GKSL master equation for a state ρ is given by

$$\frac{d}{dt}\rho = -\mathrm{i}[H,\rho] + \sum_{i,j}\left\{L_{ij}\rho L_{ij}^* - \frac{1}{2}\left(L_{ij}^* L_{ij}\rho + \rho L_{ij}^* L_{ij}\right)\right\}$$

where

$$L_{ij} = \sqrt{r_{ij}}\,|i\rangle\langle j|\,, i,j \in \{0,1\}\,, r_{ij} \geq 0.$$

Here, we put the Hamiltonian $H = 0$ and

$$\rho_0 = \begin{pmatrix} \rho_{00} & \rho_{01} \\ \rho_{10} & \rho_{11} \end{pmatrix}, \frac{d}{dt}\rho = \begin{pmatrix} \dot{\rho}_{00} & \dot{\rho}_{01} \\ \dot{\rho}_{10} & \dot{\rho}_{11} \end{pmatrix}.$$

One can calculate

$$\rho(t) = \begin{pmatrix} \frac{r_{01}}{r_{01}+r_{10}} - C_1 e^{-(r_{01}+r_{10})t} & C_2 e^{-\frac{1}{2}(r_{00}+r_{01}+r_{10}+r_{11})t} \\ \bar{C}_2 e^{-\frac{1}{2}(r_{00}+r_{01}+r_{10}+r_{11})t} & \frac{r_{10}}{r_{01}+r_{10}} + C_1 e^{-(r_{01}+r_{10})t} \end{pmatrix}$$

where C_1 and C_2 are arbitrary.

We put $p = \rho_{00}$

Lemma 2.1. *Therefore, the condition of entropy decreasing becomes*

$$p > \frac{r_{01}}{r_{01}+r_{10}} + \frac{r_{10}-r_{01}}{2(r_{01}+r_{10})} e^{(r_{01}+r_{10})t}.$$

In the case of $r_{01} = r_{10}$

$$p > \frac{1}{2}.$$

It means that for all time t the von Neumann entropy decreases for an initial parameter $p > \frac{1}{2}$.

In the case of $r_{01} < r_{10}$

$$\frac{r_{10} - r_{01}}{2\left(r_{01} + r_{10}\right)} e^{\left(r_{01}+r_{10}\right)t} > 0.$$

If the time very large, the initial parameter p does not satisfy the condition. In the case of $r_{01} > r_{10}$

$$\frac{r_{10} - r_{01}}{2\left(r_{01} + r_{10}\right)} e^{\left(r_{01}+r_{10}\right)t} < 0$$

then, after long time t, all p satisfies the condition.

2.2. *Amplification in two-qubit system*

Here we discuss about how to achieve the amplification system on two qubit system. Let $\mathcal{E} : \mathfrak{S}(\mathcal{H}) \to \mathfrak{S}(\mathcal{H}) \otimes \mathfrak{S}(\mathcal{H})$ be a lifting and

$$\Lambda : \mathfrak{S}(\mathcal{H}) \to \mathfrak{S}(\mathcal{H})$$

a channel. For a state $\rho = (1-p) \left|0\right\rangle \left\langle 0\right| + p \left|1\right\rangle \left\langle 1\right|$, the lifting $\mathcal{E}$ is defined by

$$\mathcal{E}_1(\rho) = \rho \otimes \rho.$$

This lifting is given by two different trials of same quantum algorithm because the final state ρ is prepared by the von Neumann (unread) measurement at the last step.

We consider the following problem.

Problem 2.1. Does there exist parameters of the channel Λ to achieve the Chaos amplifier such that

$$tr_2 \Lambda \mathcal{E}_1(\rho) = \Lambda_{CA}(\rho)?$$

In the paper [5], we obtain the following conditions.

$$\begin{cases} r_{02} + r_{12} - r_{21} - r_{31} = \frac{a}{2} \\ r_{03} + r_{13} = \frac{1}{2} \\ r_{20} + r_{30} = \frac{1}{2} \\ r_{ij} > 0, \text{otherwise} \end{cases}$$

where $a \in [0,4]$. Under the conditions, the quantum channel $\mathrm{tr}_2\Lambda\mathcal{E}$ works as the Chaos amplifier:

$$\Lambda_{CA}(\rho) = \frac{(I + g_a(\rho)\,\sigma_3)}{2}.$$

We also provide the alternative solution for amplification. In the case of

$$r_{01} = r_{02} = r_{03} = r_{13} = r_{20} = r_{30} = 0$$

$$r_{10} + r_{23} = 1$$

$$r_{31} + r_{32} = 2r_{10}$$

$$r_{10} + r_{21} - r_{12} - r_{32} = 1$$

this process increases the success probability more rapidly without chaotic behavior[5].

$$\mathrm{tr}_2\Lambda_r\mathcal{E}_1(\rho) = (1-p)^2\,|0\rangle\langle 0| + (1-(1-p)^2)\,|1\rangle\langle 1|\,.$$

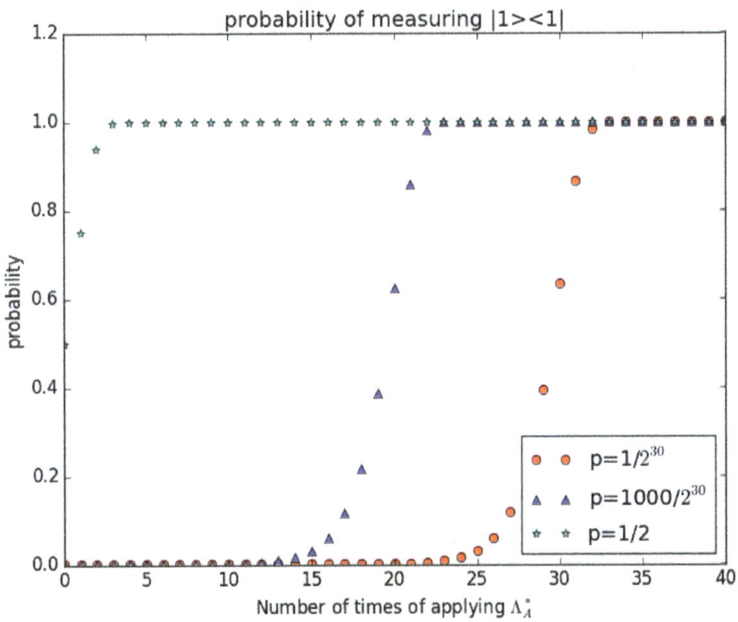

3. Combined Quantum Algorithm for Searching Problem

In this section, we discuss about a combined quantum algorithm to solve the search problem. We propose the quantum algorithm which contains some different trials using the results of previous trials. This quantum algorithm cannot be written by a single one. Let m be a positive integer which can be written by a polynomial in n. Let $\mathcal{H} = \left(\mathbb{C}^2\right)^{\otimes n+m+1}$ be a Hilbert space. We construct the following quantum algorithm $M_Q^{(1)}$ to solve the problem. Let $\left|\psi_{in}^{(1)}\right\rangle = |0^n\rangle \otimes |0^m\rangle \otimes |0\rangle \in \mathcal{H}$ be an initial vector for $M_Q^{(1)}$. The last qubit of $\left|\psi_{in}^{(1)}\right\rangle$ is for the answer, namely "yes" or "no". If the answer is "yes", then the last qubit becomes $|1\rangle$, otherwise $|0\rangle$. We start $M_Q^{(1)}$ with $\varepsilon_1 = 0$.

Step 1. Apply Hadamard gates from the 2nd qubit to the n-th qubit.

$$\begin{aligned} & I \otimes U_H^{\otimes n-1} \otimes I^{m+1} \left|\psi_{in}^{(1)}\right\rangle \\ &= \frac{1}{\sqrt{2^{n-1}}} \left|\varepsilon_1 \left(=0\right)\right\rangle \otimes \left(\sum_{i=0}^{2^{n-1}-1} |e_i\rangle\right) \otimes |0^m\rangle \otimes |0\rangle \qquad (1) \\ &= \left|\psi_1^{(1)}\right\rangle, \qquad (2) \end{aligned}$$

where $|e_i\rangle$ are

$$\begin{aligned} |e_0\rangle &= |0\cdots 0\rangle \\ |e_1\rangle &= |1\cdots 0\rangle \\ &\vdots \\ |e_{2^{n-1}-1}\rangle &= |1\cdots 1\rangle . \end{aligned} \qquad (3)$$

Step 2. Let U_f be the unitary operator on $\mathcal{H} = \left(\mathbb{C}^2\right)^{\otimes n+m+1}$ to compute f, defined by

$$U_f |x\rangle \otimes |0^m\rangle \otimes |0\rangle = |x\rangle \otimes |z_x\rangle \otimes |f(x)\rangle , \qquad (4)$$

where z_x is the dust qubit produced by the computation. Apply the unitary operator U_f to the state. The state becomes

$$\begin{aligned} U_f \left|\psi_1^{(1)}\right\rangle &= \frac{1}{\sqrt{2^{n-1}}} |0\rangle \otimes \left(\sum_{i=0}^{2^{n-1}-1} |e_i\rangle \otimes |z_i\rangle \otimes |f(0e_i)\rangle\right) \\ &= \left|\psi_2^{(1)}\right\rangle, \qquad (5) \end{aligned}$$

where z_i is the dust qubits depending on e_i.

Step 3. Take the last qubit by unread measurement from the final state $\left|\psi_2^{(1)}\right\rangle$ such that

$$(1-p)\left|0\right\rangle\left\langle 0\right| + p\left|1\right\rangle\left\langle 1\right|, \tag{6}$$

where $p = card\left\{x|f(x)=1, x=0\varepsilon_2\cdots\varepsilon_n\right\}/2^{n-1}$.

Step 4. Apply the Chaos Amplifier to check whether the last qubit is in the state $\left|1\right\rangle\left\langle 1\right|$.

After this algorithm, we know that if $\varepsilon_1 = 0$ or 1, then the last qubit is 1 or 0, respectively. We put this process as $M_Q^{(1)}(0^n) = \varepsilon_1$ where 0^n means the initial vector.

For $i = 1, 2, \cdots$, we write the algorithm $M_Q^{(i)}\left(\varepsilon_1, \varepsilon_2, \cdots, \varepsilon_{i-1}, 0^{n-i+1}\right)$ for an initial vector $\left|\psi_{in}^{(i)}\right\rangle = \left|\varepsilon_1, \varepsilon_2, \cdots, \varepsilon_{i-1}, 0^{n-i+1}\right\rangle \otimes \left|0^m\right\rangle \otimes \left|0\right\rangle$ as the following:

Step 1. Apply Hadamard gates from $i+1$-th to n-th qubits.

$$\begin{aligned}
&I^{\otimes i} \otimes U_H^{\otimes n-i} \otimes I^{m+1}\left|\psi_{in}^{(i)}\right\rangle \\
&= \frac{1}{\sqrt{2^{n-i}}}\left|\varepsilon_1, \varepsilon_2, \cdots, \varepsilon_{i-1}\right\rangle \otimes \left(\sum_{k=0}^{2^{n-i}-1}\left|e_k\right\rangle\right) \otimes \left|0^m\right\rangle \otimes \left|0\right\rangle \\
&= \left|\psi_1^{(i)}\right\rangle.
\end{aligned} \tag{7}$$

Step 2. Apply the unitary gate to compute f for the superposition made in the Step 1, and store the result in $n+m+1$-th qubit.

$$\begin{aligned}
&U_f\left|\psi_1^{(i)}\right\rangle \\
&= \frac{1}{\sqrt{2^{n-i}}}\left|\varepsilon_1, \varepsilon_2, \cdots, \varepsilon_{i-1}\right\rangle \\
&\otimes \left(\sum_{k=0}^{2^{n-1}-1}\left|e_k\right\rangle \otimes \left|z_k\right\rangle \otimes \left|f\left(\varepsilon_1, \varepsilon_2, \cdots, \varepsilon_{i-1}, e_k\right)\right\rangle\right) \\
&= \left|\psi_2^{(i)}\right\rangle.
\end{aligned} \tag{8}$$

Step 3. Take the last qubit by the projection from the final state $\left|\psi_2^{(i)}\right\rangle$ such that

$$(1-p)\left|0\right\rangle\left\langle 0\right| + p\left|1\right\rangle\left\langle 1\right|. \tag{9}$$

Step 4. Apply the Chaos Amplifier to the amplitude p, so that we can easily find that the last qubit is $|1\rangle\langle 1|$.

After this algorithm $M_Q^{(i)}\left(\varepsilon_1, \varepsilon_2, \cdots, \varepsilon_{i-1}, 0^{n-i+1}\right)$, we know the bit ε_i such that $f(x) = 1$. Each $M_Q^{(i)}$, $i \geq 2$ use the result of all $M_Q^{(j)}$ $(j < i)$ as an initial vector. We run this algorithm $M_Q^{(i)}$ for each i $(i = 1, \cdots, n)$.

The computational complexity of this combined quantum algorithm is estimated as follows[4].

Theorem 3.1. *The number of unitary gates T in the quantum algorithm for searching problem becomes*

$$T = \frac{13}{8}n^2 - \frac{9}{4}n + nT\left(U_f\right)$$

where $T(U_f)$ is a number of unitary gates for the calculation of given objective function f.

References

1. M. Ohya and I.V. Volovich, J. opt. B, 5, No. 6 639–642, 2003.
2. M. Ohya and I.V. Volovich, Rep. Math. Phys., 52, No. 1, 25–33, 2003.
3. L. Accardi and M. Ohya, Open Systems and Information dynamics, 11, 1–16, 2004.
4. S. Iriyama, M. Ohya, I.V. Volovich, Open Systems and Information Dynamics, Vol. 22, No. 3, 2015.
5. K. Furusho, S. Iriyama, Open Systems and Information Dynamics Vol. 24, No. 02, 1750008 (2017).
6. M. Ohya and I.V. Volovich, Springer, 2011.

Infinite Dimensional Analysis, Quantum Probability and Related Topics

pp. 113–121

ON EFFECTIVE METHODS OF INVESTIGATION OF NONPOSITIVE MAPS

ANDRZEJ JAMIOŁKOWSKI

Institute of Physics, Nicolaus Copernicus University,
87-100 Torun, Poland
E-mail: jam@fizyka.umk.pl

The paper is dedicated to Prof. L. Accardi, Prof. I. V. Volovich and late Prof. T. Hida.

The main aim of the paper is to give an effective procedure which allows us to distinguish between positive and nonpositive maps inside the set of all hermiticity preserving linear transformations. This procedure is important because we do not have any effective method of characterizing positive maps which are not completely positive. The fundamental role in our considerations is played by the well-known in literature matrix CJ. We analyze signs of eigenvalues of this matrix for a given superoperator Φ.

1. Introduction

During the past fifty years positive and completely positive maps represented one of the most active and fertile research topics in matrix algebra and, more generally, in C^*-algebra theory. Very often motivations for such investigations come from physics, in particular from quantum information theory, because positive maps can be used for description of evolution of quantum states [1,2]. Moreover, positive but not completely positive maps can be used to characterize entangled states of quantum systems.

Presently it is already known that constraints imposed by the initial system–environment correlations can lead to nonpositive transformations. Therefore, it is highly desirable to characterize the set of positive maps and their completion (nonpositive maps) in the set of all linear maps on the set of matrices. It appears, that at least for some types of systems, hermicity preserving map can arise as a legitimate subsystem dynamical map from a joint evolution of a system coupled to the bath, cf. e.g. [3,4].

It is now a commonly accepted point of view that central to quantum information theory is the analysis of quantum operations and quantum channels [5]. Mathematically, a quantum operation is given by completely positive trace nonincreasing linear map Φ which acts on the set of operators on a finite-dimensional Hilbert space $\mathcal{H}$. If Φ also preserves trace it is said to be a channel. In this paper we will discuss some *effective methods* (*effective procedures*) of study of certain properties of positive and nonpositive maps, and their signs. In particular, we will analyse the spectra of such maps. In our study of maps we will use the natural identification

$$M_k(\mathbb{C}) \otimes B(\mathcal{H}) \cong M_k(B(\mathcal{H})), \tag{1}$$

where $\mathcal{H}$ denotes a finite-dimensional Hilbert space, $B(\mathcal{H})$ represents the set of all linear operators on $\mathcal{H}$ and $M_k(\mathcal{A})$ for $k = 2, 3, \ldots$ denotes $k \times k$ matrices with elements in an algebra $\mathcal{A}$. On the other hand we have also the identification

$$M_k(\mathbb{C}) \otimes B(\mathcal{H}) \cong B\left(\bigoplus_{i=1}^{k} \mathcal{H}\right). \tag{2}$$

Now we explain what it means that we are interested in effective criteria to check certain properties of quantum maps. By effective criteria we will understand such procedures which employ only a finite number of arithmetic operations on given mathematical data. A beautiful example of such a procedure is given by Sylvester's criterion to determine whether a Hermitian matrix is positive-definite. According to usually used definition, a Hermitian operator A is positive-definite iff $(Ax, x) > 0$ for all $x \in \mathcal{H}$. This means that we have an infinite number of conditions and we are not able to check whether A satisfies all these inequalities. Another possible definition (equivalent to the former one) is given by the condition that all eigenvalues of A should be positive. However, it is well known that, in general, eigenvalues of an operator (a matrix) on $\mathcal{H} \cong \mathbb{C}^n$ cannot be calculated in radicals if $n \geq 5$. Fortunately, there exist Sylvesters' conditions — a Hermitian matrix is positive-definite if and only if all of the leading principal minors of A are positive.

Our aim in this paper is to formulate an effective method which allows us to check whether a given map on $B(\mathcal{H}) \cong M_n(\mathbb{C})$ is nonpositive. We will use the so-called Descartes' rule of signs and we will concentrate on spectra of considered maps. One has to say that our methods can give definite results only for some subsets of nonpositive maps.

In the algebraic formulation of quantum mechanics, a fixed quantum mechanical system is represented by an algebra $\mathcal{A}$ of operators acting on some Hilbert space $\mathcal{H}$. In this approach, the observables of the system are identified with self-adjoint elements in $\mathcal{A}$ and the physical states are given by positive unital functionals on $\mathcal{A}$. We will consider the case when $\mathcal{H}$ is finite-dimensional, then the set of states can be identified with the set of $\mathcal{S}(\mathcal{H})$ of density operators. The evolutions of the system are described by transformations on the set $\mathcal{S}(\mathcal{H})$, or more generally, by linear maps between $\mathcal{S}(\mathcal{H})$ and $\mathcal{S}(\mathcal{K})$, where $\mathcal{H}$ and $\mathcal{K}$ represent two finite-dimensional Hilbert spaces.

In this paper, generalizations of completely positive maps will be explored to answer some questions connected with properties of positive and nonpositive maps. In Section 2 some required definitions and preliminaries will be presented. Section 3 will introduce the reader to the so-called Descartes rule of signs and applications of this rule to hermitian matrices. In Section 4 we will discuss some applications of the Descartes rule to dynamical transformations.

2. Preliminaries and some quantum notions

Quantum theory, in its nonrelativistic formulation, is built on the theory of Hilbert spaces and operators. Assume that $\mathcal{H}$ is a fixed Hilbert space associated with a given quantum system S. We denote by $B(\mathcal{H})$ the set of all linear continuous operators on $\mathcal{H}$. Then the set of states of the system S is, by definition, represented by all semipositive elements of $B(\mathcal{H})$ with trace equal to one.

In the beginning of the seventies it appeared that some natural questions connected with fundamentals of quantum mechanics (more precisely, with the theory of open quantum systems) lead to investigations of linear maps in a real Banach space of self-adjoint operators on a fixed Hilbert space (for details we refer to [1,6,7]). The concept of a Banach space with the partial order defined by a specific cone, namely, the cone of positive semi-definite operators, constitutes a basic idea in the description of open quantum systems and in the quantum information theory [1].

In this paper we consider finite-dimensional Hilbert spaces — such spaces are the current focus in quantum computing and quantum information theory for experimental reasons. This means we assume $\mathcal{H} \cong \mathbb{C}^n$ and $B(\mathcal{H}) \cong \mathbb{M}_n(\mathbb{C})$. It should be stressed that to describe all possible changes of quantum states one has to consider some specific linear operators in $B(\mathcal{H})$. Very often, at least in physical literature, they are called *superoperators*. The general form of such maps is well known. Namely, for a given superoperator $\Phi : B(\mathcal{H}) \longrightarrow B(\mathcal{H})$, $\dim \mathcal{H} < \infty$, there always exists an operator-sum representation given by

$$\Phi(X) = \sum_{i=1}^{s} A_i X B_i \,, \tag{3}$$

where A_i, B_i are elements of $B(\mathcal{H})$, for all $i = 1, ..., s$. As it was mentioned in Section 1, a particular class of such maps, the so-called *completely positive maps* (or in physical terminology *quantum operations*), plays a prominent role in formulations of evolution of open quantum systems and in the theory of quantum measurements.

It is obvious from the above considerations that properties of superoperators $\Phi : B(\mathcal{H}) \longrightarrow B(\mathcal{H})$ are connected with properties of the sets $\{A_1, ..., A_s\}$ and $\{B_1, ..., B_s\}$ of matrices. In particular, in case of completely positive maps, we have $B_i = A_i^*$ for $i = 1, ..., s$, where A_i^* denotes matrix adjoint to A_i, and thus quantum operations are described by maps of the form

$$\Phi(X) = \sum_{i=1}^{s} A_i X A_i^* . \tag{4}$$

The matrices A_i occurring in the expression above are called the *Kraus coefficients* of Φ and the $*$-matrix algebra generated by them is called the *interaction algebra*.

The so-called Choi–Jamiołkowski isomorphism (CJ) is an isomorphism between linear maps from $M_n(\mathbb{C})$ to $M_n(\mathbb{C})$ and operators living in the tensor product space $M_n(\mathbb{C}) \otimes M_n(\mathbb{C})$, where we use representation (1) with $B(\mathcal{H}) \cong M_n(\mathbb{C})$. The matrix

$$J(\Phi) := \sum_{i,j=1}^{n} |i\rangle\langle j| \otimes \Phi(|i\rangle\langle j|) \tag{5}$$

is customarily called CJ matrix of the map Φ. In the formula (5) $\{|i\rangle\}$ denotes the natural basis in the space $\mathcal{H}$.

It is known that the set of maps which preserve hermiticity, that is maps Φ such that $\Phi(X)$ is hermitian whenever X is hermitian, corresponds according to (5) to the set of hermitian operators. Now, using the spectral decomposition of $J(\Phi)$, it follows that if Φ preserves the hermicity, then there are real constants $\{\alpha_i\}$ such that

$$\Phi(X) = \sum_{i=1}^{r} \alpha_i A_i X A_i^* . \tag{6}$$

When α_i for $i = 1, \ldots, p$ are positive and the rest of α_i is negative then the above formula can be rewritten in the form

$$\Phi(X) = \sum_{i=1}^{r-p} \alpha_i A_i X A_i^* - \sum_{j=r-p+1}^{r} |\alpha_j| A_j X A_j^* , \tag{7}$$

which means that each map preserving hermiticity can be represented as difference of two completely positive maps.[8] If we absorb the magnitude of the constant α_i into the operator A_i then we can write (6) in the form (7) where each $\alpha_i = \pm 1$.

One says that a linear map Φ is positive if $\Phi(X)$ is positive semidefinite whenever X is semipositive. In this case the associated operator $J(\Phi)$ defined by (6) satisfies the inequality

$$\Big((|\Psi_1\rangle \otimes |\Psi_2\rangle), J(\Phi)(|\Psi_1\rangle \otimes |\Psi_2\rangle) \Big) \geq 0 \tag{8}$$

for all $|\Psi_1\rangle$ and $|\Psi_2\rangle$.

In quantum information (cf. e.g. [9]) these maps that are positive on separable states are called block-positive operators. Unfortunately, we do not have here any effective way of testing whether or not a matrix $J(\Phi)$ is block-positive, i.e. whether or not a linear map Φ is positive.

However, also in this case we know that any positive map can be written as the difference of two completely positive maps (similar to equation (7)) [8]. Therefore we have the following situation. All maps which preserve hermiticity constitute the real vector space. We will denote the space by $M_n^*(\mathbb{C})$. As a part of this space we can consider all maps which preserve positivity. All these maps constitute the convex subset of the space $M_n^*(\mathbb{C})$. The natural question arises. How to recognize that a given map Φ belongs to $M_n^*(\mathbb{C})$ but does not belong to the set of positive maps? We discuss the question in the next section.

3. Maps which are not positive

In this section we will discuss the question, how to distinguish between linear maps (superoperators) $\Phi : M_n^*(\mathbb{C}) \longrightarrow M_n^*(\mathbb{C})$ which are positive or nonpositive. As we know, all positive maps are, in general all maps which preserve hermiticity, and have similar representations — they are differences of some completely positive maps. Unfortunately, in the case of positive maps we do not have any universal, effective method of verification if Φ is positive or not. We would like to have a method which would be similar to the one known for completely positive maps. We recall that a map Φ is completely positive if and only if the corresponding CJ matrix $J(\Phi)$ is semipositive [10]. To this end, we can analyse some properties of the spectrum of CJ matrix and to formulate effective criteria of non-positivity of maps $\Phi : M_n^*(\mathbb{C}) \longrightarrow M_n^*(\mathbb{C})$. It should be stressed that for our analysis we do not need the values of eigenvalues of $J(\Phi)$ but only their signs.

It is known since the eighties that if two Hilbert spaces $\mathcal{H}_1$ and $\mathcal{H}_2$ are finite-dimensional and Ω is a block-semipositive operator on $\mathcal{H}_1 \otimes \mathcal{H}_2$ then the number η of negative eigenvalues of Ω satisfies the inequality

$$\eta \leq (m-1)(n-1)\,, \tag{9}$$

where $m = \dim \mathcal{H}_1$ and $n = \dim \mathcal{H}_2$. This implies that every $n^2 \times n^2$ complex matrix that is block-semipositive on $\mathbb{C}^n \otimes \mathbb{C}^n$ has at least $2n-1$ nonnegative eigenvalues. In case, if the matrix Ω does not have zero as an eigenvalue, then it has at least $2n-1$ positive eigenvalues among all n^2 eigenvalues.

In the last decade the number of positive and negative eigenvalues of the CJ matrix $J(\Phi)$ was discussed in [11,12]. One can generalize the definition of a positive map to a k-positive map such that Φ is k-positive when the map $I_k \otimes \Phi$ is positive. The isomorphism (5) allows us to describe properties of k-positive maps. It appears that if we represent Φ in the so-called Kraus form

$$\Phi(X) = \sum_{i=1}^{\omega} A_i X A_i^* - \sum_{j=1}^{\eta} B_j X B_j^*\,, \tag{10}$$

and matrices A_i, B_j are linearly independent for $i = 1, \ldots, \omega$ and $j = 1, \ldots, \eta$, then for Φ k-positive we obtain

$$\eta \leq (m-k)(n-k)\,. \tag{11}$$

As a consequence of the inequality (9) one can say that if the CJ matrix $J(\Phi)$ has the number η of the negative eigenvalues larger than $(n-1)^2$ then the map Φ which preserves hermiticity cannot be positive.

This observation immediately suggests the question: can we recognize the number of positive or negative eigenvalues for a given $J(\Phi)$ (we assume that Φ is defined by (3) that is by sets of matrices $A_1, \ldots, A_s$ and $B_1, \ldots, B_s$ or as a difference of two completely positive maps). The answer to this question can be formulated on the basis of general Descartes rule of signs.

We recall it in the form of theorem.

Descartes theorem.

For a given polynomial $p(x)$, the number of all positive roots (a root of multiplicity k being counted as k roots) is equal to the number of variations of sign in the sequence of coefficients of this polynomial or is less by an even number.

In our consideration the polynomial $p(x)$ is realized by the characteristic polynomial of the matrix $J(\Phi)$, which for a given form (3) can be calculated without problems. Moreover, because of preservation of hermiticity by Φ, the matrix $J(\Phi)$ is hermitian. In this case the Descartes rule admits a certain refinement.

Now we come to the main point of our analysis. Namely, if all the roots of the polynomial $p(x)$ are real — in our case we have such situation because $J(\Phi)$ is hermitian — the number η of negative roots is equal to the number of variations in sign in the sequence of coefficients of the polynomial $p(-x)$. In other words, in our case, it is enough to count variations of sign in the characteristic polynomial of the matrix $J(\Phi)$ with the negative argument: $\chi(-x) = \det(J(\Phi) + xI)$.

It is known that when we look for the coefficients of the characteristic polynomial of any matrix Ω we can represent the coefficients in terms of traces of powers of Ω.[13] Indeed, for any complex matrix Ω its characteristic polynomial $\chi(\Omega)$ factors into the form

$$\chi(\Omega) = \prod_{i=1}^{n} (x - \lambda_i)\,, \tag{12}$$

where λ_i are the eigenvalues of the matrix Ω. The coefficients $c_1, \ldots, c_n$ of the characteristic polynomial $x^n + c_1 x^{n-1} + \cdots + c_n$ can be represented by symmetric powers of the eigenvalues

$$T_k := \sum_{i=1}^{n} \lambda_i^k = \mathrm{Tr}(\Omega^k)\,, \tag{13}$$

where the summation is over all eigenvalues of Ω (cf. e.g. [13,14]).

The relations between c_i and T_j are given by

$$kc_k + c_{k-1}T_1 + c_{k-2}T_2 + \cdots + c_1 T_{k-1} + T_k = 0\,, \tag{14}$$

where $k = 1, \ldots, n$. The above formula allows us to represent recursively $c_1, \ldots, c_n$ as functions of $T_1, \ldots, T_n$. The formula (14), although not promising short computation, nevertheless makes it possible to compute effectively the coefficients $c_1, \ldots, c_n$ independently as combinations of $T_1, \ldots, T_n$. One obtains [14]

$$c_k = \sum_{(p_1, p_2, \ldots, p_k) \in S_k} \prod_{m=1}^{k} \frac{1}{p_m!} \left[-\frac{T_m}{m} \right]^{p_m}, \quad k = 1, 2, \ldots, n, \tag{15}$$

with T_m standing for $\mathrm{Tr}(\Omega^m)$ and S_k being a set including all nonnegative-integer solutions $\{(p_1, p_2, \ldots, p_k)\}$ of the equation $p_1 + 2p_2 + \cdots kp_k = k$.

In our case it means that the sequence of traces of powers of the CJ matrix $J(\Phi)$

$$\mathrm{Tr}\, J(\Phi),\ \mathrm{Tr}[J(\Phi)]^2,\ \ldots,\ \mathrm{Tr}[J(\Phi)]^n \tag{16}$$

can play a key role to distinguish between the positive and nonpositive hermitian maps Φ.

4. Conclusions

The aim to characterize positive maps which are not completely positive is very difficult. A partial information about a given map Φ, which preserves hermiticity, can be obtained by analysing the signs of the roots of characteristic polynomial of the CJ matrix $J(\Phi)$. It appears that for the study of these signs one can use *effectively* the Descartes rule of signs because in the considered situation the matrix $J(\Phi)$ has only real roots.

Acknowledgement

The author acknowledges the support by the Polish National Science Center (NCN) under the grant number 2015/19/B/ST1/03095.

References

1. A. Kossakowski: Rep. Math. Phys. **3** (1972) 247.
2. A. Jamiołkowski: Rep. Math. Phys. **3** (1972) 275.
3. J. M. Dominy, D. A. Lidar: Quantum Inf. Process **15** (2016) 1349.
4. K. Modi *et al.*: Phys. Rev. A **86** (2012) 064102.
5. R. Horodecki *et al.*: Rev. Mod. Phys. **81** (2009) 865.
6. A. Jamiołkowski: in Open Systems, Entanglement and Quantum Optics, InTech (2013) 3.

7. D. Chruściński: in Open Systems, Entanglement and Quantum Optics, In-Tech (2013) 21.
8. T. Jordan *et al.*: Phys. Rev. A **70** (2004) 052110.
9. I. Bengtsson, K. Życzkowski: Geometry of Quantum States, Cambridge Univ. Press 2006.
10. M. D. Choi: Linear Alg. Appl. **12** (1975) 95.
11. G. Sarbicki: J. Phys. A **41** (2008) 375303.
12. D. Chruściński, G. Sarbicki: J. Phys. A **47** (2014) 483001.
13. R. R. Silva: J. Math. Phys. **39** (1998) 6206.
14. Hong-Hao Zhang *et al.*: Commun. Theor. Phys. **49** (2008) 801.

Infinite Dimensional Analysis, Quantum Probability and Related Topics

pp. 123–128

QUANTUM-LIKE MODELING OF THE ORDER EFFECT IN DECISION MAKING: POVM VIEWPOINT ON THE WANG-BUSEMEYER QQ-EQUALITY

A. LEBEDEV and A. KHRENNIKOV

International Center for Mathematical Modeling in Physics, Engineering, Economics, and Cognitive Science, Linnaeus University, Växjö, S-35195, Sweden

In recent years, quantum theory has been actively used in areas outside of physics, such as psychology, sociology, theory of decision-making, game theory, and others. In particular, quantum formalism (especially its probabilistic counterpart) is used to explain the paradoxes arising in cognitive psychology and decision making. Wang and Busemeyer invented a quantum model and approach as well as non-parametric equality (so-called QQ-equality), explaining the questions order effect. The primary objective of this note is to test the possibility to expand the Wang-Busemeyer model by considering questions which are mathematically represented by positive operator valued measures. We found that, for such observables, the QQ-equality can be violated. But, we also showed that, in principle, it is possible to reduce the expanded model to the original Wang-Busemeyer model by expanding the context of the questions.

Keywords: Quantum probability, psychology, decision making, quantum-like models, QQ-equality, open quantum systems, Neumark dilation theorem, positive operator valued measures.

1. Introduction

Quantum mechanics was originally created to explain the paradoxes arising in classical physics. At the same time, a powerful mathematical apparatus of quantum probability theory was created, which was later effectively used to explain the paradoxes not only in physics, but also in other fields, such as cognitive psychology, decision making, and social sciences, see, for example, monographs [1] - [8] and a few recent representative papers [9] - [18]. (Such modeling is known as *quantum-like* — to distinguish it from genuine quantum models of cognition.) In particular, Wang and Busemeyer [14] used the quantum formalism and methodology of experiment to explain order effects in question answering.

Wang and Busemeyer established a non-parametric inequality (known as QQ-inequality) to which the probabilities of an experiment must satisfy in order for a quantum model to exist for them, as follows:

$$p(AyBn) + p(AnBy) - p(ByAn) - p(BnAy) = 0,$$

where A and B correspond to questions with two possible outcomes "Yes" and "No". The joint probabilities are the probabilities of receiving given answers to questions A and B in the same order as they appear, e.g. $P(AyBn)$ means the probability to obtain negative answer to the question B before obtaining affirmative answer to the question A. The quantum-like model assumes that questions are represented by Hermitian operators; therefore the answers Ay, An, By, Bn are represented by orthogonal projectors.

The following questions naturally arise:

(1) Is it imperative to require these operators to be projectors?
(2) Is it possible to expand the context of the questions in such a way that, although the original operators are not projectors, the extended questions would already correspond to the projectors?

The paper presents examples of measurement operators corresponding to POVM for which the QQ-inequality does not hold. The dependence of the left part of the QQ-inequality on the state is discussed. Further, using the Neumark theorem[19], lifting of such operators are constructed for which this equality does hold.

2. The example of violation QQ-inequality

POVM stands for 'Positive operator-valued measure'. More precisely, it is the measurement operators $\{E_a\}$ which form a complete set of Hermitian non-negative operators. It means that it has the following properties:

(1) $E_a = E^\dagger$;
(2) $\langle\phi|E_a|\phi\rangle \geq 0$ for all vectors $|\phi\rangle$
(3) $\sum_a E_a = 1$.

POVM is the most general formulation of the measurement in physics. For atomic POVMs which can be represented as $E_a = M_a M_a^\dagger$, the state after a measurement is given by the formula

$$\frac{M_a \rho M_a^\dagger}{tr(E_a \rho)}.$$

The operators corresponding to two measurement procedures are

$$Q_1 + Q_2 = 1$$

$$P_1 + P_2 = 1.$$

First:

$$\begin{pmatrix} 5/6 & 1/\sqrt{12} \\ 1/\sqrt{12} & 1/2 \end{pmatrix} + \begin{pmatrix} 1/6 & -1/\sqrt{12} \\ -1/\sqrt{12} & 1/2 \end{pmatrix} = 1.$$

Moreover, the second:

$$\begin{pmatrix} 1/6 & 1/\sqrt{12} \\ 1/\sqrt{12} & 1/2 \end{pmatrix} + \begin{pmatrix} 5/6 & -1/\sqrt{12} \\ -1/\sqrt{12} & 1/2 \end{pmatrix} = 1.$$

We need to calculate the following quantity:

$$p(AyBn) + p(AnBy) - p(ByAn) - p(BnAy)$$

where A represents the first measurement and B represents the second one. Which is, in our terms,

$$Tr(P_2Q_1\rho Q_1P_2 + P_1Q_2\rho Q_2P_1 - Q_2P_1\rho P_1Q_2 - Q_1P_2\rho P_2Q_1)$$

$$= Tr(P_2Q_1\rho Q_1P_2 + P_1Q_2\rho Q_2 - Q_2P_1\rho P_1 - Q_1P_2\rho P_2Q_1)$$

$$= Tr(Q_1P_2P_2Q_1\rho + Q_2P_1Q_2\rho - P_1Q_2P_1\rho - P_2Q_1Q_1P_2\rho).$$

We calculate the products and obtain

$$= Tr\rho \begin{pmatrix} 0 & \frac{2}{27\sqrt{3}} \\ \frac{2}{27\sqrt{3}} & 0 \end{pmatrix} \tag{1}$$

which is equal to

$$\frac{2}{27\sqrt{3}} \neq 0$$

even in the case when the pure state ϕ is represented by the density matrix:

$$\rho = \frac{1}{2}\begin{pmatrix} 1 & 1 \\ 1 & 1 \end{pmatrix}.$$

3. State dependence of QQ

We can check for which pure states the left-hand side of the QQ-equality is not equal to zero:

$$\rho = \frac{1}{|\alpha|^2 + |\beta|^2} \begin{pmatrix} |\alpha|^2 & \bar{\alpha}\beta \\ \bar{\beta}\alpha & |\beta|^2 \end{pmatrix}$$

(1) became ($c \neq 0$)

$$c\,Tr\left(\begin{pmatrix} \bar{\alpha}\beta & |\alpha|^2 \\ |\beta|^2 & \bar{\beta}\alpha \end{pmatrix}\right) = c(\bar{\alpha}\beta + \bar{\beta}\alpha).$$

We can now find for which pure states the previous is equal to 0:

$$\alpha = x_1 + ix_2,\ \beta = y_1 + iy_2$$

$$(x_1 - ix_2)(y_1 + iy_2) + (x_1 + ix_2)(y_1 - iy_2) = 0$$

$$\Leftrightarrow x_1 y_1 + x_2 y_2 = 0$$

which means that

$$\alpha = x_2 i\beta,$$

where x_2 is a real number.

Examples of such pure states are the following:

$$\beta = 1,\ \alpha = i,\ x_2 = 1,\ \rho = \frac{1}{2}\begin{pmatrix} 1 & -i \\ i & 1 \end{pmatrix}$$

$$\beta = 1,\ \alpha = 2i,\ x_2 = 2,\ \rho = \frac{1}{5}\begin{pmatrix} 4 & -2i \\ 2i & 1 \end{pmatrix}.$$

The sum of these operators with coefficients $\frac{1}{2}$ gives the example of mixed state of that kind:

$$\rho = \frac{1}{4}\begin{pmatrix} 1 & -i \\ i & 1 \end{pmatrix} + \frac{1}{10}\begin{pmatrix} 4 & -2i \\ 2i & 1 \end{pmatrix}.$$

4. Lifting

Neumark's theorem states that, for any given POVM the Hilbert space, can be extended to a larger space that POVM can be realized as performing orthogonal measurements in that larger Hilbert space. We denote this process of obtaining the mapping from POVM to orthogonal measurements as 'lifting'. We can now consider 3-dimensional space and new operators

$$Q'_1 + Q'_2 = 1$$

$$P'_1 + P'_2 = 1.$$

The first:

$$\begin{pmatrix} \frac{5}{6} & \frac{1}{2\sqrt{3}} & \frac{1}{3\sqrt{2}} \\ \frac{1}{2\sqrt{3}} & \frac{1}{2} & -\frac{1}{\sqrt{6}} \\ \frac{1}{3\sqrt{2}} & -\frac{1}{\sqrt{6}} & \frac{2}{3} \end{pmatrix} + \begin{pmatrix} \frac{1}{\sqrt{6}} & -\frac{1}{2\sqrt{3}} & -\frac{1}{3\sqrt{2}} \\ -\frac{1}{2\sqrt{3}} & \frac{1}{2} & \frac{1}{\sqrt{6}} \\ -\frac{1}{3\sqrt{2}} & \frac{1}{\sqrt{6}} & \frac{1}{3} \end{pmatrix} = 1.$$

The second:

$$\begin{pmatrix} \frac{1}{\sqrt{6}} & \frac{1}{2\sqrt{3}} & -\frac{1}{3\sqrt{2}} \\ \frac{1}{2\sqrt{3}} & \frac{1}{2} & -\frac{1}{\sqrt{6}} \\ -\frac{1}{3\sqrt{2}} & -\frac{1}{\sqrt{6}} & \frac{1}{3} \end{pmatrix} + \begin{pmatrix} \frac{5}{6} & -\frac{1}{2\sqrt{3}} & \frac{1}{3\sqrt{2}} \\ -\frac{1}{2\sqrt{3}} & \frac{1}{2} & \frac{1}{\sqrt{6}} \\ \frac{1}{3\sqrt{2}} & \frac{1}{\sqrt{6}} & \frac{2}{3} \end{pmatrix} = 1.$$

Note that these operators are self-adjoint. Furthermore, they are projectors.

It is easy to check

$$Q'_1P'_2P'_2Q'_1 + Q'_2P'_1Q'_2 - P'_1Q'_2P'_1 - P'_2Q'_1Q'_1P'_2 = 0.$$

Which means that QQ holds in this case. The reason is that a multiplication of matrix involves summation over three dimensions.

Acknowledgment

This study was performed on the basis of the EU-project "Quantum Information Access and Retrieval Theory" (QUARTZ), Grant No. 721321.

References

1. A. Khrennikov *Information dynamics in cognitive, psychological, social, and anomalous phenomena* (Kluwer, Dordreht, 2004).
2. P. Pylkkänen, *P. Mind, Matter and the Implicate Order* (Springer, Heidelberg and New York, 2007).
3. A. Khrennikov, *Ubiquitous quantum structure: from psychology to finances* (Springer, Berlin-Heidelberg-New York, 2010).

4. J. R. Busemeyer and P. D. Bruza, Quantum models of cognition and decision, (Cambridge University Press, 2012).
5. F. Bagarello, Quantum dynamics for classical systems: with applications of the Number operator, (Wiley Ed., New York, 2012).
6. E. Haven, A. Khrennikov, Quantum social science (Cambridge University Press, 2013).
7. M. Asano, A. Khrennikov, M. Ohya, Y. Tanaka, I. Yamato, *Quantum adaptivity in biology: from genetics to cognition* (Springer, Heidelberg-Berlin-New York, 2015).
8. E. Haven, A. Khrennikov, T. Robinson, *Quantum Methods in Social Science: A First Course* (WSP, Singapore, 2017).
9. E. M. Pothos, J. R. Busemeyer, A quantum probability explanation for violation of rational decision theory. *Proc. Royal. Soc.* **276** (2009).
10. A. Brandenburger, The Relationship between quantum and classical correlation in games. *Game Econ. Behav.* **69** (2010).
11. J. R. Busemeyer, E. M. Pothos, R. Franco, J. Trueblood, A quantum theoretical explanation for probability judgment errors. *Psych. Rev.* **118** (2011).
12. E. M. Pothos, J. R. Busemeyer, Can quantum probability provide a new direction for cognitive modeling? *Behav. Brain Sc.* **36** (2013).
13. F. Bagarello, F. A. Oliveri, Phenomenological operator description of interactions between populations with applications to migration. *Mathematical Models and Methods in Applied Sciences* **23**(03) (2013).
14. Z. Wang and J. R. Busemeyer, A quantum question order model supported by empirical tests of an a priori and precise prediction. *Top Cogn Sci.* **5**(4), (2013).
15. A. Plotnitsky, Are quantum-mechanical-like models possible, or necessary, outside quantum physics? *Phys. Scripta* **T163** 014011, 33, (2014).
16. T. Boyer-Kassem, S. Duchene, E. Guerci, Quantum-like models cannot account for the conjunction fallacy. *Theor. Decis.* **10** (2015).
17. J. Broekaert, I. Basieva, P. Blasiak, E. M. Pothos, Quantum dynamics applied to cognition: A consideration of available options. *Phil. Trans. R. Soc.* **375** (2106), (2017).
18. A. Khrennikov, Quantum Bayesianism as the basis of general theory of decision-making. *Phil. Trans. R. Soc.* **374** 20150245, (2016).
19. J. Preskill, Lecture notes quantum information and computation, http://www.theory.caltech.edu/people/preskill/ph229/notes/chap3.pdf (1998).

Infinite Dimensional Analysis, Quantum Probability and Related Topics

pp. 129–141

ENTANGLEMENT AND ITS CONDITIONALITY

TAKASHI MATSUOKA

Suwa University of Science,
Toyohira 5001, Chino City, Nagano 391-0292, Japan
E-mail: matsuoka@rs.sus.ac.jp

It is shown that quantum states on composite system may be represented by a composition of two types of CP maps, especially one of them is called by *quantum conditionality*, and such construction can be regarded as the quantum analogue of Bayes formula between a classical joint probability and a conditional probability.

1. Introduction

In classical information theory[1,2,3] the mutual entropy $I\,(X \otimes Y)$ on a classical composite system $X \otimes Y$ measures the correlation of the joint probability distribution of two random variables X and Y which is regarded as a common information encoded into X and Y. The mutual entropy can be also regarded as the information gain from Y when we observe X. This interpretation is due to the fact that the mutual entropy can be represented by the conditional entropy $S(Y|X)$ which measures how uncertain we are of Y on the average when we know X. Then Shannon applied the mutual entropy to a communication system in which it can be regarded as a proper measure of the amount of information transmitted from an input system $(X,\, p)$ to an output system $(Y,\, q)$ through a (classical) channel Λ_{c}^* given as a transition probability matrix corresponding to a conditional probability distributions, i.e. Λ_{c}^* changes an input state (or a probability) p into an output state $q = \Lambda_{\mathrm{c}}^*\,(p)$. In this scheme the mutual entropy $I\,(X \otimes Y)$ may be denoted by $I\,(p,\, \Lambda_{\mathrm{c}}^*)$ and the validity of this interpretation is caused by the following inequality which is called the Shannon fundamental inequality

$$0 \leq I(p,\, \Lambda_{\mathrm{c}}^*) \leq \min\left\{S(p), S(\Lambda_{\mathrm{c}}^*\rho)\right\}, \tag{1}$$

where $S\,(p)$ is the Shannon entropy of the probability p. Such Shannon's construction is based on the Bayes formula between joint probabilities and conditional probabilities.

For a quantum composite system $\mathcal{G} \otimes \mathcal{H}$ given by Hilbert space $\mathcal{G}$ and $\mathcal{H}$, we can define a quantum analogue of mutual entropy $I(\mathcal{G} \otimes \mathcal{H})$ by using the von Neumann entropy for density matrices of composite system and its marginal systems:

$$I\left(\mathcal{G} \otimes \mathcal{H}\right) \equiv S\left(\rho_{\mathcal{G}}\right) + S\left(\rho_{\mathcal{H}}\right) - S\left(\rho_{\mathcal{G} \otimes \mathcal{H}}\right), \tag{2}$$

where $S\left(\rho_{\mathcal{G}}\right) = -\mathrm{tr}\rho_{\mathcal{G}}\log\rho_{\mathcal{G}}$ is the von Neumann entropy of a density matrix $\rho_{\mathcal{G}}$ on the quantum system $\mathcal{G}$. The quantum analogue can be also a measure of the correlation encoded into quantum composite system and it is also called *information correlation*[5]. Note, however, that the notion of conditional probability is not unique defined in quantum theory and hence, unlike the classical one, it is not clear how to relate the quantum analogue (2) to the associated channel. Moreover it is well known that $I\left(\mathcal{G} \otimes \mathcal{H}\right)$ in general violates the Shannon fundamental inequality (1). Simple example is provided by the pure entangled state $\rho_{\mathcal{G} \otimes \mathcal{H}} = \left|\Phi\right\rangle\left\langle\Phi\right|$, $\Phi \in \mathcal{G} \otimes \mathcal{H}$. In this case $I(\mathcal{G} \otimes \mathcal{H}) = 2S\left(\rho_{\mathcal{G}}\right) = 2S\left(\rho_{\mathcal{H}}\right)$. The quantum picture of information theory is much more complicated because of the existence of entangled states[3,4,5,6].

There have been mathematical studies of the entanglements in Werner (1989, 1998)[7,8], Bennett *et al.* (1996)[9] and Schumacher (1993a, b)[10,11] in which the entangled state is defined by a compound state which can not be written as convex combination of separable states. Horodeckis (1996)[12] and Peres (1996)[13] gave the another classification of quantum compound states by using the so-called PPT condition. There have been so many works about the classification of entanglements on PPT condition.

On the other hand the notion of quantum conditional probability were introduced by some authors. Cref and Adami (1997)[14] introduced the conditional amplitude operator by the Trotter decomposition of compound density matrix (see (30) in the Section 3.1.). Then they defined the quantum analogue of conditional entropy (see (31)). The another one was proposed by Asory-Kossakowski-Marmo-Sudarshan (2005)[15]. They introduced the construction method of operator which can be regarded as the quantum analogue of conditional probability given by applying CP maps to Choi-Jamiołkowski isomorphism[18,19], and they define the compound state by the product of the marginal density and the quantum analogue of conditional operator. See also the references [20,21].

Belavkin-Ohya (2002)[22] and Belavkin (2011)[23]: they gave also the method of construction of compound state by using CP maps. Both of above methods (Asory *et al.* and Belavkin-Ohya) are based on the duality between quantum maps and quantum states on composite system, and the both characterize the PPT conditions (see Refs.[15,16]).

In this paper we will review their approaches and make clear their relations. Using the relation it will be shown that quantum states on composite system may be represented by the composition of two types of CP maps, especially one of them may be called by *quantum conditionality*, and such construction can be regarded as the quantum analogue of Bayes formula in classical probability theory.

2. Quantum states on entanglement

2.1. *Preliminaries*

Consider a quantum system living in a Hilbert space $\mathcal{H}$ (resp. $\mathcal{G}$). Denote a set of all bounded operators and a set of trace class operators in $\mathcal{H}$ by $\mathcal{B}(\mathcal{H})$ and by $\mathcal{T}(\mathcal{H})$ respectively. Normally a quantum state ς on $\mathcal{B}(\mathcal{H})$ is defined with a linear functional

$$\varsigma(b) := \mathrm{tr} b\sigma, \tag{3}$$

giving the expectation of $b \in \mathcal{B}(\mathcal{H})$. Here σ is a trace class operator, so called a density operator, satisfying the positivity and normalized condition which is in one-to-one correspondence $\varsigma \mapsto \sigma$ with the state ς. Let $\mathcal{S}(\mathcal{H})$ be a set of density operators corresponding to normal states

$$\mathcal{S}(\mathcal{H}) = \{\sigma \in \mathcal{T}(\mathcal{H});\, \sigma \geq 0,\, \mathrm{tr}\sigma = 1\}\,. \tag{4}$$

For simplicity in the subsequent argument we call the linear functional ς and its density σ the same name "state" if without confusion.

Let us recall the definition of completely positive map. A linear map $\chi : \mathcal{B}(\mathcal{H}) \to \mathcal{B}(\mathcal{G})$ is said to be completely positive (CP) if the map

$$\chi_n \,:\, M_n(\mathcal{B}(\mathcal{H})) \to M_n(\mathcal{B}(\mathcal{G}));\, [\,a_{ij}] \to [\chi\,(a_{ij})] \tag{5}$$

is positive for all $n \in \mathbb{N}$, where $M_n(\mathcal{B}(\mathcal{H}))$ stands for $n \times n$ matrices with entries from $\mathcal{B}(\mathcal{H})$. A linear map $\chi : \mathcal{B}(\mathcal{H}) \to \mathcal{B}(\mathcal{G})$ is said to be co-completely positive (CCP) if its composition with transposition τ, i.e. $\chi \circ \tau$ is CP.

Here let us give the definition of transpose operation τ on $\mathcal{B}(\mathcal{H})$ or $\mathcal{B}(\mathcal{G})$. For any element $b \in \mathcal{B}(\mathcal{H})$ let $\tau_{\mathcal{H}}$ be a transposition on a some (fixed) complete orthonormal system (CONS) $\{e_i\} \subseteq \mathcal{H}$, for example one can choice the canonical basis, defined by

$$b^{\mathrm{t}} := \tau_{\mathcal{H}}(b) \equiv J_{\mathcal{H}} b^* J_{\mathcal{H}}, \tag{6}$$

where $J_{\mathcal{H}}$ denotes an isometric complex conjugation defined on $\{e_i\}$ as follows:

$$J_{\mathcal{H}}y = J_{\mathcal{H}}\sum \langle e_i, y\rangle e_i \equiv \sum \overline{\langle e_i, y\rangle} e_i = \sum \langle y, e_i\rangle e_i =: \overline{y}. \tag{7}$$

In the matrix elements representation on the basis $\{e_i\}$ we have

$$\begin{aligned} b^{\mathrm{t}} &= \sum |e_i\rangle \langle e_i, J_{\mathcal{H}} b^* J_{\mathcal{H}} e_j\rangle \langle e_j| \\ &= \sum |e_i\rangle \langle b^* e_j, e_i\rangle \langle e_j| = \sum |e_i\rangle \langle e_j, b e_i\rangle \langle e_j| . \end{aligned}$$

A transposition $\tau_{\mathcal{G}}$ on $\mathcal{B}(\mathcal{G})$ with an isometric complex conjugation $J_{\mathcal{G}}$ are defined analogously on a some (fixed) CONS in $\mathcal{G}$.

2.2. *Duality between quantum states and linear maps, and their classification*

In this subsection we recall the construction methods of compound state given in[22,23] in a little bit modified way.

Now let us consider a state $\theta \in \mathcal{S}(\mathcal{H} \otimes \mathcal{K})$ and let $\phi : \mathcal{B}(\mathcal{H}) \to \mathcal{T}(\mathcal{G})$ (or $\mathcal{B}(\mathcal{H})_*$, i.e. it is the pre-dual space of $\mathcal{B}(\mathcal{H})$) be a linear map defined by

$$\phi(b) \equiv \mathrm{tr}_{\mathcal{H}}(I_{\mathcal{G}} \otimes b)\theta \tag{8}$$

for any $b \in \mathcal{B}(\mathcal{H})$, and its dual map $\phi^* : \mathcal{B}(\mathcal{G}) \to \mathcal{T}(\mathcal{K})$ reads

$$\phi^*(a) \equiv \mathrm{tr}_{\mathcal{G}}(a \otimes I_{\mathcal{H}})\theta \tag{9}$$

for any $a \in \mathcal{B}(\mathcal{G})$, where $I_{\mathcal{G}}$ and $I_{\mathcal{H}}$ are identity operators respectively. Then the compound state ϑ and its expectation $\vartheta(a \otimes b)$ may be rewritten as follows;

$$\vartheta(a \otimes b) = \mathrm{tr}(a \otimes b)\theta = \mathrm{tr}_{\mathcal{G}} a\phi(b) = \mathrm{tr}_{\mathcal{H}}\phi^*(a)b. \tag{10}$$

It is clear that the marginal states read

$$\rho = \phi(I_{\mathcal{H}}) = \mathrm{tr}_{\mathcal{H}}\theta \in \mathcal{T}(\mathcal{G}), \qquad \sigma = \phi^*(I_{\mathcal{G}}) = \mathrm{tr}_{\mathcal{G}}\theta \in \mathcal{T}(\mathcal{H}). \tag{11}$$

Belavkin and Ohya observed that if $\theta \in S(\mathcal{H} \otimes \mathcal{K})$, then both ϕ and ϕ^* are always CCP but not always CP[22]. From the (physical) operational points of view, however, it is suitable to deal with these maps as CP operations. According to this motivation Belavkin[23] introduced the notion of *contra-variant densities* (see Remark 2.2) and gave the another map by reformulating (8) as follows:

$$\phi(b) \equiv \mathrm{tr}_{\mathcal{H}}(I_{\mathcal{G}} \otimes b^{\mathrm{t}})\theta^{\mathrm{t}}. \tag{12}$$

Using (12) we have

$$\begin{aligned}\vartheta(a\otimes b)&=\mathrm{tr}(a\otimes b)\theta\\&=\mathrm{tr}(a^{\mathrm{t}}\otimes b^{\mathrm{t}})\theta^{\mathrm{t}}\\&=\mathrm{tr}_{\mathcal{G}}a^{\mathrm{t}}\phi(b)=\mathrm{tr}_{\mathcal{H}}\phi^{*}(a)b^{\mathrm{t}}.\end{aligned}\tag{13}$$

Actually we can show that ϕ may be a CP map and the complete positivity of ϕ gives the classification of composite system. Let us check this fact in the case of pure compound states, i.e. pure entangled states.

Let ϑ_{P} be a pure state on $\mathcal{B}(\mathcal{H}\otimes\mathcal{K})$ given by

$$\vartheta_{\mathrm{P}}\left(a\otimes b\right)=\left\langle\Phi,\,\left(a\otimes b\right)\Phi\right\rangle=\mathrm{tr}\left(a\otimes b\right)\left|\Phi\right\rangle\left\langle\Phi\right|,\tag{14}$$

where Φ is an entangled vector state, i.e. its Schmidt decomposition is given by $\Phi=\sum\lambda_i^{\frac{1}{2}}g_i\otimes h_i$, $\{g_i\}\subset\mathcal{G}$, $\{h_i\}\subset\mathcal{H}$ are orthonormal systems corresponding to Schatten decompositions, i.e. the spectral decomposition into one-dimensional orthogonal projectors, of marginal state ρ and σ respectively, and $\{\lambda_n\}$ denotes a probability distribution: $\lambda_i\geq 0$ and $\sum\lambda_i=1$. In this case the explicit formulas of the map $\phi:\mathcal{B}(\mathcal{H})\to\mathcal{T}(\mathcal{G})$ is given by the following simple computation:

$$\begin{aligned}\phi(b)&=\mathrm{tr}_{\mathcal{H}}(I_{\mathcal{G}}\otimes b^{\mathrm{t}})\left(\left|\Phi\right\rangle\left\langle\Phi\right|\right)^{\mathrm{t}}\\&=\sum\left|\bar{g}_i\right\rangle\lambda_i^{\frac{1}{2}}\lambda_j^{\frac{1}{2}}\left\langle\bar{g}_j\right|\left\langle\bar{h}_j,\,b^{\mathrm{t}}\bar{h}_i\right\rangle\\&=\sum\left|\bar{g}_i\right\rangle\lambda_i^{\frac{1}{2}}\left\langle h_i,\,bh_j\right\rangle\lambda_j^{\frac{1}{2}}\left\langle\bar{g}_j\right|=\kappa^{*}b\kappa,\end{aligned}\tag{15}$$

where $\kappa\equiv\sum\left|h_i\right\rangle\lambda_i^{\frac{1}{2}}\left\langle\bar{g}_i\right|$ which is the Hilbert-Schmidt operator from $\mathcal{G}$ to $\mathcal{H}$, and $\phi\left(\cdot\right)\equiv\kappa^{*}\cdot\kappa$ is a CP map given in the Steinspring form. However ϕ is not always CCP, because the transposition τ is positive but not CP unless the algebra on which τ is defined is Abelian. In the pure state case, it can be shown that ϕ is not CCP unless θ_{P} is given as a simple product state, i.e., θ_{P} is a separable state (see[16]). Namely the non separability of θ_{P} is characterized by the non co-complete positivity of its associated map ϕ.

In the same way its dual map ϕ^{*} is given by

$$\phi^{*}(a)\equiv\mathrm{tr}_{\mathcal{G}}(a^{\mathrm{t}}\otimes I_{\mathcal{H}})\left|\Phi\right\rangle\left\langle\Phi\right|^{\mathrm{t}}=\kappa^{\mathrm{t}*}a\kappa^{\mathrm{t}}\tag{16}$$

where $\kappa^{\mathrm{t}}:=J_{\mathcal{H}}\kappa^{*}J_{\mathcal{H}}=\sum\left|g_i\right\rangle\lambda_i^{\frac{1}{2}}\left\langle\bar{h}_i\right|$ from $\mathcal{H}$ to $\mathcal{G}$. Furthermore the entangled vector Φ may be represented by κ as follows:

$$\left\langle\zeta\otimes\eta,\,\Phi\right\rangle=\left\langle\eta,\,\kappa J_{\mathcal{G}}\zeta\right\rangle,\tag{17}$$

where $\zeta \in \mathcal{G}$, $\eta \in \mathcal{H}$. Then $\vartheta_{\mathrm{P}}(a \otimes b)$ is represented by

$$\vartheta_{\mathrm{P}}(a \otimes b) = \mathrm{tr}_{\mathcal{G}} a^{\mathrm{t}} \kappa^* b \kappa = \mathrm{tr}_{\mathcal{G}} a^{\mathrm{t}} \phi(b) \tag{18}$$

$$= \mathrm{tr}_{\mathcal{H}} \kappa^{\mathrm{t}*} a \kappa^{\mathrm{t}} b^{\mathrm{t}} = \mathrm{tr}_{\mathcal{H}} \phi^*(a)\, b^{\mathrm{t}} \tag{19}$$

Following[22] we call κ the *entangling operator*. Its marginals are obtained by the entangling operator κ:

$$\rho^{\mathrm{t}} = \phi(I_{\mathcal{H}}) = \kappa^* \kappa = \sum |\bar{g}_i\rangle \lambda_i \langle \bar{g}_i|, \tag{20}$$

$$\sigma^{\mathrm{t}} = \phi^*(I_{\mathcal{G}}) = \kappa^{\mathrm{t}*} \kappa^{\mathrm{t}} = \sum |\bar{h}_i\rangle \lambda_i \langle \bar{h}_i|. \tag{21}$$

Notice that the map ϕ may be recognized as the CP map form $\mathcal{B}(\mathcal{H})$ to the transposed algebra $\widetilde{\mathcal{B}(\mathcal{G})} \equiv \mathcal{B}(\mathcal{G})^{\mathrm{t}} = \{a^{\mathrm{t}}\,;\, a \in \mathcal{B}(\mathcal{G})\}$. In usual argument we may assume that the transposition leaves invariant the sub-algebra $\mathcal{A} \subseteq \mathcal{B}(\mathcal{G})$ such that $\mathcal{A} = \mathcal{A}^{\mathrm{t}}$, however from the notation and operational points of view, it is preferable to distinguish the algebra $\mathcal{A}$ from the transposed algebra $\tilde{\mathcal{A}}$.

We can extend the above construction to mixture states. In general the map ϕ can be given by using a set of stochastic entangling operators $\kappa_i : \mathcal{G} \to \mathcal{H}$. Let $\theta \in \mathcal{S}(\mathcal{G} \otimes \mathcal{H})$ be a mixed compound state with the Schatten decomposition $\theta = \sum |\Phi_i\rangle p_i \langle \Phi_i| = \sum |\Psi_i\rangle \langle \Psi_i|$ where $\Psi_i = p_i^{\frac{1}{2}} \Phi_i$. Applying (15) to the density θ the expectation $\vartheta(a \otimes b)$ may be represented by ϕ or ϕ^* as follows:

$$\vartheta(a \otimes b) = \mathrm{tr}(a \otimes b)^{\mathrm{t}} \theta^{\mathrm{t}} = \sum \mathrm{tr}\left(a^{\mathrm{t}} \otimes b^{\mathrm{t}}\right) \left(|\Psi_i\rangle \langle \Psi_i|\right)^{\mathrm{t}}$$

$$= \sum \mathrm{tr}_{\mathcal{G}} a^{\mathrm{t}} \kappa_i^* b \kappa_i = \mathrm{tr}_{\mathcal{G}} a^{\mathrm{t}} \phi(b) \tag{22}$$

$$= \sum \mathrm{tr}_{\mathcal{H}} \kappa_i^{\mathrm{t}*} a \kappa_i^{\mathrm{t}} b^{\mathrm{t}} = \mathrm{tr}_{\mathcal{H}} \phi^*(a)\, b^{\mathrm{t}}, \tag{23}$$

where

$$\phi(\cdot) = \sum \kappa_i^* \cdot \kappa_i, \quad \phi^*(\cdot) = \sum \kappa_i^{\mathrm{t}*} \cdot \kappa_i^{\mathrm{t}}, \tag{24}$$

which are CP maps given in the Kraus-Sudarshan form, and the stochastic entangling operator κ_i is derived from the stochastic (unnormalized) vector state Ψ_i. Then ϕ and ϕ^* satisfy the following marginal normalized conditions:

$$\rho^{\mathrm{t}} = \phi(I_{\mathcal{H}}) = \sum \kappa_i^* \kappa_i, \quad \sigma^{\mathrm{t}} = \phi^*(I_{\mathcal{G}}) = \sum \kappa_i^{\mathrm{t}*} \kappa_i^{\mathrm{t}}. \tag{25}$$

Notice that the map given by (8) or (9) are both always CCP but not always CP as mentioned above. It is easy to read

$$\tilde{\phi}(b) := \mathrm{tr}_{\mathcal{H}}(I_{\mathcal{G}} \otimes b)\theta = \tau_{\mathcal{G}} \circ \phi(b) = \phi \circ \tau_{\mathcal{H}}(b) \tag{26}$$

which is equivalent to $\phi = \tilde{\phi} \circ \tau_{\mathcal{H}}$, i.e. $\tilde{\phi}$ is always CCP.

Now let us consider the partial positive transpose of compound state $\theta \in \mathcal{S}(\mathcal{G} \otimes \mathcal{H})$. Recall that θ is a PPT state if $\theta_{\mathrm{PPT}} := (1_{\mathcal{G}} \otimes \tau_{\mathcal{H}})\theta \in \mathcal{S}(\mathcal{G} \otimes \mathcal{H})$. This means that its associated map ϕ_{PPT} is CP. This condition is equivalent to the co-complete positivity of ϕ, because it can be shown that the following equality holds by simple computation:

$$\phi_{\mathrm{PPT}}(b) = \mathrm{tr}_{\mathcal{H}}(I_{\mathcal{G}} \otimes b^{\mathrm{t}})\theta_{\mathrm{PPT}}^{\mathrm{t}} = \phi \circ \tau_{\mathcal{H}}(b) \,. \tag{27}$$

Corollary 2.1. [16]*PPT states are completely characterized by ϕ which are both CP and CCP.*

Let us summarize the above observation as the following definition according to Belavkin[23]:

Definition 2.1. A CP map (or a CP operation) $\phi : \mathcal{B}(\mathcal{H}) \to \widetilde{\mathcal{B}(\mathcal{G})}$ normalized to a state density $\rho^{\mathrm{t}} = \phi(I_{\mathcal{H}})$ is called *generalized entanglement* of the state $\varsigma(b) = \mathrm{tr}_{\mathcal{H}} b^{\mathrm{t}} \sigma^{\mathrm{t}}$ to the state $\varrho(a) = \mathrm{tr}_{\mathcal{G}} a^{\mathrm{t}} \rho^{\mathrm{t}}$.

The generalized entanglement defines the *generalized compound state*ϑ:

$$\mathrm{tr}_{\mathcal{G}} a^{\mathrm{t}} \phi(b) = \vartheta(a \otimes b) = \mathrm{tr}_{\mathcal{H}} \phi^{*}(a)\, b^{\mathrm{t}}. \tag{28}$$

Remark 2.1. Belavkin and Ohya (BO for short)[22] called the CCP map $\tilde{\phi} : \mathcal{B}(\mathcal{H}) \to \mathcal{T}(\mathcal{G})$ *normal coupling* of the state $\varsigma(b) = \mathrm{tr}_{\mathcal{H}} b\sigma$ to the state $\varrho(a) = \mathrm{tr}_{\mathcal{G}} a\rho$. On the other hand the CP map $\phi^{*} : \mathcal{B}(\mathcal{G}) \to \widetilde{\mathcal{T}(\mathcal{H})}$ normalized to the density $\sigma^{\mathrm{t}} = \phi^{*}(I_{\mathcal{G}})$ of state ς was called *normal entanglement* of the system $(\mathcal{B}(\mathcal{G}), \varrho)$ with the density $\rho^{\mathrm{t}} = \phi(I_{\mathcal{H}})$ to $(\mathcal{B}(\mathcal{H}), \varsigma)$. Using $\tilde{\phi}$ and ϕ BO gave the several types of quantum encoding corresponding to several types of entanglement ϕ and discussed the channel capacity on each encoding by using the mutual entropy defined for such entanglements ϕ.

Remark 2.2. On the base of results in[22] Belavkin disclosed an operational meaning of entanglement as specifically quantum encoding by only using the CP map ϕ[23]. He introduced the notion of *contra-variant density* in order to avoid the consideration of the unphysical operations and to deal only with CP operation. He described the normal state ς not using by the usual density operator σ, called *covariant density,* but using by the transposed density $\tilde{\sigma} := \sigma^{\mathrm{t}}$ such that

$$\varsigma(b) := \langle b, \tilde{\sigma} \rangle \equiv \mathrm{tr}_{\mathcal{H}} b^{\mathrm{t}} \tilde{\sigma} =: b_{ij} \tilde{\sigma}^{ij}, \tag{29}$$

where b_{ij} is the matrix element of b, $\tilde{\sigma}$ is called a *contra-variant density* of the state ς with respect to the natural tensor paring (29) (see also Ref. [24]).

Remark 2.3. Corollary 2.1 was shown by Majewski, Ohya and T. M. [16]. They applied the isometric isomorphism between linear maps and quantum states given by Størmer [17] which is a reformulation of Choi result [18] (see also Ref. [19]). Using the Størmer isomorphism it is shown that representation of $\vartheta(a \otimes b)$ in (10) gives the criterion of co-complete positivity of ϕ (or ϕ^*).

3. On quantum analogue of Bayes formula

3.1. *Quantum compound state vs classical joint probability*

As mentioned in the introduction Cerf and Adami [14] introduced the following definition of the conditional amplitude operator

$$\rho_{\mathcal{H}|\mathcal{G}} = \lim_{n\to\infty} [\rho_{\mathcal{G}\otimes\mathcal{H}}^{1/n} \cdot \rho_{\mathcal{G}}^{-1/n} \otimes I_{\mathcal{H}}]^n, \tag{30}$$

and shown that such definition reproduces the formula for the conditional von Neumann entropy $S(\mathcal{H}|\mathcal{G}) = S(\rho_{\mathcal{G}\otimes\mathcal{H}}) - S(\rho_{\mathcal{H}})$, that is,

$$S(\mathcal{H}|\mathcal{G}) = -\mathrm{Tr}(\rho_{\mathcal{G}\otimes\mathcal{H}} \log \rho_{\mathcal{H}|\mathcal{G}}). \tag{31}$$

Asorey, Kossakowski, Marmo and Sudarshan, 2005: They introduced the another approach of notion of *quantum conditional probability* operator (QCP for short) [15] (see also Refs. [20,21]). In this section it will be shown that the entanglement ϕ has a similar structure as the classical Bayes formula via the QCP. For this purpose, at first, we recall the notion of *standard entanglement* given by BO [22].

The entanglement ϕ in contra-variant form is called *standard* for the state ς on $\mathcal{B} := \mathcal{B}(\mathcal{H})$ if it couples σ to the transposed state $\rho = \sigma^{\mathrm{t}}$ on $\mathcal{A} := \tilde{\mathcal{B}} \equiv \{b^{\mathrm{t}} : b \in \mathcal{B}(\mathcal{H})\}$ by

$$\phi(b) = \rho^{\mathrm{t}\frac{1}{2}} b \rho^{\mathrm{t}\frac{1}{2}} =: \rho(b), \; \forall b \in \mathcal{B} = \tilde{\mathcal{A}}. \tag{32}$$

The standard entanglement ρ may be derived from the pure entangled state ϑ_{P} with marginal density ρ in the case of $\mathcal{G} = \mathcal{H}$. Obviously

$$\rho^*(a) = \sigma^{\mathrm{t}\frac{1}{2}} a \sigma^{\mathrm{t}\frac{1}{2}} =: \sigma(a), \; \forall a \in \mathcal{A} = \tilde{\mathcal{B}}. \tag{33}$$

In (32) and (33) we use the same symbols ρ and σ for both the density operators and their entanglements.

Given ς, the *standard entangled state* ϑ_{s} is defined by

$$\mathrm{tr}_{\mathcal{H}} a^t \rho(b) = \vartheta_{\mathrm{s}}(a \otimes b) = \mathrm{tr}_{\mathcal{H}} b^t \sigma(a) . \tag{34}$$

Beralvkin & Dai[25] showed that every entanglement ϕ can be decomposed to composition of standard entanglement with a CP map π. For convenient of readers we show its proof.

Theorem 3.1. [25] *Every entanglement ϕ on $\mathcal{B}(\mathcal{H})$ to the state $\rho^{\mathrm{t}} \in \widetilde{\mathcal{T}(\mathcal{G})}$ has a decomposition,*

$$\phi(b) = \rho^{\mathrm{t}^{\frac{1}{2}}} \pi(b) \rho^{\mathrm{t}^{\frac{1}{2}}}, \tag{35}$$

where π is normal CP contraction $\mathcal{B}(\mathcal{H}) \to \widetilde{\mathcal{B}(\mathcal{G})}$ such that $I_{\mathcal{G}} \geq \pi(I_{\mathcal{H}}) \geq E_{\rho}^{\mathrm{t}}$ on the minimal orthoprojector denoted by $E_{\rho}^{\mathrm{t}} \in \widetilde{\mathcal{B}(\mathcal{G})}$ supporting the density operator ρ^{t}. This decomposition is unique by the normalization condition $\pi(I_{\mathcal{H}}) = E_{\rho}^{\mathrm{t}}$.

Proof. π can be found as the solution to the equation

$$\rho^{\mathrm{t}^{\frac{1}{2}}} \pi(b) \rho^{\mathrm{t}^{\frac{1}{2}}} = \phi(b) \ \forall b \in \mathcal{B}(\mathcal{H}) \tag{36}$$

which is unique if ρ^{t} is non-degenerate:

$$\pi(b) = \rho^{\mathrm{t}^{-\frac{1}{2}}} \phi(b) \rho^{\mathrm{t}^{-\frac{1}{2}}} \tag{37}$$

If ρ^{t} is degenerate, one should reduce $\mathcal{G}$ to the Hilbert subspace $\mathcal{G}_{\rho} = E_{\rho}^{\mathrm{t}} \mathcal{G}$ given by E_{ρ}^{t}. □

Let us recall the notion of QCP[15]. Given a compound density operator $\theta \in \mathcal{S}(\mathcal{G} \otimes \mathcal{H})$ with its marginal $\rho = \mathrm{tr}_{\mathcal{H}} \theta$, one can define the operator $\Pi \in \mathcal{T}(\mathcal{G} \otimes \mathcal{H})$ as

$$\Pi \equiv \left(\rho^{-\frac{1}{2}} \otimes I\right) \theta \left(\rho^{-\frac{1}{2}} \otimes I\right) \tag{38}$$

which has the properties

$$\Pi \geq 0, \tag{39}$$

$$\mathrm{tr}_{\mathcal{H}} \Pi = I_{\mathcal{G}}. \tag{40}$$

Notice that if ρ is degenerate, one should reduce $\mathcal{G}$ to the Hilbert subspace $\mathcal{G}_{\rho} = E_{\rho} \mathcal{G}$, then $\mathrm{tr}_{\mathcal{H}} \Pi = I_{\mathcal{G}_{\rho}}$. It is clear that

$$\theta = \left(\rho^{\frac{1}{2}} \otimes I\right) \Pi \left(\rho^{\frac{1}{2}} \otimes I\right). \tag{41}$$

In the sense of (39), (40) the operator Π can be regarded as the quantum analogue of conditional probability. In the same sense the decomposition (41) may be corresponding to the classical construction of joint probability by conditional probability on Bayes formula. Kossakowski *et al.* called the operator $\Pi \in \mathcal{T}(\mathcal{G} \otimes \mathcal{H})$ satisfying (39) and (40) the *quantum conditional probability* (QCP). In [20] it is called *conditional quantum state.* Using the transposed density θ^{t} in (41) we can represent the compound state ϑ by the entanglement map in contra-variant form corresponding to the transposed QCP as follows:

$$\begin{aligned}\vartheta(a \otimes b) &= \operatorname{tr}\left(a^{\mathrm{t}} \otimes b^{\mathrm{t}}\right) \theta^{\mathrm{t}} \\ &= \operatorname{tr}\left(a^{\mathrm{t}} \otimes b^{\mathrm{t}}\right)\left(\rho^{\mathrm{t}\frac{1}{2}} \otimes I_{\mathcal{H}}\right) \Pi^{\mathrm{t}}\left(\rho^{\mathrm{t}\frac{1}{2}} \otimes I_{\mathcal{H}}\right) \qquad (42)\\ &= \operatorname{tr}_{\mathcal{G}} a^{\mathrm{t}} \rho^{\mathrm{t}\frac{1}{2}} \pi_{\Pi}(b) \rho^{\mathrm{t}\frac{1}{2}} = \operatorname{tr}_{\mathcal{G}} a^{\mathrm{t}} \phi_{\Pi}(b), \qquad (43)\end{aligned}$$

where $\pi_{\Pi}(b) \equiv \operatorname{tr}_{\mathcal{H}}\left(I_{\mathcal{G}} \otimes b^{\mathrm{t}}\right) \Pi^{\mathrm{t}}$ and $\phi_{\Pi}(b) = \rho^{\mathrm{t}\frac{1}{2}} \pi_{\Pi}(b) \rho^{\mathrm{t}\frac{1}{2}} = \rho \circ \pi_{\Pi}(b)$. Comparing (42) with the decomposition of ϕ in (36), due to the Theorem 3.1, it is clear that $\phi = \phi_{\Pi}$, $\pi = \pi_{\Pi}$ (see also Ref. [26]).

Definition 3.1. A normal unital CP map π from $\mathcal{B}(\mathcal{H})$ to $\widetilde{\mathcal{B}(\mathcal{G})}$ is called a *quantum conditionality.*

Then a generalized entanglement ϕ is given by the composition of a standard entanglement ρ and a quantum conditionality π:

$$\phi = \rho \circ \pi. \qquad (44)$$

In classical probability theory a conditional probability gives a classical channel denoted by Λ_{c}^{*} defined as the transition probability matrix. A quantum conditionality π defines also the quantum channel Λ_{q}^{*} on the $\mathcal{T}(\mathcal{G})$ into $\widetilde{\mathcal{T}(\mathcal{H})}$ as its dual map $\pi^{*} : \mathcal{B}(\mathcal{G}) \to \widetilde{\mathcal{B}(\mathcal{H})}$ given by the following duality:

$$\varsigma(b) = \operatorname{tr}_{\mathcal{G}} \rho^{\mathrm{t}} \pi(b) = \operatorname{tr}_{\mathcal{H}} \pi^{*}(\rho) b^{\mathrm{t}} = \operatorname{tr}_{\mathcal{H}} \sigma^{\mathrm{t}} b^{\mathrm{t}}, \qquad (45)$$

i.e. $\Lambda_{\mathrm{q}}^{*} \equiv \pi^{*}|_{\mathcal{T}(\mathcal{G})}$. This duality is given by the same dual relation between entanglements ϕ and ϕ^{*} in (22) and (23).

From the above observation we can say that there exists the similar structure between quantum entanglements ϕ and classical joint probability distributions $r = (r_{ij})$ with its marginals $p = (p_i)$, $q = (q_j)$ as follows:

$$\text{Compound measure}: (r_{ij}) = \left(p_i p_{j|i}\right) \longleftrightarrow \phi = \rho \circ \pi \qquad (46)$$

$$\text{Conditional measure}: \left(p_{j|i}\right) = \left(\frac{r_{ij}}{p_i}\right) \longleftrightarrow \pi = \rho^{-1} \circ \phi \tag{47}$$

$$\text{Channel}: \left(q_j\right) = \Lambda_{\mathrm{c}}^{*}(p) = \left(\sum_i p_i p_{j|i}\right) \longleftrightarrow \sigma^{\mathrm{t}} = \Lambda_{\mathrm{q}}^{*}(\rho) = \pi^{*}(\rho). \tag{48}$$

We call the above relations (46), (47), (48) the *quantum Bayes relations.*

3.2. *Compound state via Choi-Jamiołkowski isomorphism*

Let d be a dimensionality of $\mathcal{G}$ and $\mathcal{H}$, i.e., $d = \dim\mathcal{G} = \dim\mathcal{H} < \infty$, and $\{g_i\} \subset \mathcal{G}$ and $\{h_i\} \subset \mathcal{H}$ be a CONS on which the transpositions $\tau_{\mathcal{G}}$ and $\tau_{\mathcal{H}}$ are defined respectively. Given entanglement ϕ (or ϕ^*) normalized to $\rho^{\mathrm{t}} = \phi(I_{\mathcal{H}})$ (or $\sigma^{\mathrm{t}} = \phi^*(I_{\mathcal{G}})$) we can construct a compound state θ_ϕ in $\mathcal{B}(\mathcal{G} \otimes \mathcal{H})$ via Choi-Jamiołkowski isomorphism [18,19] as follows:

$$\theta_\phi \equiv (\phi \otimes I_{\mathcal{H}})P_{\mathcal{H}} = \sum \phi(|h_i\rangle\langle h_j|) \otimes |h_i\rangle\langle h_j| \tag{49}$$

$$= (I_{\mathcal{G}} \otimes \phi^*)P_{\mathcal{G}} = \sum |g_i\rangle\langle g_j| \otimes \phi^*(|g_i\rangle\langle g_j|), \tag{50}$$

where $P_{\mathcal{H}} = \sum |h_i\rangle\langle h_j| \otimes |h_i\rangle\langle h_j|$, $P_{\mathcal{G}} = \sum |g_i\rangle\langle g_j| \otimes |g_i\rangle\langle g_j|$. Due to the duality given by Choi-Jamiołkowski isomorphism we know that the complete positivity of ϕ leads the positivity of θ_ϕ, and it is clear that $\rho^{\mathrm{t}} = \mathrm{tr}_{\mathcal{G}}\theta_\phi$, $\sigma^{\mathrm{t}} = \mathrm{tr}_{\mathcal{H}}\theta_\phi$ which means the normality condition $\mathrm{tr}\theta_\phi = 1$, i.e. $\theta_\phi \in \mathcal{S}(\mathcal{G} \otimes \mathcal{H})$. Then the expectation $\vartheta_\phi(a \otimes b)$ is given by

$$\vartheta_\phi(a \otimes b) = \mathrm{tr}\left(a^{\mathrm{t}} \otimes b^{\mathrm{t}}\right)\theta_\phi$$

$$= \mathrm{tr}_{\mathcal{G}} a^{\mathrm{t}} \sum \phi(|h_i\rangle\langle h_j|)\langle h_j, b^{\mathrm{t}} h_i\rangle = \mathrm{tr}_{\mathcal{G}} a^{\mathrm{t}} \phi(b) \tag{51}$$

$$= \mathrm{tr}_{\mathcal{H}} \sum \langle g_j, a^{\mathrm{t}} g_i\rangle \phi^*(|g_i\rangle\langle g_j|) b^{\mathrm{t}} = \mathrm{tr}_{\mathcal{H}} \phi^*(a) b^{\mathrm{t}} \tag{52}$$

for any $a \in \mathcal{B}(\mathcal{G})$ and $b \in \mathcal{B}(\mathcal{H})$. If its partial transpose $(1_{\mathcal{G}} \otimes \tau_{\mathcal{H}})\theta_\phi$ is positive, i.e. $\theta_\phi \in \mathcal{S}_{\mathrm{PPT}}$, then it can be represented by

$$(1_{\mathcal{G}} \otimes \tau_{\mathcal{H}})\theta_\phi = (\phi \otimes \tau_{\mathcal{H}})P_{\mathcal{H}}$$

$$= \sum \phi(|h_i\rangle\langle h_j|) \otimes \tau_{\mathcal{H}}(|h_i\rangle\langle h_j|)$$

$$= \sum \phi(|h_i\rangle\langle h_j|) \otimes |h_j\rangle\langle h_i|$$

$$= \sum \phi \circ \tau_{\mathcal{H}}(|h_j\rangle\langle h_i|) \otimes |h_j\rangle\langle h_i| > 0. \tag{53}$$

Therefore the condition $\theta_\phi \in \mathcal{S}_{\mathrm{PPT}}$ is equivalent to the co-complete positivity of ϕ. In finite dimensional case, via the Choi-Jamiołkowski criterion of CP map, we reach the characterization of $\mathcal{S}_{\mathrm{PPT}}$ corresponding to Corollary 2.1 in the previous section.

Applying the decomposition $\phi = \rho \circ \pi$, due to the duality of quantum conditionality π, the compound state θ_ϕ may be represented by its channel π^* or QCP $\Pi_{\mathcal{H}|\mathcal{G}}$ as follows:

$$\theta_\phi = \sum \rho \circ \pi(|h_i\rangle \langle h_j|) \otimes |h_i\rangle \langle h_j| \tag{54}$$

$$= \sum \rho(|g_i\rangle \langle g_j|) \otimes \pi^* (|g_i\rangle \langle g_j|) = \left(\rho^{\mathrm{t}\frac{1}{2}} \otimes I_{\mathcal{H}}\right) \Pi_{\mathcal{H}|\mathcal{G}} \left(\rho^{\mathrm{t}\frac{1}{2}} \otimes I_{\mathcal{H}}\right) \tag{55}$$

where $\Pi_{\mathcal{H}|\mathcal{G}} = \sum |g_i\rangle \langle g_j|) \otimes \pi^* (|g_i\rangle \langle g_j|)$. The channel Λ_ϕ^* may be corresponding to the partial trace operation:

$$\sigma^{\mathrm{t}} = \mathrm{tr}_{\mathcal{G}} \theta_\phi = \sum \left\langle g_j, \rho^{\mathrm{t}} g_i \right\rangle \pi^* (|g_i\rangle \langle g_j|) = \pi^* (\rho) =: \Lambda_\phi^* (\rho) . \tag{56}$$

4. Summary and forthcoming problem

In classical information theory, as mentioned in the introduction, Shannon's mutual information has three types of interpretation at the same time: (I) the common information (or correlation), (II) the information gain by observation, (III) the transmitted information via channel. Such interpretations can be done on the base of Bayes formula of probabilities.

In this paper we showed that quantum states also have a similar structure as quantum analogue of Bayes formula in ((46), (47), (48)) or ((54), (56)). Notice that, using the decomposition $\phi = \rho \circ \pi$, one can redefine the quantum analogue of mutual entropy $I(\mathcal{G} \otimes \mathcal{H})$ in (2) as follows:

$$I(\mathcal{G} \otimes \mathcal{H}) =: I_\phi(\mathcal{G} \otimes \mathcal{H}) = S(\rho) + S(\pi^* (\rho)) - S(\theta_\phi). \tag{57}$$

In (57) we can express the mutual information by using the associated channel π^*, however it is not clear how we can relate (57) to the interpretation (II) and (III). Moreover, in quantum theory, we have several types of mutual information, for example see Refs.[27,28,29,30,31], and it is not clear the relations between some of them.

In the forthcoming paper[32] we will try to answer the above questions by studying the relations between another types of quantum mutual information on the base of channel representation of compound state in (54), which is given by the duality of quantum conditionality.

Acknowledgements

The author is grateful to the late Professor V.P. Belavkin for fruitful discussions on entanglement maps. The author also wish to thank Professor D. Chruściński for good suggestions to improve the original manuscript.

References

1. C. Shannon, *Mathematical theory of communication*, The Bell System Technical Journal, **27**, 379–423, 623–656, (1948).
2. T. Cover and J. A. Thomas, *Elements of information theory* (2nd ed.), New York, Wiley-Interscience (2006).
3. M. Wilde, *Quantum Information Theory*, Cambridge, CUP, 2013.
4. M.A. Nielsen and I.L. Chuang, *Quantum Computation and Quantum Information*, Cambridge Univ. Press, Cambridge, 2000.
5. A. Holevo, *Quantum Systems, Channels, Information: A Mathematical Introduction*, De Gruyter Studies in Mathematical Physics **16**, 2012.
6. M. Ohya and I. Volovich, *Mathematical Foundations of Quantum Information and Computation and Its Applications to Nano- and Bio-systems*, Dordrecht-Heidelberg-London-New York, Springer, 2011
7. R.F. Werner, Lett. Math. Phys. **17**, 359 (1989).
8. R.F. Werner, Phys. Rev. **A58**, 1827 (1998).
9. C.H. Bennett, G. Brassard, S. Popescu, B. Schumacher, J.A. Smolin and W.K. Wootters, Phys. Rev. Lett. **76**, 722 (1996).
10. B. Schumacher, Phys. Rev. **A51**, 2614 (1993a).
11. B. Schumacher, Phys. Rev. **A51**, 2738 (1993b).
12. M. Horodecki, P. Horodecki and R. Horodecki, Phys. Lett. **A223**, 1 (1996).
13. A. Peres, Phys. Rev. Lett. **77**, 1413 (1996).
14. N.J. Cerf and C. Adami, Phys. Rev. Lett. **79**, 5194 (1997).
15. M. Asorey, A. Kossakowski, G. Marmo, E.C.G. Sudarshan, Open Syst. Inf. Dyn. **12**, 319 (2005).
16. W.A. Majewski, T. Matsuoka, and M. Ohya, J. Math. Phys. **50**, 113509 (2009).
17. E. Størmer, Contemp. Math. **62**, 345 (1987).
18. M.D. Choi, Linear Algebra Appl. **10**, 285 (1975).
19. A. Jamiołkowski, Rep. Math. Phys. **3**, 275 (1972).
20. M.S. Leifer, Phys. Rev. A **74**, 042310 (2006).
21. D. Chruściński, Phys. Lett. A **377**, 606 (2013).
22. V.P. Belavkin, and M. Ohya, Proc. Royal. Soc. London A **458**, 209 (2002).
23. V.P. Belavkin, in Quantum Bio-Informatics IV, Eds. L. Accardi *et al.*, Singapore, World Scientific 2011.
24. V. P. Belavkin, Rep. Math. Phys. **55**, 61 (2005).
25. V. P. Belavkin, and X. Dai, International Journal of Quantum Inforamtion, **6**, 357 (2008).
26. D. Chruściński, A. Kossakowski, T. Matsuoka, and M. Ohya, in Quantum Bio-Informatics IV, Eds. L. Accardi *et al.*, Singapore, World Scientific 2011.
27. A. Holevo, Problems of Information Transmission, **9**, 177 (1973).
28. M. Ohya, IEEE T. Inform. Theory **29**, 770 (1983).
29. B. Schumacher, Phys. Rev. A **54**, 2614 (1996).
30. L. Henderson and V. Vedral, J. Phys. A: Math. Gen. **34**, 6899 (2001).
31. H. Ollivier and W. H. Żurek, Phys. Rev. Lett. **88**, 017901 (2001).
32. D. Chruściński and T. Matsuoka, in preparation.

Infinite Dimensional Analysis, Quantum Probability and Related Topics

pp. 143–151

STABILITY AND PERTURBATIONS OF C_0-MARKOV SEMIGROUPS ON ABSTRACT STATE SPACES

NAZIFE ERKURUN ÖZCAN

Department of Mathematics,
Faculty of Science,
Hacettepe University,
Ankara, 06800, Turkey
E-mail: erkursun.ozcan@hacettepe.edu.tr

FARRUKH MUKHAMEDOV

Department of Mathematical Sciences,
College of Science,
United Arab Emirates University,
15551, Al-Ain, United Arab Emirates
E-mail: far75m@yandex.ru; farrukh.m@uaeu.ac.ae

Our investigation's aim is to explore stability and perturbation bounds of positive C_0-semigroups on abstract state spaces by means of the Dobrushin's ergodicity coefficient. We obtain a linear relation between the stability of the semigroup and the sensitivity of its fixed point with respect to perturbations of C_0-Markov semigroups.

Keywords: Uniform asymptotically stable; C_0-Markov semigroup; ergodicity coefficient; perturbation bound.

1. Introduction

If we consider a classical Markov process on a general state space $(E, \mathcal{F}, \mu)$ given by the transition probabilities $P(t, x, A)$, then one can define a semigroup as follows: $T_t f(x) = \int f(y)P(t, x, dy)$, which acts on $L^1(E, \mu)$-space (see [10] for more details). Hence, the theory of perturbations [9,12] can be reformulated in terms of the defined semigroup defined on $L^1(E, \mu)$-space which is an ordered Banach space (lattice) w.r.t. the standard ordering. There are many papers devoted to the investigation of stability and ergodic properties of semigroups defined on Banach lattices [3,10,11].

In the present paper, we investigate stability and perturbation bounds of positive C_0-semigroups defined on abstract state spaces (i.e. ordered Banach spaces). Roughly speaking, an abstract state space is an ordered Banach space with additivity property on the cone of positive elements. Such spaces contain all classical L^1-spaces and the space of density operators acting on some Hilbert space [1]. We refer the reader to [4] (see also [2]) for more information about the structure of positive semigroups defined ordered spaces.

2. Preliminaries

In this section we recall some necessary definitions and fact about ordered Banach spaces.

2.1. *Ordered Banach spaces*

Let X be an ordered vector space with a cone $X_+ = \{x \in X : x \geq 0\}$. A subset $\mathcal{K}$ is called a *base* for X, if one has $\mathcal{K} = \{x \in X_+ : f(x) = 1\}$ for some strictly positive (i.e. $f(x) > 0$ for $x > 0$) linear functional f on X. An ordered vector space X with generating cone X_+ (i.e. $X = X_+ - X_+$) and a fixed base $\mathcal{K}$, defined by a functional f, is called *an ordered vector space with a base* [1]. In what follows, we denote it as $(X, X_+, \mathcal{K}, f)$. Let U be the convex hull of the set $\mathcal{K} \cup (-\mathcal{K})$, and let

$$\|x\|_{\mathcal{K}} = \inf\{\lambda \in \mathbb{R}_+ : x \in \lambda U\}.$$

Then one can see that $\|\cdot\|_{\mathcal{K}}$ is a seminorm on X. Moreover, one has $\mathcal{K} = \{x \in X_+ : \|x\|_{\mathcal{K}} = 1\}$, $f(x) = \|x\|_{\mathcal{K}}$ for $x \in X_+$. In what follows, we always assume that the set U is *radially compact* [1], (i.e. $\ell \cap U$ is closed and bounded segment for every line ℓ though the origin of X), which implies that $\|\cdot\|_{\mathcal{K}}$ is a norm, and in this case $(X, X_+, \mathcal{K}, f)$ is called *base-norm space*. When X is complete with respect to the norm $\|\cdot\|_{\mathcal{K}}$ and the cone X_+ is closed, then $(X, X_+, \mathcal{K}, f)$ is called *Banach base norm space* or *abstract state spaces*. In the sequel, for the sake of simplicity instead of $\|\cdot\|_{\mathcal{K}}$ we will use usual notation $\|\cdot\|$.

We notice that the dual space of an ordered normed space with order unit is an example of an abstract state space. The Hermitian part of the predual space of a von Neumann algebra is another example of an abstract state spaces, the base being the family of all normal states on this algebra (see [16,10] for more examples).

Let $(X, X_+, \mathcal{K}, f)$ be an abstract state space. It is well-known (see [1]) that every element x of X admits a decomposition $x = y - z$, where $y, z \geq 0$ and $\|x\| = \|y\| + \|z\|$.

2.2. *Markov operators and Dobrushin's ergodicity coefficient*

Let $(X, X_+, \mathcal{K}, f)$ be an abstract state space. A linear operator $T : X \to X$ is called positive, if $Tx \geq 0$ whenever $x \geq 0$. A positive linear operator $T : X \to X$ is called *Markov*, if $T(\mathcal{K}) \subset \mathcal{K}$. It is clear that $\|T\| = 1$, and its adjoint mapping $T^* : X^* \to X^*$ acts in ordered Banach space X^* with unit f, and moreover, one has $T^* f = f$. For each $y \in X$ we define a linear operator $T_y : X \to X$ by $T_y(x) = f(x)y$.

The main object in this paper is a strongly continuous semigroup or C_0*-semigroup* on a Banach space X is a map $T_t : \mathbb{R}_+ \to L(X)$ such that

(i) $T_0 = I$, identity operator on X;
(ii) $\forall t, s \geq 0 : \ T_{t+s} = T_t T_s$;
(iii) $\forall x_0 \in X : \ \|T_t x_0 - x_0\| \to 0$, as $t \downarrow 0$.

The first two axioms are algebraic, and state that $\mathcal{T} = (T_t)$ is a representation of the semigroup $(\mathbb{R}_+, +)$; the last is topological, and states that the map $\mathcal{T}$ is continuous in the strong operator topology.

Let $\mathcal{T} = (T_t)_{t\geq 0}$ be a C_0-semigroup. If for each $t \in \mathbb{R}_+$ the operator T_t is Markov, then $\mathcal{T} = (T_t)_{t\geq 0}$ is called a C_0*-Markov semigroup.* An element $x_0 \in X$ is called a *fixed point* or *stationary point* of $\mathcal{T} = (T_t)_{t\geq 0}$ if one has $T_t x_0 = x_0$ for all $t \in \mathbb{R}_+$. In [4] a basic structure of positive semigroups defined on ordered Banach spaces is given. We refer the reader to [6,5], for more concrete examples of Markov semigroups associated with certain physical systems.

The Cesaro means C_0-semigroup is denoted by

$$A_t(\mathcal{T}) = \frac{1}{t}\int_0^t T_s ds, \quad t \in \mathbb{R}_+.$$

The integral above is taken with respect to the strong operator topology.

Definition 2.1. A $\mathcal{T} = (T_t)_{t\geq 0}$ be a C_0-Markov semigroup defined on X is called

(i) *uniform asymptotically stable* if there exist an element $x_0 \in \mathcal{K}$ such that

$$\lim_{t\to\infty} \|T_t - T_{x_0}\| = 0;$$

(ii) *uniform mean ergodic* if there exist an element $x_0 \in \mathcal{K}$ such that

$$\lim_{t\to\infty} \left\| A_t(\mathcal{T}) - T_{x_0} \right\| = 0;$$

(iii) *weakly ergodic* if one has

$$\lim_{t\to\infty} \sup_{x,y\in\mathcal{K}} \|T_t x - T_t y\| = 0;$$

(iv) *weakly mean ergodic* if one has

$$\lim_{t\to\infty} \sup_{x,y\in\mathcal{K}} \|A_t(\mathcal{T})x - A_t(\mathcal{T})y\| = 0.$$

The uniform asymptotical stability implies uniform mean ergodicity. Moreover, if $\mathcal{T}$ is uniform mean ergodic, then x_0 corresponding to T_{x_0}, is a fixed point of $\mathcal{T}$. Indeed, $T_t T_{x_0} = T_{x_0}$, which yields $T_t x_0 = x_0$ for all $t \in \mathbb{R}_+$.

Let $(X, X_+, \mathcal{K}, f)$ be an abstract state space and $T : X \to X$ be a Markov operator. Letting

$$N = \{x \in X : \ f(x) = 0\}, \tag{1}$$

we define

$$\delta(T) = \sup_{x\in N,\ x\neq 0} \frac{\|Tx\|}{\|x\|}. \tag{2}$$

The quantity $\delta(T)$ is called *Dobrushin's ergodicity coefficient* of T (see [16]).

The next result establishes several properties of the Dobrushin's ergodicity coefficient.

Theorem 2.1. [16] *Let $(X, X_+, \mathcal{K}, f)$ be an abstract state space and $T, S : X \to X$ be Markov operators. The following assertions hold:*

(i) $0 \leq \delta(T) \leq 1$;
(ii) $|\delta(T) - \delta(S)| \leq \delta(T - S) \leq \|T - S\|$;
(iii) $\delta(TS) \leq \delta(T)\delta(S)$;
(iv) if $H : X \to X$ *is a linear bounded operator such that* $H^*(f) = 0$, *then* $\|TH\| \leq \delta(T)\, \|H\|$;
(v) one has

$$\delta(T) = \frac{1}{2} \sup_{u,v\in\mathcal{K}} \|Tu - Tv\|\,;$$

(vi) if $\delta(T) = 0$, *then there exists* $y_0 \in X_+$ *such that* $T = T_{y_0}$.

Note that taking into account Theorem 2.1(v) we obtain that the weak ergodicity is equivalent to the condition $\delta(T_t) \to 0$ as $t \to \infty$.

3. Uniform asymptotic stability and perturbation bounds of C_0-Markov semigroups

In this section, we formulate the uniform asymptotic stability of C_0-Markov semigroups in terms of the Dobrushin's ergodicity coefficient. This allows us to establish perturbation bounds for the semigroups using the ergodicity coefficient. The following result is an continuous time analogue of Theorem 3.5 [16].

Theorem 3.1. *Let $(X, X_+, \mathcal{K}, f)$ be an abstract state space and $\mathcal{T} = (T_t)_{t\geq 0}$ be a C_0-Markov semigroup. Then the following statements are equivalent:*

(i) *$\mathcal{T}$ is weakly ergodic;*
(ii) *there exists $t_0 \in \mathbb{R}$ such that $\delta(T_{t_0}) < 1$;*
(iii) *$\mathcal{T}$ is uniformly asymptotical stable. Moreover, there are positive constants $C, \alpha, t_0 \in \mathbb{R}$ and $x_0 \in \mathcal{K}$ such that*

$$\|T_t - T_{x_0}\| \leq Ce^{-\alpha t}, \quad \forall t \geq t_0. \tag{3}$$

We can conclude that if at least one operator T_{t_0} (for some $t_0 \in \mathbb{R}_+$) of a C_0-Markov semigroup $\mathcal{T} = (T_t)_{t\geq 0}$ is uniformly asymptotical stable, then whole semigroup $\mathcal{T}$ is uniform asymptotically stable. In the classical setting, i.e. if $X = L^1(E, \mu)$, a similar kind of result has been established in [13].

To establish perturbation bounds for C_0-semigroups, we use the following well-known fact (see [7] for the proofs).

Theorem 3.2. *Let $(A, D(A))$ be the generator of a C_0-semigroup $\mathcal{T} = (T_t)_{t\geq 0}$ on a Banach space X satisfying*

$$\|T_t\| \leq Me^{wt} \quad \forall t \geq 0$$

and some $w \in \mathbb{R}, M \geq 1$. If $B \in L(X)$, then $C := A + B$ with $D(C) := D(A)$ generates a C_0-semigroup $\mathcal{S} = (S_t)_{t\geq 0}$ satisfying

$$\|S_t\| \leq Me^{(w+M\|B\|)t} \quad \forall t \geq 0.$$

Moreover, for a representation formula for this new semigroup, it satisfies an integral equation.

$$S_t x = T_t x + \int_0^t T_{t-s} B S_s x ds$$

holds for every $t \geq 0$ *and* $x \in X$.

Our first result about perturbation bounds is the following one.

Theorem 3.3. *Let* $(X, X_+, \mathcal{K}, f)$ *be an abstract state space and* $\mathcal{S} = (S_t)_{t\geq 0}$, $\mathcal{T} = (T_t)_{t\geq 0}$ *be two* C_0*-Markov semigroups on* X *with generators* A *and* C, *respectively, satisfying* $B := C - A$ *is bounded. If there exists* $\rho \in [0, 1)$ *and* $t_0 \in \mathbb{R}$ *such that* $\delta(T_{t_0}) \leq \rho$, *then one has*

$$\|T_t x - S_t z\| \leq \tag{4}$$
$$\begin{cases} \|x - z\| + t\|B\|, & \forall t \leq t_0, \\ \rho^{[t/t_0]}\|x - z\| + \Big(\dfrac{t_0(1 - \rho^{[t/t_0]})}{1 - \rho} + \rho^{[t/t_0]}(t - t_0[t/t_0])\Big)\|B\|, & \forall t > t_0 \end{cases}$$

for every $x, z \in \mathcal{K}$.

Remark 3.1. Note that this result extends all existing results (see [14,15,17]) for general spaces. In particularly, if we take as X a predual of any von Neumann algebra, then we extend the result of [17] (where it was proved similar result for matrix algebras) for arbitrary von Neumann algebras.

Corollary 3.1. *Let the conditions of Theorem 3.3 be satisfied. Then for every* $x, y \in \mathcal{K}$ *one has*

$$\sup_{t\geq 0}\|T_t x - S_t z\| \leq \|x - z\| + \frac{t_0}{1-\rho}\|B\|. \tag{5}$$

In addition, if $\mathcal{S}$ *is uniformly asymptotical stable to* S_{z_0} *then*

$$\|T_{x_0} - S_{z_0}\| \leq \frac{t_0}{1-\rho}\|B\|. \tag{6}$$

The following theorem gives an alternative method of obtaining perturbation bounds in terms of $\delta(T_{t_0})$.

Theorem 3.4. *Let the conditions of Theorem 3.3 be satisfied. Then for every* $x, z \in \mathcal{K}$ *one has*

$$\|T_t x - S_t z\| \leq \delta(T_{t_0})^{\lfloor t/t_0 \rfloor}(\|x - z\| + \sup_{0<t<t_0}\|T_t - S_t\|) \tag{7}$$
$$+\frac{1 - \delta(T_{t_0})^{\lfloor t/t_0 \rfloor}}{1 - \delta(T_{t_0})}\|T_{t_0} - S_{t_0}\|, \quad t \in \mathbb{R}_+.$$

Theorem 3.5. *Let $(X, X_+, \mathcal{K}, f)$ be an abstract state space and $\mathcal{T} = (T_t)_{t\geq 0}$ be C_0-Markov semigroups on X such that $\delta(T_{t_0}) < 1$ for some $t_0 \in \mathbb{R}_+$. Then every C_0-Markov semigroup $\mathcal{S} = (S_t)_{t\geq 0}$ satisfying $\|S_{t_0} - T_{t_0}\| < 1 - \delta(T_{t_0})$ is uniformly asymptotical stable and has a unique fixed point $z_0 \in \mathcal{K}$ such that*

$$\|x_0 - z_0\| \leq \frac{\|S_{t_0} - T_{t_0}\|}{1 - \delta(T_{t_0}) - \|S_{t_0} - T_{t_0}\|}. \tag{8}$$

4. Uniform mean ergodicity of Cesaro means of C_0-Markov semigroups

In this section, we are going to establish an analogue of Theorem 3.1 for uniform mean ergodic Markov semigroups, and provide its application.

Theorem 4.1. *Let $(X, X_+, \mathcal{K}, f)$ be an abstract state space and $\mathcal{T} = (T_t)_{t\geq 0}$ be a C_0-Markov semigroup. Then the following statements are equivalent:*

(i) $\mathcal{T}$ is weakly mean ergodic;
(ii) There exists $t_0 \in \mathbb{R}_+$ such that $\delta(A_{t_0}(T)) < 1$;
(iii) $\mathcal{T}$ is uniformly mean ergodic.

We notice that in [8] an analogous result for discrete semigroup of Markov operators has been proved.

Corollary 4.1. *Let $(X, X_+, \mathcal{K}, f)$ be an abstract state space and $\mathcal{T} = (T_t)_{t\geq 0}$ be a C_0-Markov semigroup. If there exist $\rho \in [0, 1)$ and $t_0 \in \mathbb{R}_+$ such that $\delta(A_{t_0}(T)) \leq \rho$, then one has*

$$\sup_{x\in\mathcal{K}} \|A_t(\mathcal{T})x - x_0\| \leq \frac{2t_0}{t(1-\rho)},$$

where x_0 is a unique fixed point of $\mathcal{T}$.

This result allows to establish an analogue of Theorem 3.5 for $A_t(\mathcal{T})$.

Theorem 4.2. *Let $(X, X_+, \mathcal{K}, f)$ be an abstract state space and $\mathcal{T} = (T_t)_{t\geq 0}$ be a C_0-Markov semigroup. If $\delta(A_{t_0}(T)) < 1$ for some $t_0 > 0$, then every C_0-Markov semigroup $\mathcal{S} = (S_t)_{t\geq 0}$ satisfying $\|A_{t_0}(\mathcal{S}) - A_{t_0}(\mathcal{T})\| < 1-\delta(A_{t_0}(\mathcal{T}))$, is uniformly mean ergodic and has a unique fixed point $z_0 \in \mathcal{K}$ such that*

$$\|x_0 - z_0\| \leq \frac{\|A_{t_0}(\mathcal{S}) - A_{t_0}(\mathcal{T})\|}{1 - \delta(A_{t_0}(\mathcal{T})) - \|A_{t_0}(\mathcal{S}) - A_{t_0}(\mathcal{T})\|}, \tag{9}$$

here as before, x_0 is a unique fixed point of $\mathcal{T}$.

Corollary 4.2. *Let the conditions of Theorem 4.2 be satisfied. Then for every $x, z \in \mathcal{K}$ one has*

$$\sup_{x,z\in\mathcal{K}} \|A_{t_0}(\mathcal{T})x - A_{t_0}(\mathcal{S})z\| \leq \frac{2t_0}{t(1-\delta(A_{t_0}(\mathcal{T})))} + \frac{2t_0}{t(1-\delta(A_{t_0}(\mathcal{T})) - \|A_{t_0}(\mathcal{S}) - A_{t_0}(\mathcal{T})\|)} + \frac{\|A_{t_0}(\mathcal{S}) - A_{t_0}(\mathcal{T})\|}{1-\delta(A_{t_0}(\mathcal{T})) - \|A_{t_0}(\mathcal{S}) - A_{t_0}(\mathcal{T})\|}.$$

5. Conclusion

We note that all obtained results are new even if one takes X as predual of either a von Neumann algebra or a JBW-algebra. Moreover, they give new insight between the rate of convergence to stationary points of perturbed uniform/weak mean ergodic C_0-Markov semigroups.

Moreover also all obtained results can be adopted to arbitrary Banach spaces by considering a construction given by $\tilde{X} = \mathbb{R} \oplus X$ where X is an arbitrary Banach space over $\mathbb{R}$. Let (T_t) be a C_0-semigroup of contractions of X (i.e. $\|T_t x\| \leq \|x\|$ for all $x \in X$, $t \in \mathbb{R}_+$), then the equality $\tilde{T}_t(\alpha, x) = (\alpha, T_t x)$ defines a C_0-Markov semigroup on $\tilde{X}$. It is clear that $A_t(\tilde{T})(\alpha, x) = (\alpha, A_t(T)x)$. Hence, all obtained results can be applied to the C_0-Markov semigroup $(\tilde{T}_t)$, which allows to produce new types of theorems for the semigroup (T_t).

Acknowledgments

The second named author (F.M.) would like to thank Professor Noboru Watanabe for his kind invitation, hospitality and support during the conference.

References

1. E.M. Alfsen, *Compact convex sets and booundary integrals*, Springer-Verlag, Berlin, (1971).
2. L. Arlotti, B. Lods, M. Mokhtar-Kharroubi, On perturbed stochastic semigroups on abstract state spaces, *Z. Anal. Anwend.* **30** (2011), 457–495.
3. W. Bartoszek, Asymptotic properties of iterates of stochastic operators on (AL) Banach lattices, *Anal. Polon. Math.* **52** (1990), 165–173.
4. C.J.K. Batty, D. W. Robinson, Positive one-parameter semigroups on ordered Banach spaces, *Acta Appl. Math.* **1** (1984), 221–296.

5. N. Crawford, W. De Roeck, M. Schutz, Uniqueness regime for Markov dynamics on quantum lattice spin systems, *J. Phys. A: Math. Theor.* **48** (2015), no. 42, 425203.
6. T. S. Cubitt, A. Lucia, S. Michalakis, D. Perez-Garcia, Stability of Local Quantum Dissipative Systems, *Comm. Math. Phys.* **337** (2015), no. 3, 1275–1315.
7. K. J. Engel, R. Nagel, *A short course on operator semigroups*, Springer, New York, 2006.
8. N. Erkursun-Ozcan, F. Mukhamedov, Uniform ergodicities of Lotz - Räbiger nets of Markov operators on ordered Banach spaces, *Results Math.* **73** (2018), no. 1, 35.
9. N.V. Kartashov, Inequalities in theorems of ergodicity and stability for Markov chains with common Phase space, I, *Probab. Theor. Appl.* **30**(1986), 247–259.
10. U. Krengel, *Ergodic Theorems*, Walter de Gruyter, Berlin-New York, 1985.
11. A. Lasota, T.Y. Li, J.A. Yorke, Asymptotic periodicity of the iterates of Markov operator, *Trans. Amer. Math. Soc.* **286** (1984), 751–764.
12. A. Mitrophanov, Sensitivity and convergence of uniform ergodic Markov chains, *J. Appl. Probab.* **42** (2005), 1003–1014.
13. A. Mitrophanov, Stability estimates for finite homogeneous continuous-time Markov chains, *Theory Probab. Appl.* **50** (2006), no. 2, 319–326.
14. A. Mitrophanov, Ergodicity coefficient and perturbation bounds for continuous-time Markov chains, *Math. Inequal. Appl.* **8** (2005), no. 1, 159–168.
15. A. Mitrophanov, Stability and exponential convergence of continuous-time Markov chains, *J. Appl. Probab.* **40** (2003), no. 4, 970–979.
16. F. Mukhamedov, Ergodic properties of nonhomogeneous Markov chains defined on ordered Banach spaces with a base, *Acta. Math. Hungar.* **147** (2015), 294–323.
17. O. Szehr, M.M. Wolf, Perturbation bounds for quantum Markov processes and their fixed points, *J. Math. Phys.* **54** (2013), 032203.

Infinite Dimensional Analysis, Quantum Probability and Related Topics

pp. 153–167

DECOHERENCE IN AN EXCITATION ENERGY TRANSPORT MODEL

J. AGREDO*

Department of Mathematics, Colombian School of Engineering Julio Garavito, Autopista Norte AK 45 No. 205-59, Bogotá, D.C., Colombia

J. C. GARCIA-CORTE** and R. QUEZADA†

Department of Mathematics,
Autonomous Metropolitan University-Iztapalapa Campus,
Mexico City, Iztapalapa 09340, Mexico
**E-mail: jaagredoe@gmail.com*
***E-mail: jcgc@xanum.uam.mx*
†E-mail: roqb@xanum.uam.mx

We study decoherence-free subspaces of weak coupling limit type Markov generators and make a full description of the decoherence-free sub-algebra of the excitation energy transport AKV's model studied in Ref.[13].

Keywords: Quantum Markov semigroup, energy transport, decoherence.

1. Introduction

Quantum Markov Semigroups (QMSs) are a non-commutative extension of Markov semigroups defined in classical probability, they represent an evolution without memory of a microscopic system in accordance with the laws of quantum physics and fit into the framework of open quantum systems. In Heisenberg's picture the semigroup $\mathcal{T}$ acts on observables: given any observable x, $\mathcal{T}_t(x)$ describes its evolution at time t. Evolution of a density matrix ρ, is described in Schrodinger's picture (or pre-dual representation) of the semigroup $\mathcal{T}_{*t}(\rho)$, where $tr(\rho\mathcal{T}_t(x)) = tr(\mathcal{T}_{*t}(\rho)x)$.

Decoherence occurs when a quantum system interacts with its environment in an irreversible way. Decoherence and noise typically affect quantum features of a state over its time evolution, however it may be possible to find states with a unitary evolution in some "good" portion of a system.

Two main approaches to decoherence of open quantum systems have been proposed in the literature; both are based on quantum Markov semigroups. Blanchard and Olkiewicz[4], starting from an algebraic setting, defined environment induced decoherence and found many physical models where the system algebra decomposes as the direct sum of two pieces: a subalgebra, called the decoherence-free algebra, where the semigroup acts homomorphically, and a Banach subspace where the semigroup action is purely dissipative (see e.g. papers[5,6] and the references therein) and vanishing as time tends to infinity. The decoherence-free subalgebra, that was used several years before by Frigerio and Verri[11,12] to study of the asymptotic behavior of QMS's, was characterised in Ref.[8] as the commutant (or generalised commutant for unbounded operators) of the Kraus operators arising in a GKSL (Gorini-Kossakowski-Sudarshan-Lindblad) representation of the generator.

In the approach to decoherence proposed by Lidar *et al.*[14] registers of a quantum computer are modeled by a quantum open system on a finite-dimensional Hilbert space h. The time evolution of states is described by a pre-dual semigroup $\mathcal{T}_*$ on the Banach space of trace class operators on h. Given that, for a quantum computer to execute a quantum algorithm efficiently, it needs to evolve by unitary quantum dynamics, in this approach, is defined a particular sector where a quantum algorithm is executable: the decoherence-free subspaces.

A subspace h_f of h is *decoherence-free* if the time evolution of states ρ supported in h_f is given by $\rho \to e^{-itK}\rho e^{itK}$ for some self-adjoint operator K on h_f.

Decoherence-free subspaces were identified in Ref.[7] (see also Ref.[17]) under some physical (somewhat implicit) assumptions, we refer to Ref.[14] for an introduction to the theory of decoherence-free subspaces with a lot of examples and applications to protection of quantum information.

The structure of the paper is as follows. In Section 2 we introduce Markov generators of stochastic limit type with respect to an Hamiltonian H. Section 3 contains the definition of decoherence-free subspaces and we study decoherence-free subspace related with generators of stochastic limit type. Finally, in Sections 4 and 5 we make a complete description of the decoherence-free subalgebra of an excitation energy transport model due to I. Arefeva, S. Kozyrev and I. Volovich,[3] whose invariant states where characterized in Ref.[13]

2. Quantum Markov semigroups of weak coupling limit type

We are interested in a special class of quantum Markov semigroups (QMS). This class arises in the weak coupling limit as well as in the stochastic limit of a Hamiltonian system S interacting with a reservoir (see Ref.[1]). We consider a separable Hilbert space h and a positive self-adjoint operator (Hamiltonian) with a pure point spectrum $H_S \in \mathcal{B}(\mathsf{h})$, where $\mathcal{B}(\mathsf{h})$ is the von Neumann algebra of all bounded operators on h. Assume that the spectral decomposition of H_S is

$$H_S = \sum_{\epsilon_m \in Spect(H_s)} \epsilon_m P_{\epsilon_m} \tag{1}$$

where P_{ϵ_m} is the orthogonal projection onto the nullspace of $H_S - \epsilon_m 1\!\!1$, where $1\!\!1$ denotes the identity operator on h. We call Bohr frequencies the differences $\omega = \epsilon_n - \epsilon_m$ with $\epsilon_n > \epsilon_m$. Moreover,we consider an interaction operator $D \in \mathcal{B}(\mathsf{h})$ and define

$$D_\omega := \sum_{\epsilon_n - \epsilon_m = \omega} P_{\epsilon_m} D P_{\epsilon_n}. \tag{2}$$

QMS's in our class are generated by a linear operator $\mathcal{L}$ acting on $\mathcal{B}(\mathsf{h})$. This generator has the form:

$$\mathcal{L}(x) = \sum_\omega \mathcal{L}_\omega(x), \quad x \in \mathcal{B}(\mathsf{h}) \tag{3}$$

where

$$\begin{aligned}\mathcal{L}_\omega(x) :=& i[\Delta_\omega, x] - \frac{\Gamma_{-,\omega}}{2}(D_\omega^* D_\omega x - 2D_\omega^* x D_\omega + x D_\omega^* D_\omega) \\ &- \frac{\Gamma_{+,\omega}}{2}(D_\omega D_\omega^* x - 2D_\omega x D_\omega^* + x D_\omega D_\omega^*)\end{aligned}$$

with $\Gamma_{-,\omega} > 0, \; \Gamma_{+,\omega} > 0, \; \Delta_\omega = \Delta_\omega^* \in \{H_S\}'$ for all ω.

Henceforth, we will always assume that $\mathcal{L}$ generates a uniformly continuous quantum Markov semigroup $\mathcal{T}$. We will call weak coupling limit type (WCLT) QMS to any semigroup in the above class.

3. Decoherence-free subspaces

We recall that a state ρ on $\mathcal{B}(\mathsf{h})$ is a positive, trace-one, operator on h. The support $supp(\rho)$ of ρ is the closed subspace of h generated by eigenvectors with strictly positive eigenvalues.

Definition 3.1. A subspace h_f of h is called *decoherence-free* (DF) if there exists a self-adjoint operator K on h_f such that for all state ρ with support in h_f we have

$$\mathcal{T}_{*t}(\rho) = e^{-\mathrm{i}tK}\,\rho\, e^{\mathrm{i}tK} \tag{4}$$

for all $t \geq 0$. Where $\mathcal{T}_{*t}$ is the predual semigroup defined by means of the relation $\operatorname{tr}\big(\rho\mathcal{T}_t(x)\big) = \operatorname{tr}\big(\mathcal{T}_{*t}(\rho)x\big), \ \ \forall x \in \mathcal{B}(\mathsf{h}), \ \ t \geq 0$.

Notice that a self-adjoint operator K on h_f can always be extended to the whole Hilbert space h, therefore DF subspaces could be defined in an equivalent way with a self-adjoint operator K on h leaving the subspace h_f invariant. In a more precise way, for an unbounded K, this means that $e^{-\mathrm{i}tK}(\mathsf{h}_f) \subseteq \mathsf{h}_f$ for all $t \in \mathbb{R}$. Moreover, notice that a DF subspace is necessarily closed with respect to the norm topology of h.

Using the GKSL structure of the generator of a uniformly continuous semigroup

$$\begin{aligned}\mathcal{L}(x) &= i[H,x] - \frac{1}{2}\sum j(L_j^*L_jx + xL_j^*L_j - 2L_j^*xL_j) \\ &= G^*x + xG + \sum_j L_j^*xL_j\end{aligned} \tag{5}$$

where H is a self-adjoint operator, called *the effective Hamiltonian* and

$$G = -iH - \frac{1}{2}\sum_j L_j^*L_j \tag{6}$$

it is possible to give necessary and sufficient conditions for a subspace to be DF. The following is a general result due to F. Fagnola, R. Rebolledo and J. Agredo, see Proposition 7 in Ref.[2]

Theorem 3.1. *A subspace h_f is a DF subspace with associated self-adjoint operator K, if and only if in any GKSL representation of $\mathcal{L}$ by means of operators L_ℓ, G there exist complex numbers λ_ℓ $(\ell \geq 1)$ and a real number r such that $\sum_{\ell\geq 1}|\lambda_\ell|^2 < \infty$ and*

(1) $L_\ell\, u = \lambda_\ell\, u$ for all $u \in \mathsf{h}_f$ and $\ell \geq 1$,
(2) $(G + \mathrm{i}K)u = -(\frac{1}{2}\sum_{\ell\geq 1}|\lambda_\ell|^2 + \mathrm{i}r)u$ for all $u \in \mathsf{h}_f$

where G is given by (6).

A description of decoherence-free subspaces for our class of semigroups is obtained by applying Theorem 3.1.

Proposition 3.1. *If $H = \sum_{\omega} \Delta_{\omega}$, then DF subspace h_f of MGWLT QMSs is given by*

$$\mathsf{h}_f = \bigcap_{\ell} Ker(L_{\ell})$$

with $K = H$ associated self-adjoint operator.

Proof. First notice that the only eigenvalue of a non-null operator D_{ω}^* is 0. Indeed, by Lemma 3.1 in Ref.[1], condition $(\epsilon_n, \epsilon_m), (\epsilon_n, \epsilon_{m'}) \in B_{+,\omega}$ implies that $\epsilon_m = \epsilon_{m'}$. Hence for each n such that $(\epsilon_n, \epsilon_m) \in B_{+,\omega}$, the spectral projection P_n appears in (2) on the right of only one summand. This implies that the operators $P_n D P_m$, $(\epsilon_n, \epsilon_m) \in B_{+,\omega}$ have mutually orthogonal ranges. Moreover, one can order the set $B_{\omega} := \{\epsilon_m : \epsilon_m + \omega \in \text{Spect}(H_S)\}$ so that $B_{\omega} = \{\epsilon_0(\omega) < \epsilon_1(\omega) < \cdots\}$. From now on we write simply ϵ_k instead $\epsilon_k(\omega)$. Now, if we suppose that exist u eigenvector of D_{ω}^* with eigenvalue associated $\lambda \neq 0$ then

$$D_{\omega}^* u = \sum_{\{\epsilon_n \in B_{\omega} : n \geq 1\}} P_{\epsilon_n} D P_{\epsilon_{n-1}} u = \lambda \sum_{\epsilon_k \in \text{Spect}(H_S)} P_{\epsilon_k} u. \tag{7}$$

Therefore, due to orthogonality of ranges, we get that $\lambda P_{\epsilon_k} u = 0$ for all $\epsilon_k \notin B_{+,\omega}$; moreover, for every $\epsilon_n \in B_{+,\omega}$ we have that $\lambda P_{\epsilon_0} u = 0$ and $P_{\epsilon_n} D P_{\epsilon_{n-1}} u = \lambda P_{\epsilon_n} u$. Clearly, with $n = 1$ the last two relations imply $P_{\epsilon_1} u = 0$ and by induction we get $P_{\epsilon_n} u = 0$ for all $\epsilon_n \in B_{+,\omega}$. Therefore $P_{\epsilon_k} u = 0$ for all $\epsilon_k \in \text{Spect}(H_S)$ and, consequently, $u = 0$. This contradiction implies that $\lambda = 0$.

In a similar way one can show that the only eigenvalue of a non-null D_{ω} is zero. The result follows from Theorem 3.1 with $K = H$ and $r = 0$, since, by (6),

$$(G + iK)u = (G + iH)u = -\sum_j L_j^* L_j u = 0, \quad \forall u \in \cap_{\ell} \ker L_{\ell}.$$

□

4. Decoherence-free subalgebra

The decoherence-free (DF) subalgebra of $\mathcal{T}$ is denoted by $\mathcal{N}(\mathcal{T})$ and defined by

$$\{x \in \mathcal{B}(\mathsf{h}); \mathcal{T}_t(x^* x) = \mathcal{T}_t(x)^* \mathcal{T}_t(x), \mathcal{T}_t(x x^*) = \mathcal{T}_t(x) \mathcal{T}_t(x)^*, \forall t \geq 0\}.$$

In particular $\mathcal{N}(\mathcal{T})$ is a $\mathcal{T}_t$-invariant for all $t \geq 0$, von Neumann subalgebra of $\mathcal{B}(\mathsf{h})$ (see Ref.[9], Theorem 3.1 and Ref.[8], Proposition 2.1). In addition, this subalgebra is characterized in terms of operators H, L_ℓ in any GKSL representation of $\mathcal{L}$ (see Ref.[8]). First define iterated commutators $\delta_H^n(X)$ recursively by

$$\delta_H^0(X) = X, \quad \delta_H^1(X) = [H, X], \quad \delta_H^{n+1}(X) = [H, \delta_H^n(X)].$$

Proposition 4.1 (A. Dahri, F. Fagnola and R. Rebolledo). *The decoherence-free subalgebra $\mathcal{N}(\mathcal{T})$ is the commutant of the set of operators*

$$\{\delta_H^n(L_\ell), \delta_H^n(L_\ell^*); n \geq 0, \ell \geq 1\}$$

and for all self-adjoint H in a GKSL representation of the $\mathcal{L}$ we have

$$\mathcal{N}(\mathcal{T}) \subset \{x \in \mathcal{B}(\mathsf{h}); \mathcal{T}_t(x) = e^{itH} x e^{-itH} \quad \forall t \geq 0\}.$$

See Ref.[8], Theorem 3.2 and Theorem 4.1 for a proof.

We now briefly recall some results on the set

$$\mathcal{F}(\mathcal{T}) = \{x \in \mathcal{B}(\mathsf{h}); \mathcal{T}_t(x) = x \quad \forall t \geq 0\}$$

of fixed points for the linear maps $\mathcal{T}_t$ that will be useful. Clearly, $\mathcal{F}(\mathcal{T})$ is a linear subspace of operators. Unlike $\mathcal{N}(\mathcal{T})$, it may not be an algebra and not necessarily $\mathcal{F}(\mathcal{T})$ is contanined in $\mathcal{N}(\mathcal{T})$ (see Ref.[6], Section 4 for examples).

Some useful properties are collected in the following proposition.

Proposition 4.2. *Let $\mathcal{T}$ be a QMS on $\mathcal{B}(\mathsf{h})$, then*

(1) $\mathcal{F}(\mathcal{T})$ is a $*$-algebra if and only if it is contained in $\mathcal{N}(\mathcal{T})$,
(2) if the QMS $\mathcal{T}$ has a faithful normal invariant state, then $\mathcal{F}(\mathcal{T})$ is a von Neumann subalgebra of $\mathcal{B}(\mathsf{h})$,
(3) if $\mathcal{F}(\mathcal{T})$ is a von Neumann subalgebra of $\mathcal{B}(\mathsf{h})$, then it coincides with the commutant of the set of operators $\{L_\ell, L_\ell^*, H; \ell \geq 1\}$.

See Refs.[9,10] for a proof.

5. Decoherence in the excitation energy transport AKV's model

5.1. *Quantum transport AKV's model*

The invariant states of a quantum transport model due to Aref'eva, Kozyrev and Volovich[3] were studied in Ref.[13] The corresponding generator belongs to the class of WCLT and the associated reference Hamiltonian is given by

$$H_S = 0P_0 + \epsilon_1 P_1 + \epsilon_2 P_2 + \epsilon_3 P_3$$

with $P_0 = |e_0\rangle\langle e_0|$, $P_1 = |e_1\rangle\langle e_1|$, $P_2 = \sum_{j=2}^{N} |e_j\rangle\langle e_j|$, $P_3 = \sum_{j=N+1}^{N+M} |e_j\rangle\langle e_j|$ and the eigenvalues satisfy the inequalities $0 < \epsilon_3 < \epsilon_2 < \epsilon_1$. The interaction operator is

$$D = |Te_{N+1}\rangle\langle Te_1| + Z + |Te_0\rangle\langle Te_{N-1}|$$

where Z is an interference operator performing transitions between levels ϵ_2 and ϵ_3

$$Z : \mathrm{span}\{e_2, \cdots, e_N\} \mapsto \mathrm{span}\{e_{N+1}, \cdots, e_{N+M}\}$$

defined in terms of some interference coefficients g_{ab} (see Remark 2.2 in Ref. [13]) as

$$Z = \frac{1}{\sqrt{N-1}} \sum_{b=N+1}^{N+M} \sum_{a=2}^{N} g_{ab} |b\rangle\langle a|$$

and $T = P_0 + P_1 + Z + Z^*$. The maximally entangled states are defined, up to normalization, as $|\psi\rangle\langle\psi|$ with $\psi = \sum_{a=2}^{N} |a\rangle$ and $|\psi'\rangle\langle\psi'|$ with $\psi' = \sum_{b=N+1}^{N+M} |b\rangle$. Writing simply T_a instead $T|a\rangle$, one can observe that $\sqrt{N-1}T_{N-1} = \psi'$ and $\sqrt{N-1}T_{N+1} = \psi$.

Remark 5.1. The relations $T = T^*$, $Z^2 = 0$ and $Z^{*2} = 0$ immediately follow. Moreover, by direct computation one can show that $|Z|$ is an orthogonal projection, $TP_2T = P_3 = ZZ^*$ and $TP_3T = |Z|$, see Corollary 2.1 in Ref. [13]

There exist six Bohr frequencies: $\omega_1 = \epsilon_1 - \epsilon_2$, $\omega_2 = \epsilon_2 - \epsilon_3$, $\omega_3 = \epsilon_3$, $\omega_4 = \epsilon_1$, $\omega_5 = \epsilon_2$, $\omega_6 = \epsilon_1 - \epsilon_3$. But only three of them have associated non-trivial generators, so that

$$\mathcal{L} = \mathcal{L}_{\omega_1} + \mathcal{L}_{\omega_2} + \mathcal{L}_{\omega_3}$$

and $\mathcal{L}_{\omega_4} = \mathcal{L}_{\omega_5} = \mathcal{L}_{\omega_6} = 0$.

One can see that

$$\begin{aligned} D_{\omega_1} &= |\psi\rangle\langle e_1| \\ D_{\omega_2} &= Z \\ D_{\omega_3} &= |e_0\rangle\langle\psi'|. \end{aligned} \tag{8}$$

Hence

$$\begin{aligned}
L_1^{\omega_1} &= \sqrt{2(N-1)\Gamma_{Re,-,\omega_1}}\,|\psi\rangle\langle e_1| \\
L_1^{\omega_2} &= \sqrt{2(N-1)\Gamma_{Re,-,\omega_2}}\,Z \\
L_1^{\omega_3} &= \sqrt{2(N-1)\Gamma_{Re,-,\omega_3}}\,|e_0\rangle\langle\psi'| \\
L_2^{\omega_1} &= \sqrt{2(N-1)\Gamma_{Re,+,\omega_1}}\,|e_1\rangle\langle\psi| \\
L_2^{\omega_2} &= \sqrt{2(N-1)\Gamma_{Re,+,\omega_2}}\,Z^* \\
L_2^{\omega_3} &= 0
\end{aligned} \tag{9}$$

where

$$\begin{aligned}
&\Gamma_{-,\omega_1} = (N-1)\Gamma_{Re,-,\omega_1} > 0, \quad \Gamma_{-,\omega_2} = (N-1)\Gamma_{Re,-,\omega_2} > 0 \\
&\Gamma_{-,\omega_3} = \Gamma_{Re,-,\omega_3} > 0, \quad \Gamma_{+,\omega_1} = (N-1)\Gamma_{Re,+,\omega_1} > 0 \\
&\Gamma_{+,\omega_2} = \Gamma_{Re,+,\omega_2} > 0, \quad \Gamma_{+,\omega_3} = 0.
\end{aligned} \tag{10}$$

The effective Hamiltonian is give by

$$\begin{aligned}
H_{eff} = {} & \Gamma_{Im,+,\omega_1}|\psi\rangle\langle\psi| - (N-1)\Gamma_{Im,-,\omega_1}|e_1\rangle\langle e_1| \\
& + (N-1)\Gamma_{Im,+,\omega_2}P_3 - (N-1)\Gamma_{Im,-,\omega_2}|Z| \\
& - \Gamma_{Im,-,\omega_3}|\psi'\rangle\langle\psi'|
\end{aligned} \tag{11}$$

with all coefficients $\Gamma_{Im,\pm,\omega_l}$, $1 \leq l \leq 3$, positive.

Likewise, a direct computation using the result in previous section shows that a decoherence-free subspace is

$$\begin{aligned}
\mathsf{h}_f &= \{e_1\}^\perp \cap \ker Z \cap \ker Z^* \\
&= \{e_1\}^\perp \cap \{Z^*e_j : j = N+1, \cdots N+M\}^\perp \cap \{Ze_j : j = 2, \cdots N\}^\perp.
\end{aligned} \tag{12}$$

5.2. *Decoherence-free subalgebra*

Proposition 5.1. *With the previous notations, for all $n \in \mathbb{N}$ and $X, Y \in \mathcal{B}(h)$ we have,*

(1) If $Y^* = Y$ then

$$\delta_Y^n(X^*) = (-1)^n(\delta_Y^n(X))^* \tag{13}$$

(2)

$$\delta_Y^n(X) = \sum_{j=0}^{n}(-1)^j\binom{n}{j}Y^{n-j}XY^j. \tag{14}$$

Proof. Direct computations using Newton's binomial expansion and induction. □

The following is an immediate consequence of results in Refs.[18,8]

Theorem 5.1. *$\mathcal{N}(\mathcal{T})$ is the commutant of the family of operators,*

$$\mathcal{C}_n := \{\delta^n_{H_{eff}}(D_{\omega_\ell}), \delta^n_{H_{eff}}(D^*_{\omega_\ell}), \delta^n_{H_{eff}}(D_{\omega_3}) : \ 1 \le \ell \le 2\}$$

where H_{eff} is the effective Hamiltonian in (11).

By (13), $\delta^n_{H_{eff}}(X^*) = (-1)^n \delta^n_{H_{eff}}(X)^*$. Hence each $\mathcal{C}_{2n}$ is a self-adjoint family. In particular,

$$\begin{aligned} \mathcal{C}_0 &= \{D_{\omega_\ell}, D^*_{\omega_\ell} : \ 1 \le \ell \le 3\} \\ &= \{|\psi\rangle\langle e_1|, |e_1\rangle\langle\psi|, Z, Z^*, |e_0\rangle\langle\psi'|, |\psi'\rangle\langle e_0|\} \end{aligned} \tag{15}$$

consists of the Kraus operators of the model along with their adjoints.

Theorem 5.2. *Let $\mathcal{D}_1 = \mathcal{C}_0 \cup \{H_{eff}\}$. Then*

(1) $\mathcal{C}_0' = \mathcal{D}_1' \subset \mathcal{F}(\mathcal{T})$
(2) $\mathcal{C}_0' \subset \cap_{n\ge 1}\mathcal{C}_n'$
(3) $\mathcal{N}(\mathcal{T}) = \mathcal{C}_0' \subset \mathcal{F}(\mathcal{T})$.

Proof.

(1) To prove that $\mathcal{C}_0' = D_1'$, notice that each summand in H_{eff} is a product of elements in $\mathcal{C}_0$. Indeed,

$$\begin{aligned} |\psi\rangle\langle\psi| &= |\psi\rangle\langle e_1||e_1\rangle\langle\psi| \\ |e_1\rangle\langle e_1| &= |e_1\rangle\langle\psi||\psi\rangle\langle e_1| \\ |Z| &= |Z|^2 = Z^*Z \\ P_3 &= ZZ^* \\ |\psi'\rangle\langle\psi'| &= |\psi'\rangle\langle e_0||e_0\rangle\langle\psi'|. \end{aligned} \tag{16}$$

Therefore any operator in $\mathcal{C}_0'$ commutes with H_{eff}. This proves that $D_1' \subset \mathcal{C}_0'$, the opposite inequality is immediate.

To prove $\mathcal{D}_1' \subset \mathcal{F}(\mathcal{T})$, let $x \in \mathcal{D}_1'$ then x commutes with all Krauss operators D_{ω_ℓ}'s in the Lindblad generator and also commutes with the effective Hamiltonian. Hence, it readily follows that $\mathcal{L}(x) = 0$, i.e., $x \in \mathcal{F}(\mathcal{T})$.

(2) Due to (14), any $x \in \mathcal{D}_1'$ commutes with any linear combination of products of powers of H_{eff} and the D_{ω_ℓ}, $1 \le \ell \le 3$. Thus, it commutes with all the $\delta^n_{H_{eff}}(D_{\omega_\ell})$, $1 \le \ell \le 3$. This proves that $x \in \mathcal{C}_n'$, $n \ge 1$. So, along with item (1), it shows that $\mathcal{C}_0' \subset \cap_{n\ge1}\mathcal{C}_n'$.

(3) By Theorem 5.1 it is clear that, $\mathcal{N}(\mathcal{T}) = \cap_{n=0}^{\infty}\mathcal{C}_n'$. Therefore, by items (1) and (2), $\mathcal{N}(\mathcal{T}) = \mathcal{C}_0' \subset \mathcal{F}(\mathcal{T})$.

□

The following is a simple but very useful result.

Lemma 5.1. *Given two non-zero vectors $u, v \in \mathsf{h}$ and $a \in \mathcal{B}(\mathsf{h})$, operators a and $|u\rangle\langle v|$ commute if and only if there exists $\lambda \in \mathbb{C}$ such that $au = \lambda u$ and $a^*v = \bar{\lambda}v$.*

Proof. If a and $|u\rangle\langle v|$ commute, then $\langle v, w\rangle au = \langle v, aw\rangle u, \quad \forall \quad w \in \mathsf{h}$. Hence, with $w = v$ we get $au = \lambda u$ with $\lambda = \frac{\langle v, av\rangle}{\|v\|^2}$. In particular, since $|v\rangle\langle u|$ commutes with a^*, there exits $\mu \in \mathbb{C}$ such that $a^*v = \mu v$. But

$$\lambda|u\rangle\langle v| = a|u\rangle\langle v| = |u\rangle\langle v|a = \bar{\mu}|u\rangle\langle v|$$

hence $\mu = \bar{\lambda}$.

Conversely, if $\lambda \in \mathbb{C}$ is such that $au = \lambda u$ and $a^*v = \bar{\lambda}v$, then

$$a|u\rangle\langle v| = \lambda|u\rangle\langle v| = |u\rangle\langle a^*v| = |u\rangle\langle v|a$$

This finishes the proof. □

Theorem 5.3. *Let a be any element in $\mathcal{B}(\mathsf{h})$, then: $a \in \mathcal{C}_0' = \mathcal{N}(\mathcal{T})$ if and only if a, a^* commute with Z and Z^*, and there exists a complex number λ_0 such that, $\{e_0, e_1, \psi, \psi'^*\psi'\}$ are common eigenvectors of a and a^*, associated with the eigenvalues λ_0 and $\bar{\lambda}_0$, respectively.*

Proof. $\mathcal{C}_0'$ is a self-adjoint sub-algebra, therefore $a \in \mathcal{C}_0'$ if and only if $a^* \in \mathcal{C}_0'$. Consequently both operators commute with Z and Z^*.

If $a \in \mathcal{C}_0'$, an application of Lemma 5.1 to the rank-one operators $|e_1\rangle\langle\psi|$ and $|\psi\rangle\langle e_1|$ yields the existence of $\lambda_1, \lambda_\psi \in \mathbb{C}$ such that

$$ae_1 = \lambda_1 e_1, \quad a^*\psi = \bar{\lambda}_1\psi. \tag{17}$$

Now, applying the same lemma with the rank-one operator $|\psi\rangle\langle\psi| = |\psi\rangle e_1||e_1\rangle\langle\psi| \in \mathcal{C}_0'$ one gets that

$$a\psi = \lambda_\psi\psi, \quad \text{and} \quad a^*\psi = \bar{\lambda}_\psi\psi.$$

It readily follows that $\lambda_\psi = \lambda_1$ and, consequently, e_1, ψ are eigenvectors of a and a^* with eigenvalues λ_1 and $\bar{\lambda}_1$, respectively. But a direct computation shows that $aZ\psi = Za\psi = \lambda_1 Z\psi$ and similarly $a^*Z\psi = \bar{\lambda}_1 Z\psi$. This proves that $\{e_1, \psi, Z\psi\}$ are common eigenvectors of a and a^* with eigenvalues λ_1 and $\bar{\lambda}_1$, respectively. Similar computations show that span$\{e_0, \psi'^*\psi'\}$ are common eigenvectors of a and a^* with eigenvalues λ_0 and $\bar{\lambda}_0$.

To prove (i) it remains to show that $\lambda_0 = \lambda_1$. Clearly $a \in \mathcal{C}_0'$ if and only if the self-adjoint operators $a_1 = \frac{1}{2}(a + a^*)$ and $a_2 = \frac{1}{2i}(a - a^*)$ belong to $\mathcal{C}_0'$. Let λ_0 and λ_1 be the eigenvalues of a as above. Then a_1 and a_2, have the same eigenvectors of a with eigenvalues $\Re e\lambda_0$, $\Re e\lambda_1$ and $\Im\lambda_0$, $\Im\lambda_1$, respectively. Assume $\Re e\lambda_0 \neq \Re e\lambda_1$, then due to the self-adjointness of a_1, corresponding eigenvectors should be mutually orthogonal. But a simple computation shows that

$$\langle Z^*\psi', \psi\rangle = \langle \psi', Z\psi\rangle = \langle \psi', e_{N+1}\rangle = 1.$$

This shows that $\Re e\lambda_0 = \Re e\lambda_1$. In a similar way one can see that $\Im\lambda_0 = \Im\lambda_1$. We can conclude that $\lambda_0 = \lambda_1$.

Conversely, if a and a^* commute with Z and Z^* and have common eigenvectors $\{e_0, e_1, \psi, \psi'^*\psi'\}$ with eigenvalues λ_0, $\bar{\lambda}_0$, then due to Lemma 5.1 they belong to $\mathcal{C}_0'$. This finishes the proof. □

Theorem 5.4. *Each element $a \in \mathcal{C}_0'$ can be written, in the form*

$$a = \sigma + Z\sigma Z^* \tag{18}$$

where, $\sigma = P_3^\perp a$ is a linear operator such that,

(i) $\sigma u = \lambda_0 u$ for all $\in$ span$\{e_0, e_1, \psi, Z^*\psi'\}$
(ii) $\sigma(Im|Z| \cap \{\psi, Z^*\psi'^\perp) \subset Im|Z| \cap \{\psi, Z^*\psi'^\perp$
(iii) $\sigma(ker|Z|) \subset ker|Z|$.

Proof. Clearly, if $a \in \mathcal{C}_0'$, then it commutes with P_0, P_1. Moreover it commutes with Z and Z^*, hence $a(\mathrm{Im}|Z|) \subset \mathrm{Im}|Z|$ and $a(\ker|Z|) \subset \ker|Z|$. The identity $P_2 = \mathrm{Im}|Z| \oplus \ker|Z|$ implies that a commutes with P_2, consequently commutes with P_3 and with H_S.

Now, every element $a \in \mathcal{C}_0'$ can be written in the form $a = P_3^\perp a + P_3 a$. Let $\sigma = P_3^\perp a$, hence, $Z\sigma Z^* = ZP_3^\perp aZ^* = ZP_2Z^*a = P_3 a$, consequently, $a = \sigma + Z\sigma Z^*$.

Now, observe that

$$\sigma\psi = P_3^\perp a\psi = \lambda_0 P_3^\perp\psi = \lambda_0\psi$$

and

$$\sigma Z^*\psi' = P_3^{\perp} Z^* a\psi' = \lambda_0 P_3^{\perp} Z^*\psi' = \lambda_0 Z^*\psi'.$$

Similarly one sees that $\sigma e_0 = \lambda_0 e_0$ and $\sigma e_1 = \lambda_0 e_1$. Therefore, for every $u \in \mathrm{Im} Z \cap \{\psi, Z^*\psi'^{\perp}$ we have

$$\langle \sigma u, \psi \rangle = \langle u, P_3^{\perp} a^* \psi \rangle = \bar{\lambda}_0 \langle u, P_3^{\perp} \psi \rangle = \bar{\lambda}_0 \langle u, \psi \rangle = 0$$

and

$$\langle \sigma u, Z^*\psi' \rangle = \langle u, P_3^{\perp} a^* Z^*\psi' \rangle = \bar{\lambda}_0 \langle u, P_3^{\perp} Z^*\psi' \rangle = \bar{\lambda}_0 \langle u, Z^*\psi' \rangle = 0.$$

Similarly one can show that $\sigma u \in \ker |Z|$ for all $u \in \ker |Z|$. This finishes the proof. □

Theorem 5.5. *Let $\mathcal{W}$ be the space of operators σ on $P_3^{\perp}(\mathsf{h})$ fulfilling conditions (i)–(iii) in Theorem 5.3. The map $U : \mathcal{W} \to \mathcal{C}_0'$ given by*

$$U(\sigma) = \sigma + Z\sigma Z^*$$

is a one to one and onto $*$-homomorphism

Proof. Clearly U is linear and onto. If $U(\sigma) = U(\sigma')$, then $U(\sigma - \sigma') = 0$ and due to orthogonality of ranges we get that $\sigma = \sigma'$. Hence U is injective.

Now, if $\sigma = P_3^{\perp} a$ and $\tau = P_3^{\perp} b$,

$$\begin{aligned} U(\sigma)^* U(\tau) &= (\sigma^* + Z\sigma^* Z^*)(\tau + Z\tau Z^*) \\ &= \sigma^*\tau + \sigma^* Z\tau Z^* + Z\sigma^* Z^*\tau + Z\sigma^* Z^* Z\tau Z^* \\ &= \sigma^*\tau + Z\sigma^* |Z| \tau Z^* = \sigma^*\tau + Z|Z| a^* P_3^{\perp} b Z^* \\ &= \sigma^*\tau + P_3 Z\sigma^*\tau Z^* = \sigma^*\tau + Z\sigma^*\tau Z^* = U(\sigma^*\tau) \end{aligned} \tag{19}$$

where we have used that $\sigma^* Z = a^* P_3^{\perp} Z = 0 = Z\sigma^*$. □

Remark 5.2. In particular, $U(\sigma)$ is a self-adjoint operator, (respectively an orthogonal projection) if and only if σ is a self-adjoint operator (respectively an orthogonal projection).

Corollary 5.1.

(1) If $M \geq 2$ then dim $\mathcal{C}_0'^2 + (N - M - 1)^2$
(2) If $M = 1$ then dim $\mathcal{C}_0'^2$
(3) $\mathcal{C}_0'$ is non commutative if and only if $M \geq 4$ or $N \geq M + 3$.

Proof. Due to (18), $\mathcal{C}_0'$ is parametrized by the vector space $\mathcal{W}$, of operators σ on $P_3^{\perp}(\mathsf{h})$ fulfilling conditions (i)–(iii) in Theorem 5.3. Hence, it suffices to compute the dimension of $\mathcal{W}$.

But, ψ, $Z^*\psi'$ are fixed points of $|Z|$, indeed,

$$|Z|\psi = \sqrt{N-1}Z^*ZZ^*e_{N+1} = \sqrt{N-1}Z^*e_{N+1} = \psi$$

and $|Z|Z^*\psi'^*ZZ^*\psi'^*\psi'$. Consequently we have the decomposition

$$\mathrm{Im}|Z| = \mathrm{span}\{\psi, Z^*\psi'\} \oplus \mathrm{Im}|Z| \cap \{\psi, Z^*\psi'^{\perp}.$$

Therefore, since $P_2(\mathsf{h}) = \mathrm{Im}|Z| \oplus \ker|Z|$, we have,

$$\begin{aligned}
P_3^{\perp}(\mathsf{h}) &= P_0(\mathsf{h}) \oplus P_1(\mathsf{h}) \oplus P_2(\mathsf{h}) \\
&= P_0(h) \oplus P_1(h) \oplus \mathrm{span}\{\psi, Z^*\psi'\} \oplus \left(Im|Z| \cap \{\psi, Z^*\psi'\}^{\perp}\right) \oplus \ker|Z| \\
&= \mathrm{span}\{e_0, e_1, \psi, Z^*\psi'\} \oplus \left(Im|Z| \cap \{\psi, Z^*\psi'\}^{\perp}\right) \oplus ker|Z|.
\end{aligned}$$

Now, any $\sigma \in \mathcal{W}$ behaves as a constant on $\mathrm{span}\{e_0, e_1, \psi, Z^*\psi'\}$, maps $Im|Z| \cap V_1^T = Im|Z| \cap V_2^T$ into itself and maps freely $ker|Z|$ into itself. The space of constant operators has dimension one. $|Z| = \sum_{b=N+1}^{N+M} |Z^*e_b\rangle\langle Z^*e_b|$, hence, $\dim(\ker|Z|) = N-1-M$ consequently, the dimension of the space of maps of $\ker|Z|$ into itself is $(N-1-M)^2$.

(1) If $M \geq 2$ the dimension of the subspace of all operators mapping $Im|Z| \cap \{\psi, Z^*\psi'^{\perp}$ into itself is $(M-2)^2$ since $\dim(\mathrm{Im}|Z|) = M$. Thus, if $M \geq 2$ we get

$$\dim(\mathcal{C}_0'^2 + (N-M-1)^2.$$

(2) If $M = 1$, $|Z| = |\psi\rangle\langle\psi| = |Z^*\psi'^*\psi'|$, hence $\mathrm{Im}|Z| \cap \{\psi, Z^*\psi'^{\perp}) = \{0\}$ and the dimension of all of the operators mapping $Im|Z| \cap \{\psi, Z^*\psi'^{\perp}$ into itself is zero. Therefore, in this case,

$$dim\ \mathcal{C}_0'^2.$$

(3) If $M \geq 4$ then the dimension of all operators mapping $\mathrm{Im}|Z| \cap \{\psi, Z^*\psi'^{\perp}$ into itself is $(M-2)^2 \geq 4$, which clearly corresponds to a noncommutative algebra. The same is true if $N \geq M+3$ since, in this case, the dimension of all linear maps from $\ker|Z|$ into itself is $(N-1-M)^2 \geq 4$. Conversely, if $M < 4$ and $N < M+3$, then the dimension of all of the operators mapping $\mathrm{Im}|Z| \cap \{\psi, Z^*\psi'^{\perp}$ into itself is $(M-2)^2 < 4$, which implies that this dimension is 0 or 1. Dimension of all linear maps from $\ker|Z|$ into itself is $(N-1-M)^2 < 4$, which also implies that this dimension is 0 or 1. Hence, both algebras should be commutative.

□

5.3. *Examples and comments*

Corollary 5.2.

(1) If $N = 3$ and $M = 1$ then dim $\mathcal{C}_0' = 2$ and

$$\mathcal{C}_0' = \{\lambda(P_0 + P_1 + |Z| + P_3) + \mu|e_2 - e_3\rangle\langle e_2 - e_3| \, : \lambda, \mu \in \mathbb{C}\}. \quad (20)$$

(2) If $N = 3$ and $M = 2$ then all elements in $\mathcal{C}_0'$ are multiples of the identity operator. This is the unique case in which $\mathcal{C}_0'$ is trivial.

Proof.

(1) If $N = 3$ and $M = 1$ then $\dim(\mathcal{C}_0'^2 = 2$. Indeed, $\dim(\ker|Z|) = 1$, $e_2 - e_3 \in \ker|Z|$, $Z^*\psi'^* e_4 = e_2 = \psi$,

$$|Z| = |\psi\rangle\langle\psi|, \quad P_3 = |e_4\rangle\langle e_4| = |\psi'\rangle\langle\psi'|$$

$$P_0 + P_1 + |Z| + P_3 = \sigma + Z\sigma Z^*$$

where, $\sigma = P_0 + P_1 + |Z|$. The claim in this item readily follows.

Let us notice that this non-trivial decoherence free algebra is commutative. In this particular situation, $P_0 + P_1 + |Z| + P_3 \in \mathcal{C}_0'$ but in general, when $M < N - 1$, $P_0 + P_1 + |Z| + P_3 \notin \mathcal{C}_0'$. Let us recall that $P_0 + P_1 + |Z| + P_3 = I \in \mathcal{C}_0'$ if $M = N - 1$.

(2) If $N = 3$ and $M = 2$, the Hilbert space h has dimension $M + N + 1 = 6$. Furthermore, ψ and $Z^*\psi'$ belong to $P_2(h)$, they are not orthogonal but linearly independent. The same holds true for ψ' and $Z\psi$ which belong to $P_3(h)$. The set $\{e_0, e_1, \psi, \psi'^*\psi'\}$ is an algebraic basis of h, with eigenvectors of any $a \in \mathcal{C}_0'$. So a must be a multiple of the identity.

□

Acknowledgements

The financial support from CONACYT-Mexico (Grant 221873) and PRODEP Red Italia-UAM Mexico, is gratefully acknowledged. J. Agredo would like to highlight the support from Escuela colombiana de ingeniería "Julio Garavito".

References

1. L. Accardi, F. Fagnola and R. Quezada, *On three new principles in non-equilibrium statistical mechanics and Markov semigroups of weak coupling limit type*, *Infin. Dimens. Anal. Quantum Probab. Relat. Top.*, **19-2**, (2016), 9–46.

2. J. Agredo, F. Fagnola and R. Rebolledo, *Decoherence-free subspaces of a quantum Markov semigroup*, *J. Math. Phys*, **55**, (2014), 11–32.
3. Y. Aref'eva, I. Volovich and S. Kozyrev, *Stochastic Limit Method and Interference in Quantum Many-particles Systems*, *Theoretical and Mathematical Physics* **183**(3) (2015) 782–799.
4. Ph. Blanchard, and R. Olkiewicz, *Decoherence induced transition from quantum to classical dynamics*, *Rev. Math. Phys*, **15**, (2003), 217–243.
5. Ph. Blanchard, and R. Olkiewicz, *Decoherence as irreversible dynamical process in open quantum systems*, *Open quantum systems. III*, *Lectures Notes in Math.*, 1882, Springer, Berlin, (2006).
6. R. Carbone, E. Sasso, and V. Umanità, *Decoherence for Quantum Markov Semigroups on Matrix Algebras*, *Annales Henri Poincaré*, **14**, (2013), 17–37.
7. I. L. Chuang, D. A. Lidar and K. B. Whaley, *Decoherence-free Subspaces for Quantum Computation*, *Phys. Rev. Lett.* **81**, (1998), 2594.
8. A. Dhahri, F. Fagnola and R. Rebolledo, *The decoherence-free subalgebra of a quantum Markov semigroup with unbounded generator*, *Infin. Dimens. Anal. Quantum Probab. Relat.* Top., **13**, (2010), 413–433.
9. D. E. Evans, *Irreducible quantum dynamical semigroups*, *Comm. Math. Phys.* **54** (1977), 293–297.
10. F. Fagnola and R. Rebolledo, *Algebraic conditions for convergence of a quantum Markov semigroup to a steady state*, *Infin. Dimens. Anal. Quantum Probab. Relat. Top*, **11**, (2008), 467–474.
11. A. Frigerio, *Quantum dynamical semigroups and approach to equilibrium*, *Lett. Math. Phys.* **2** (1977) 79–87.
12. A. Frigerio and M. Verri, *Long-time asymptotic properties of dynamical semigroups on W^*-algebras*, *Math. Z.* **180** (1982) 275–286.
13. J. C. García, S. Gliouez, F. Guerrero and R. Quezada, *Entangled and dark stationary states of excitation energy transport models in quantum many-particles systems and photosynthesis*, Infinite Dimensional Analysis, Quantum Probab. Relat. Top. **21** (3) (2018) 1850018 (21 pages), DOI: 10.1142/S0219025718500182.
14. D. Lidar, *Review of Decoherence Free Subspaces, Noiseless Subsystems, and Dynamical Decoupling*, *Adv. Chem. Phys.* **154**, (2014), 295–354.
15. R. Rebolledo, *Decoherence of quantum Markov semigroups*, *Ann. Inst. Henri Poincaré (B) Probab. Stat*, **41**, (2005), 349–373.
16. R. Rebolledo, *A view on decoherence via master equations*, *Open Sys. Inf. Dyn*, **12**, (2005), 37–54.
17. F. Ticozzi and L. Viola, *Quantum Markovian Subsystems: Invariance, Attractivity, and Control*, *IEEE Trans. Automat Control* **53**, (2008), 2048.
18. V. Umanità, 2005. *Classification and Decomposition of Quantum Markov Semigroups.* Dottorato di Ricerca in Matematica. Università di Genova, Italia.

Infinite Dimensional Analysis, Quantum Probability and Related Topics

pp. 169–181

POWERS OF WHITE NOISE ASSOCIATED WITH THE PRODUCT OF DISTRIBUTIONS

SHINYA HIBINO* and KIMIAKI SAITO†

Department of Mathematics, Meijo University,
Tenpaku, Nagoya 468-8502, Japan
E-mails: *163429005@ccalumni.meijo-u.ac.jp; †ksaito@meijo-u.ac.jp

In this paper we introduce a new locally convex space of distributions, in which we have the product of any two distributions as a series expansion. Then we discuss higher powers of the complex white noise on the space consisting of distributions without any renormalization. We also extend the Lévy and Voltera Laplacians to operators on C^∞-functions on the locally convex space, and give an integral expression of the Lévy Laplacian. Based on those results, we obtain white noise distribution-valued stochastic differential equations for the delta distribution centered at the higher power of the infinite dimensional Brownian motion.

Keywords: Product of distributions, powers of white noise, infinite dimensional Laplacians, infinite dimensional Brownian motion.

1. Introduction

Let $\mathcal{S}'(\mathbb{R})$ be the Schwartz space of tempered distributions. The Gross Laplacian generates an infinite dimensional Brownian motion

$$\mathbb{B}(t) = \sum_{k=1}^{\infty} B_k(t) e_k,$$

as an $\mathcal{S}'(\mathbb{R})$-valued stochastic process with a sequence $\{B_k(t)\}_{k=1}^{\infty}$ of independent one-dimensional Brownian motions and an orthonormal basis $\{e_k\}_{k=1}^{\infty}$ of the real Hilbert space $L^2(\mathbb{R})$, which is contained in $\mathcal{S}(\mathbb{R})$. Obviously, we can not consider any power of $\mathbb{B}(t)$ in $\mathcal{S}'(\mathbb{R})$. However, in our previous paper [7], introducing dual spaces of some closed subspaces of a complex Hilbert space $L^2([0,1])$ based on functions $e_n(u) := e^{2\pi i n u}, n \in \mathbb{Z}$, we constructed a new locally convex space E_∞^* of distributions, in which we have the product of any two elements as series expansions. We also introduced an E_∞^*-valued Brownian motion and discussed the Itô formula for powers of the Brownian motion in the paper [7].

In this paper, based on the space E_∞^*, we can construct a space of white noise distributions in which we have powers of white noise without any renormalization. We also discuss the complex version of the Gross, Lévy and Volterra Laplacians, and an important formula among those Laplacians, with the divergent part $\delta(0)$ being realized as a distribution in E_∞^*, and give integral expressions of infinite dimensional Laplacians. Moreover, we define an infinite dimensional Brownian motion $\{\mathbf{B}_n(t) \equiv \sum_{k=-n}^{\infty} B_k(t)e_k;\ t \geq 0\}$ on E_∞^* for each $n \in \mathbb{N}$ and obtain stochastic differential equations induced from a delta distribution $\delta_{\mathbf{B}_n(t)^m}$ centered at the m-th power of $\mathbf{B}_n(t)$ for each $m \in \mathbb{N}$.

2. Basic spaces of distributions

Let $e_n(u) := e^{2\pi i n u}$ for $n \in \mathbb{Z}$. Then the set $\{e_n;\ \ n \in \mathbb{Z}\}$ is an orthonormal basis for $L^2([0,1])$. For any $n \in \mathbb{Z}$ let H_n be the closed subspace of $L^2([0,1])$ generated by the functions $e_k, k \geq -n, k \in \mathbb{Z}$. Then for any $n \in \mathbb{Z}$ and sequence $\{\ell_k\}_{k=-\infty}^{\infty}$ satisfying

$$1 = \ell_0 < \ell_1 < \ell_2 < \ell_3 < \cdots, \quad \sum_{k=0}^{\infty} \ell_k^{-2} < \infty, \quad \ell_{-k} = \ell_k,\ k \in \mathbb{N},$$

and introducing a densely defined selfadjoint operator A_n on H_n by

$$A_n\xi = \sum_{k=-n}^{\infty} \ell_k \alpha_k e_k, \quad \xi = \sum_{k=-n}^{\infty} \alpha_k e_k \in H_n,$$

we can define norms $|\cdot|_{n,p},\ \ p \in \mathbb{R}$ and spaces $E_{n,p}$ by

$$\begin{aligned}
|\xi|_{n,p}^2 &:= |A_n^p \xi|_0^2 = \sum_{k=-n}^{\infty} \ell_k^{2p} |\langle \xi, e_k \rangle_*|^2; \\
E_{n,p} &:= \{\xi \in H_n;\ \ |\xi|_{n,p} < \infty\} \ \text{ for } \ p \geq 0; \\
E_{n,p} &:= \text{completion of } \ H_n \ \text{ with respect to } \ |\cdot|_{n,p} \ \text{ for } \ p < 0,
\end{aligned}$$

where $|\cdot|_0$ is the norm of $L^2([0,1])$ and $\langle \cdot, \cdot \rangle_*$ is the conjugate bilinear form of E_n^* and E_n. Here E_n and E_n^* are the projective limit space of $E_{n,p},\ \ p \in \mathbb{R}$ and the inductive limit space of $E_{n,p},\ \ p \in \mathbb{R}$, respectively. We note that $H_n = E_{n,0}$ for $n \in \mathbb{Z}$. We also assume the following conditions:

(1) For any $\alpha > 0$, there exists $\beta > 0$ such that $(k+1)^\alpha \leq \ell_k^\beta$, $k = 1, 2, 3, \ldots$.
(2) $\rho \equiv \sup_{k \in \mathbb{N}_0} \ell_{k+1}/\ell_k < \infty$.
(3) For any $a > 0$, there exists $p \geq 1$ such that $\sum_{k=0}^{\infty} \ell_k^{-2p} \rho^{ak} < \infty$.

For any $n \in \mathbb{N}_0$ and $a > 1$, the sequence $\{(|k|+1)^n a^{|k|}\}_{k=-\infty}^{\infty}$ is an example satisfying the above three conditions.

Define spaces E_∞, H_∞ and E_∞^* by inductive limit spaces of E_n, $n \in \mathbb{Z}$, H_n, $n \in \mathbb{Z}$ and E_n^*, $n \in \mathbb{Z}$, respectively. Then we can get the product of distributions in E_∞^* as follows:

$$x \cdot y = \sum_{j=-n}^{\infty} \overline{\langle x, e_j\rangle_* e_j} \cdot \sum_{j=-n}^{\infty} \overline{\langle y, e_j\rangle_* e_j} = \sum_{j=-2n}^{\infty} \sum_{k=-n}^{j+n} \overline{\langle x, e_k\rangle_* \langle y, e_{j-k}\rangle_* e_j}$$

if $x = \sum_{j=-n}^{\infty} \overline{\langle x, e_j\rangle_* e_j}$ and $y = \sum_{j=-n}^{\infty} \overline{\langle y, e_j\rangle_* e_j}$ in E_n^* for some $n \in \mathbb{N}$.

In our previous paper [7], we obtained that for any x, y in E_∞^*, there exist $n \in \mathbb{Z}$ and $q > 0$ such that x and y are in $E_{n,-q}$. In the paper [7], we proved that for some $r > 0$ and any $p \geq 2q + r + 1$, there exists a positive constant M_n depending only on n such that $|x \cdot y|_{2n,-p}^2 \leq M_n \rho^{4n(r+2q)} |x|_{n,-q}^2 |y|_{n,-q}^2$. We can take the constant which is not depending on n instead of M_n. We give the more precise proof in the following:

Theorem 2.1. *For any x and y in E_∞^*, the above product $x \cdot y$ is also in E_∞^*. Moreover, the product operator $\cdot :\ E_\infty^* \times E_\infty^* \to E_\infty^*$ is a continuous bilinear operator.*

Proof. Let x, y be elements of E_∞^*. Then there exist $n \in \mathbb{Z}$ and $q > 0$ such that x and y are in $E_{n,-q}$. For some $r > 0$ and any $p \geq 2q + r + 1$ we can estimate the $(-p)$-norm $|x \cdot y|_{2n,-p}$ of $x \cdot y$ as follows:

$$\begin{aligned}
|x \cdot y|_{2n,-p}^2 &= \sum_{\nu=-2n}^{\infty} \ell_\nu^{-2p} \left| \sum_{j=-n}^{\nu+n} \overline{\langle x, e_j\rangle_* \langle y, e_{\nu-j}\rangle_*} \right|^2 \\
&\leq \sum_{\nu=-2n}^{\infty} \ell_\nu^{-2p} \sum_{j=-n}^{\nu+n} |e_j|_{n,q}^2 |x|_{n,-q}^2 \sum_{j=-n}^{\nu+n} |e_{\nu-j}|_{n,q}^2 |y|_{n,-q}^2 \\
&\leq \sum_{\nu=1}^{n} \ell_{-\nu}^{-2p} \ell_{-\nu+2n}^{2r} \ell_n^{4q} |x|_{n,-q}^2 |y|_{n,-q}^2 + \sum_{\nu=n+1}^{2n} \ell_{-\nu}^{-2p} \ell_{-\nu+2n}^{2r} \ell_n^{4q} |x|_{n,-q}^2 |y|_{n,-q}^2 \\
&\quad + \sum_{\nu=0}^{\infty} \ell_\nu^{-2p} \ell_{\nu+2n}^{2r} \ell_{\nu+2n}^{4q} |x|_{n,-q}^2 |y|_{n,-q}^2 (=: S_1 + S_2 + S_3).
\end{aligned}$$

Since

$$\begin{aligned}
\ell_{\nu+2n}^{2(r+2q)} &= \left(\frac{\ell_{\nu+2n}}{\ell_{\nu+2n-1}} \frac{\ell_{\nu+2n-1}}{\ell_{\nu+2n-2}} \cdot \frac{\ell_{\nu+1}}{\ell_\nu} \ell_\nu \right)^{2(r+2q)} \\
&\leq \left(\rho^{2n} \ell_\nu\right)^{2(r+2q)} = \rho^{4n(r+2q)} \ell_\nu^{2(r+2q)}
\end{aligned}$$

for some $r > 0$ and $q > 0$, we have

$$S_3 \leq \sum_{\nu=0}^{\infty} \ell_\nu^{-2(p-r-2q)} \rho^{4n(r+2q)} |x|_{n,-q}^2 |y|_{n,-q}^2$$

for any $p \geq 2q + r + 1$. By using similar method, we also get estimations:

$$S_1 \leq 2\sum_{\nu=1}^{n} \ell_\nu^{-2p} \rho^{4n(r+q)} |x|_{n,-q}^2 |y|_{n,-q}^2; \; S_2 \leq \sum_{\nu=n+1}^{2n} \ell_\nu^{-2p} \rho^{4n(r+2q)} |x|_{n,-q}^2 |y|_{n,-q}^2.$$

Here, we set $M := 2\sum_{\nu=-\infty}^{\infty} \ell_\nu^{-2}$. Then we obtain an estimation

$$|x \cdot y|_{2n,-p}^2 \leq M \rho^{4n(r+2q)} |x|_{n,-q}^2 |y|_{n,-q}^2.$$

This implies $x \cdot y \in E_{2n}^*$. The bilinearity follows from properties of the conjugate bilinear form between E_∞^* and E_∞. The proof of the continuity of the product operator is same to that in the paper [7]. □

For any $N \in \mathbb{N}_0$ and $p \geq 0$, we define a set $\mathcal{E}_{N,-p}$ by

$$\mathcal{E}_{N,-p} := \left\{ \left(\sum_{k=-n}^{\infty} c_{n,k} e_k \right)_{n\in\mathbb{Z}} ; \; \forall k \in \mathbb{Z}(c_{n,k} \in \mathbb{C}), \; \forall k \geq N(c_{n,k} = 0), \right.$$

$$\left. \forall n \in \mathbb{Z} \left(\sum_{k=-n}^{\infty} \ell_k^{-2p} |c_{n,k}|^2 < \infty \right) \right\}$$

with norm

$$\left| \left(\sum_{k=-n}^{\infty} c_{n,k} e_k \right)_{n\in\mathbb{Z}} \right|_{\mathcal{E}_{N,-p}} := \left(\sum_{k=-N}^{\infty} \ell_k^{-2p} |c_{n,k}|^2 \right)^{1/2}.$$

Then the space $\mathcal{E}_{N,-p}$ is Hilbert space for any $N \in \mathbb{N}_0$ and $p \geq 0$. We also define sets $\mathcal{E}_N^*$ and $\mathcal{E}_\infty^*$ by the inductive limit space of spaces $\mathcal{E}_{N,-p}$, $p \geq 0$, and the inductive limit space of spaces $\mathcal{E}_N^*$, $N \in \mathbb{N}_0$, respectively. A mapping

$$i_{N,-p}: \; E_{N,-p} \to \mathcal{E}_{N,-p}, \quad x \mapsto \left(\sum_{k=-n}^{\infty} \overline{\langle x, e_k \rangle}_* e_k \right)_{n\in\mathbb{Z}}$$

is an isometric linear operator. This mapping is extended to an isomorphism $i_{\mathcal{E}_\infty^*}$ from E_∞^* onto $\mathcal{E}_\infty^*$. Then we can identify E_∞^* to $\mathcal{E}_\infty^*$, and for any $\widetilde{x} = i_{\mathcal{E}_\infty^*}(x)$, $\widetilde{y} = i_{\mathcal{E}_\infty^*}(y) \in \mathcal{E}_\infty^*$, define the product $\widetilde{x}\widetilde{y}$ by $\widetilde{x}\widetilde{y} := (x_n y_n)_{n\in\mathbb{Z}}$. We can also identify the element $x \in E_\infty^*$ by an element $\widetilde{x} = \left(\sum_{k=-n}^{\infty} \overline{\langle x, e_k \rangle_*} e_k\right)_{n\in\mathbb{Z}}$ of $\mathcal{E}_\infty^*$ and denote it by $x = \left(\sum_{k=-n}^{\infty} \overline{\langle x, e_k \rangle_*} e_k\right)_{n\in\mathbb{Z}}$.

3. Powers of the delta distribution in E_∞^*

For $t \in [0,1]$ and $n \in \mathbb{N}_0$, we define $\delta_{t,n}$ by $\delta_{t,n} := \sum_{k=-n}^{\infty} \overline{e_k(t)} e_k$. Then $\delta_{t,n}$ is in E_n^* and the equality $\delta_{t,n}(\xi) = \xi(t)$ holds for $\xi \in E_n$. Here we define the derivative $\partial x \equiv x'$ of the distribution $x \in E_m^*, m \in \mathbb{N}_0$, by

$$\partial x := \sum_{\ell=-m}^{\infty} \overline{\langle x, e_\ell \rangle_*} \partial e_\ell \left(= 2\pi i \sum_{\ell=-m}^{\infty} \ell \overline{\langle x, e_\ell \rangle_*} e_\ell \right).$$

For any $n \in \mathbb{N}_0$ and $m \in \mathbb{N}$, the m-th power of $\delta_{t,n} \in E_n^*$ is given by

$$\delta_{t,n}^m = \sum_{k=-mn}^{\infty} \sum_{k_1+\cdots+k_m=k} \overline{e_k(t)} e_k = \sum_{k=-mn}^{\infty} \binom{k+mn+m-1}{m-1} \overline{e_k(t)} e_k$$

in E_{mn}^*. Then we have the following:

Proposition 3.1. [10] *For any $n \in \mathbb{N}_0$, $m \in \mathbb{N}$ and $t \in [0,1]$, we have $\delta_{t,n} \in E_\infty^*$ and the equality*

$$\delta_{t,n}^m = \frac{1}{(m-1)!} \sum_{j=0}^{m-1} \frac{1}{(2\pi i)^{m-j-1}} c_j \delta_{t,(m-j)n}^{(m-j-1)}$$

holds in E_∞^*, where constants c_j, $j = 0, 1, 2, \ldots, m$ are determined by the equality

$$(k+mn+1)(k+mn+2)\cdots(k+mn+m-1) = \sum_{j=0}^{m-1} c_j k^{m-j-1}.$$

By Theorem 3.1, if we regard $\delta_{t,n}^m \in E_\infty^*$ as

$$\widetilde{\delta_{t,n}^m} = \left(\sum_{k=-mn}^{\infty} \binom{k+mn+m-1}{m-1} \overline{e_k(t)} e_k \right)_{n\in\mathbb{Z}} \in \mathcal{E}_\infty^*,$$

then the formula also holds in $\mathcal{E}_\infty^*$. If we define the delta distribution δ_t by an element $(\delta_{t,n})_{n\in\mathbb{Z}}$ of $\mathcal{E}_\infty^*$, then we have

$$\delta_t^m = \frac{1}{(m-1)!} \sum_{j=0}^{m-1} \frac{1}{(2\pi i)^{m-j-1}} (c_{n,j})_{n\in\mathbb{Z}} \cdot \delta_t^{(m-j-1)},$$

where

$$(c_{n,j})_{n\in\mathbb{Z}} \cdot \delta_t^{(m-j-1)} = (c_{n,j})_{n\in\mathbb{Z}} \cdot (\delta_{t,n})_{n\in\mathbb{Z}}^{(m-j-1)} = \left(c_{n,j} \cdot \delta_{t,n}^{(m-j-1)}\right)_{n\in\mathbb{Z}}.$$

4. Powers of white noise

Let $e_{\mathbb{R},0},\ \ e_{\mathbb{R},2k},\ \ e_{\mathbb{R},2k-1},\ \ k=1,2,3,\ldots$ be functions on $[0,1]$ given by

$$e_{\mathbb{R},0}(u)=1,\quad e_{\mathbb{R},2k}(u):=\sqrt{2}\sin(2k\pi u),\quad e_{\mathbb{R},2k-1}(u):=\sqrt{2}\cos(2k\pi u),$$

which form an orthonormal basis for $L^2_{\mathbb{R}}([0,1])$. Then we can construct spaces $E_{n,\mathbb{R}}$, $E^*_{n,\mathbb{R}}$ using the sequence $\{\ell_k\}$ and an operator $A_{n,\mathbb{R}}$ given by

$$A_{n,\mathbb{R}}\xi=\sum_{k=0}^{2n}\ell_k\alpha_k e_{\mathbb{R},k},\quad \xi=\sum_{k=0}^{\infty}\alpha_k e_{\mathbb{R},k}\in L^2_{\mathbb{R}}([0,1])$$

as in Section 2. By the Bochner-Minlos theorem for any $n\in\mathbb{Z}$ and $\sigma>0$, there exists a probability measure $\mu_{n,\sigma}$ on the Borel field $\mathcal{B}(E^*_{n,\mathbb{R}})$ of $E^*_{n,\mathbb{R}}$ such that

$$\int_{E^*_{n,\mathbb{R}}} e^{i\langle x,\xi\rangle_n}d\mu_{n,\sigma}(x)=e^{-\frac{\sigma^2}{2}\langle\xi,\xi\rangle_n}$$

holds for all $\xi\in E_{n,\mathbb{R}}$, where $\langle\cdot,\cdot\rangle_n$ is the canonical bilinear form on $E^*_{n,\mathbb{R}}\times E_{n,\mathbb{R}}$. For any $n\in\mathbb{N}$, we define a projection p_n from $E^*_{\infty,\mathbb{R}}$ into $E^*_{n,\mathbb{R}}$ by

$$p_n(x):=\sum_{k=0}^{2n}\overline{\langle x,e_{\mathbb{R},k}\rangle_n}e_{\mathbb{R},k},\quad x\in E^*_{\infty,\mathbb{R}}.$$

Then the canonical bilinear form $\langle\cdot,\cdot\rangle$ on $E^*_{\infty,\mathbb{R}}\times E_{\infty,\mathbb{R}}$ is given by

$$\langle x,\xi\rangle:=\lim_{n\to\infty}\langle p_n(x),p_n(\xi)\rangle_n,\quad x\in E^*_{\infty,\mathbb{R}},\ \xi\in E_{\infty,\mathbb{R}}.$$

We define a probability measure $\mu_{\infty,\sigma}$ on the Borel field $\mathcal{B}(E^*_{\infty,\mathbb{R}})$ of $E^*_{\infty,\mathbb{R}}$ by $\mu_{\infty,\sigma}(B):=\sum_{n=1}^{\infty}\mu_{n,\sigma}(B\cap(E^*_{n,\mathbb{R}}\setminus E^*_{n-1,\mathbb{R}}))$ for any $B\in\mathcal{B}(E^*_{\infty,\mathbb{R}})$, where $E^*_{0,\mathbb{R}}=\phi$. Then

$$\int_{E^*_{\infty,\mathbb{R}}} e^{i\langle x,\xi\rangle}d\mu_{\infty,\sigma}(x)=e^{-\frac{\sigma^2}{2}\langle\xi,\xi\rangle}$$

holds for all $\xi\in E_{\infty,\mathbb{R}}$. Let $\mu=\mu_{\infty,1/\sqrt{2}}\times\mu_{\infty,1/\sqrt{2}}$. Then we have a probability space (E^*_∞,μ) and the equality

$$\int_{E^*_\infty} e^{\langle\overline{z},\xi\rangle+\langle z,\eta\rangle}d\mu(z)=e^{\langle\xi,\eta\rangle}$$

holds for $\xi,\eta\in E_\infty$.

Proposition 4.1. *For any $n\in\mathbb{N}_0$, the following equalities hold:*

$$E_n=E_{n,\mathbb{R}}+iE_{n,\mathbb{R}}+E_{-(n+1)},\quad E^*_n=E^*_{n,\mathbb{R}}+iE^*_{n,\mathbb{R}}+E^*_{-(n+1)}.$$

Consequently we have

$$E_\infty = E_{\infty,\mathbb{R}} + iE_{\infty,\mathbb{R}}, \quad E_\infty^* = E_{\infty,\mathbb{R}}^* + iE_{\infty,\mathbb{R}}^*.$$

For each $n \in \mathbb{N}$, we can construct spaces $\mathcal{W}_n$ of white noise test functionals and $\mathcal{W}_n^*$ of white noise distributions with $\mathcal{W}_n \subset L^2(E_\infty^*, \mu_{X_n}) \subset \mathcal{W}_n^*$, where X_n is a continuous operator from E_∞^* into E_∞^* given by $X_n(x) := x^n, \quad x \in E_\infty^*$ and μ_{X_n} is a probability measure defined by $\mu_{X_n}(B) := \mu(X_n^{-1}(B)), \quad B \in \mathcal{B}(E_\infty^*)$. Since

$$x^n(t) = \sum_{j_1,\dots,j_n=-m}^{\infty} \overline{\langle x, e_{j_1}\rangle_*} \cdots \overline{\langle x, e_{j_n}\rangle_*} e_{j_1+\cdots+j_n}(t) = x(t)^n$$

for $x \in E_m^*$ and $n \in \mathbb{N}_0$, we can define $x(t)^n$ as an element of $\mathcal{W}_n^*$ with the measure μ_{X_n}.

5. The Lévy and Volterra Laplacians and integral expressions

Let $\mathcal{L}(\mathcal{X};\mathcal{Y})$ denote the set of continuous linear operators from $\mathcal{X}$ into $\mathcal{Y}$ for locally convex linear topological spaces $\mathcal{X}$ and $\mathcal{Y}$. We denote $\mathcal{L}(\mathcal{X};\mathcal{X})$ by $\mathcal{L}(\mathcal{X})$ simply. For $p \geq 0$ let $C^p(\mathcal{X};\mathcal{Y})$ be the set of $\mathcal{Y}$-valued C^p-functions on $\mathcal{X}$ and $C^\infty(\mathcal{X};\mathcal{Y})$ the set of $\mathcal{Y}$-valued C^∞-functions on $\mathcal{X}$. We also denote $C^0(\mathcal{X};\mathcal{Y})$ and $C^p(\mathcal{X};\mathbb{C})$ simply by $C(\mathcal{X};\mathcal{Y})$ and $C^p(\mathcal{X})$, respectively.

By Theorem 2.1 we can define an *LV functional* φ as a functional in $C^\infty(E_\infty^*)$, such that for any $x \in E_\infty^*$, its second derivative $\varphi''(x)(y,z), y, z \in E_\infty^*$ is given by the form

$$\varphi''(x)(y,z) = \varphi_L''(x)(yz) + \varphi_V''(x)(y,z), \quad y, z \in E_\infty^*,$$

where $\varphi_L''(x) \in \mathcal{L}(E_\infty^*)$ and $\varphi_V''(x) \in \mathcal{L}(E_\infty^* \times E_\infty^*)$ that is a trace class operator of H_∞.

For any $n \in \mathbb{N}_0$, we define the operators $\mathcal{A}$ and $\mathcal{A}_n$ on $C^\infty(E_\infty^*)$ by

$$\mathcal{A}\varphi(x) = \sum_{k=-\infty}^{\infty} \varphi''(x)(e_k, e_k), \quad \mathcal{A}_n\varphi(x) = \sum_{k=-n}^{\infty} \varphi''(x)(e_k, e_k)$$

for $\varphi \in C^\infty(E_\infty^*)$. Since $\sum_{k=-\infty}^{\infty} e_k \otimes e_k \in E_\infty^* \otimes E_\infty^*$, for any $\varphi \in C^\infty(E_\infty^*)$, $\mathcal{A}_n\varphi$ exists and can be written by

$$\mathcal{A}_n\varphi(x) = \sum_{k=-n}^{\infty} \varphi''(x)(e_k \otimes e_k) = \varphi''(x)\left(\sum_{k=-n}^{\infty} e_k \otimes e_k\right), \quad x \in E_\infty^*$$

for each $n \in \mathbb{N}_0$. We also define the *Lévy Laplacian* $\Delta_L\varphi$ of $\varphi \in C^\infty(E_\infty^*)$ by

$$\Delta_L\varphi(x) = \lim_{N\to\infty} \frac{1}{2N+1} \sum_{n=-N}^{N} \varphi''(x)(J(e_n), e_n), \quad x \in E_\infty^*,$$

if the limit exists, where J is the conjugate operator. Then for an LV functional φ we have $\Delta_L\varphi(x) = \varphi_L''(x)(e_0)$. Any function $\varphi \in C^\infty(E_{\infty,\mathbb{R}}^*)$ can be regarded as $\widetilde{\varphi} \in C^\infty(E_\infty^*)$ given by $\widetilde{\varphi}(x) = \varphi(\mathrm{Re}(x))$, $x \in E_\infty^*$. Therefore, the Lévy Laplacian $\Delta_L\varphi$ of φ is defined by

$$\Delta_L\varphi(x_r) := \Delta_L\widetilde{\varphi}(x_r + i0), \quad x_r \in E_{\infty,\mathbb{R}}^*.$$

This means the Laplacian Δ_L acting on $C^\infty(E_\infty^*)$ is an extension of its action on $C^\infty(E_{\infty,\mathbb{R}}^*)$.

Since $\sum_{k=-\infty}^{\infty} e_{2k} \in E_\infty^*$ we denote this distribution by $\delta(0)$. This is the divergent part in the usual infinite dimensional analysis. However, we can get the part by a distribution in E_∞^*. Therefore, we can introduce an operator $\tau_{\delta(0)}$ defined on the range of the Lévy Laplacian by $\tau_{\delta(0)}\Delta_L\varphi(x) = \varphi_L''(x)(\delta(0)e_0)$. The Volterra Laplacian Δ_V for an LV functional is defined as a trace of $\varphi_V''(x)$ for $x \in E_\infty^*$:

$$\Delta_V\varphi(x) := \mathrm{trace}\ \varphi_V''(x) = \sum_{k=-\infty}^{\infty} \varphi_V''(x)(e_k, e_k).$$

The Volterra Laplacian Δ_V acting on $C^\infty(E_\infty^*)$ is also an extension of its action on $C^\infty(E_{\infty,\mathbb{R}}^*)$. Then we have an interesting formula:

$$\mathcal{A}\varphi = \tau_{\delta(0)}\Delta_L\varphi + \Delta_V\varphi$$

for any LV functional φ. Since the operator $\mathcal{A}$ means the complex version of the Gross Laplacian and $\tau_{\delta(0)}$ is implied from the divergent part $\frac{1}{dx}$, this formula is important with the setup of $\tau_{\delta(0)}\Delta_L\varphi$ as an element of $\mathcal{L}(E_\infty^*)$ instead of the formal expression $\Delta_G = \frac{1}{dx}\Delta_L + \Delta_V$. We give examples to show the calculation of the Lévy and Volterra Laplacians acting on certain functions. We note that $e^x := \sum_{\nu=0}^{\infty} \frac{x^\nu}{\nu!} \in E_\infty^*$ for any $x \in E_\infty^*$.

Example 5.1. [1] Let $\varphi(x) = x^\alpha(f)$, $f \in E_\infty$, $p \geq 2$. Then φ is in $C^\infty(E_\infty^*)$ and

$$\Delta_L\varphi(x) = \alpha(\alpha-1)x^{\alpha-2}(f), \quad \Delta_V\varphi(x) = 0.$$

[2] Let $\varphi(x) = e^{x(f)}$, $f \in E_\infty$. Then φ is in $C^\infty(E_\infty^*)$ and

$$\Delta_L\varphi(x) = 0, \quad \Delta_V\varphi(x) = \sum_{k=-\infty}^{\infty} e_k(f)^2\varphi(x).$$

[3] Let $\varphi(x) = e^{x^\alpha}(f),\ \ f \in E_\infty,\ \ \alpha \geq 1$. Then φ is in $C^\infty(E_\infty^*)$ and

$$\Delta_L \varphi(x) = (\alpha(\alpha-1)x^{\alpha-2} + \alpha^2 x^{2(\alpha-1)})\varphi(x),\ \ \Delta_V \varphi(x) = 0.$$

[4] Let $\varphi(x) = e^{iyx}(f),\ \ f \in E_\infty,\ \ y \in E_\infty^*$. Then φ is in $C^\infty(E_\infty^*)$ and

$$\Delta_L \varphi(x) = -y^2 \varphi(x),\ \ \Delta_V \varphi(x) = 0.$$

Remark 5.1. In the white noise analysis a normal monomial φ is defined by $\varphi(x) = \int_{\mathbb{R}} f(u) x^{\diamond\alpha}(u)du$ using the Wick product $\diamond$ as a white noise distribution. This is realized as part [1] in Example 5.1 without the Wick product.

For $n \in \mathbb{N}_0$ and $p \geq 1$, let $L^2(E_\infty^*; E_{n,p}^*)$ be the set of $E_{n,p}^*$-valued functions on E_∞^* with $||\Phi||_{n,-p}^2 := \int_{E_\infty^*} |\Phi(x)|_{n,-p}^2 d\mu(x) < \infty$. Let $\mathcal{P}_k(x) := x^k,\ x \in E_\infty^*$ for $k \in \mathbb{N}_0$ and denote the set of all elements Φ expressed in the form

$$\Phi = \sum_{k=0}^{n} c_k \mathcal{P}_k,\ \ c_k \in E_\infty^*,\ \ k = 0, 1, 2, \ldots, n,\ \ n \in \mathbb{N}_0$$

by $\mathrm{Poly}(E_\infty^*; E_\infty^*)$ or simply $\mathrm{Poly}(E_\infty^*)$. This is the set of all polynomials on E_∞^* with E_∞^*-coefficients and is in the space $C^\infty(E_\infty^*; E_\infty^*)$ of C^∞-functions from E_∞^* into itself. In general let $\mathrm{Poly}(\mathcal{X}; \mathcal{Y})$ be the set of all polynomials on $\mathcal{X}$ with $\mathcal{Y}$-coefficients for locally convex linear topological spaces $\mathcal{X}$ and $\mathcal{Y}$. Then $L^2(E_\infty^*; E_{n,p}^*)$ is the Hilbert space with norm $||\cdot||_{L^2(E_\infty^*; E_{n,p}^*)}$ and we have $\mathrm{Poly}(E_\infty^*) \subset C^\infty(E_\infty^*; E_\infty^*) \subset \bigcup_{n\in\mathbb{N}_0} \bigcup_{p\geq 1} L^2(E_\infty^*; E_{n,p}^*)$.

Let $\{B_k(t);\ t \geq 0\}$ be a sequence of independent one-dimensional Brownian motions on a probability space $(\Omega, \mathcal{F}, P)$ and set $\mathbf{B}_n(t) := \sum_{k=-n}^{\infty} B_k(t) e_k,\ \ t \geq 0$, for any $n \in \mathbb{N}_0$. Then for any $n \in \mathbb{N}_0$ and $t \geq 0$, we have $\mathbf{B}_n(t) \in E_n^*$ (a.e.).

We can define $\mathbf{B}(t)$ by $\mathbf{B}(t) = \sum_{k=-\infty}^{\infty} B_k(t) e_k$ as an element of E_∞^*. Then we note that $\mathbf{B}(t) = \mathbf{B}_n(t)$ on E_n for each $n \in \mathbb{N}_0$. From Theorem 2.1 we can define $\mathbf{B}_n(t)^m \in E_\infty^*$ for any $n \in \mathbb{N}_0$, $m \in \mathbb{N}$ and $t \geq 0$. The distribution $\mathbf{B}_n(t)^m$ is given by

$$\mathbf{B}_n(t)^m = \sum_{k_1, k_2, \ldots, k_m = -mn}^{\infty} B_{k_1}(t) B_{k_2}(t) \cdots B_{k_m}(t) e_{k_1 + k_2 + \cdots + k_m}.$$

Let $\mathcal{D}_{poly} := LS\{\Phi(\cdot)(f) |\ f \in E_\infty,\ \Phi \in \mathrm{Poly}(E_\infty^*)\}$, where LS means the linear span. Then we have the following:

Theorem 5.1. [10] *For any $\varphi \in \mathcal{D}_{poly}$ and $t \geq 0$, we have*

$$e^{\frac{t}{2}\mathcal{A}_n} \varphi(x) = E[\varphi(x + \mathbf{B}_n(t))],\ \ x \in E_\infty^*$$

for some $n \in \mathbb{N}_0$.

We have also an integral expression of the Lévy Laplacians as follows.

Theorem 5.2. *For any LV-functional* φ, *we have the expression of* Δ_L :

$$\Delta_L\varphi(x) = \lim_{n\to\infty}\frac{1}{2n+1}\lim_{T\to\infty}\frac{1}{2T}\int_{-T}^{T}\varphi''(x)(\delta_{t,n},\delta_{t,n})dt, \quad x\in E_\infty^*.$$

Proof. Let φ be an LV-functional. Then for any $x\in E_\infty^*$ its second variation $\varphi''(x)(y,z), y,z\in E_\infty^*$ is given by the form $\varphi''(x)(y,z) = \varphi_L''(x)(yz) + \varphi_V''(x)(y,z),\ \ y,z\in E_\infty^*$, where $\varphi_L''(x)\in\mathcal{L}(E_\infty^*)$ and $\varphi_V''(x)\in\mathcal{L}(E_\infty^*\times E_\infty^*)$ that is the trace class operator of H_∞. Therefore we have

$$\frac{1}{2T}\int_{-T}^{T}\varphi''(x)(\delta_{t,n},\delta_{t,n})dt = \frac{1}{2T}\int_{-T}^{T}\{\varphi_L''(x)(\delta_{t,n}^2)+\varphi_V''(x)(\delta_{t,n}\otimes\delta_{t,n})\}dt$$

$$= \frac{1}{2T}\int_{-T}^{T}\sum_{j=-n}^{\infty}\sum_{k=-n}^{\infty}\overline{e_{j+k}(t)}\left\{\varphi_L''(x)(e_{j+k})+\varphi_V''(x)(e_j\otimes e_k)\right\}dt.$$

Since there exists $q>0$ such that $(k+1)\le \ell_k^q,\ k=1,2,3,\ldots$, we get estimations of $\sum_{j=-n}^{\infty}\sum_{k=-n}^{\infty}|\varphi_L''(x)(e_{j+k})|$ and $\sum_{j=-n}^{\infty}\sum_{k=-n}^{\infty}|\varphi_L''(x)(e_j\otimes e_k)|$ for some $p>q+2$ and $M_x>0$ as follows:

$$\sum_{j=-n}^{\infty}\sum_{k=-n}^{\infty}|\varphi_L''(x)(e_{j+k})| \le M_x\rho^{2nq}\sum_{\nu=-\infty}^{\infty}\ell_\nu^{-2}<\infty;$$

$$\sum_{j=-n}^{\infty}\sum_{k=-n}^{\infty}|\varphi_L''(x)(e_j\otimes e_k)| \le M_x\left(\sum_{j=-n}^{\infty}\ell_j^{-p}\right)^2 < \infty.$$

Therefore, since $\lim_{T\to\infty}\frac{1}{2T}\int_{-T}^{T}\overline{e_{j+k}(t)}dt = \delta_{j+k,0}$, by the Fubini theorem and Lebesgue convergence theorem, it holds that

$$\lim_{T\to\infty}\frac{1}{2T}\int_{-T}^{T}\varphi''(x)(\delta_{t,n},\delta_{t,n})dt = (2n+1)\varphi_L''(x)(e_0)+\varphi_V''(x)\left(\sum_{j=-n}^{n}e_j\otimes e_{-j}\right).$$

Thus, by $\frac{1}{2n+1}\sum_{j=-n}^{n}e_j\otimes e_{-j}\to 0$ in $E_\infty^*\otimes E_\infty^*$, we obtain the assertion. □

Using similar method in the proof of Theorem 5.2 we have the following:

Theorem 5.3. *For any LV-functional* φ, *we have the equality*

$$\tau_{\delta_n(0)}\Delta_L\varphi(x)+\Delta_V\varphi(x) = \lim_{T\to\infty}\frac{1}{2T}\int_{-T}^{T}\varphi''(x)(\delta_{t,n},\delta_{-t,n})dt, \quad x\in E_\infty^*.$$

6. Delta distributions centered at powers of an infinite dimensional Brownian motion

For $N, n \in \mathbb{N}_0$ and $p \geq 1$, we denote by $\mathbb{A}_{n,p}^N$ the class of all entire functions φ defined on $E_{n,-p}$ such that

$$\|\varphi\|_{\mathbb{A}_{n,p}^N} := \sup_{x \in E_{n,-p}} \left\{ |\varphi(x)| \left(1 + |x|_{n,-p}^2\right)^{-N} \right\} < \infty.$$

Then for $N, n \in \mathbb{N}_0$ and $p \geq 1$, we note that $\mathbb{A}_{n,p}^N \subset \mathbb{A}_{n,p}$.

Define $\mathbb{A}_{n,\infty}^N := \bigcap_{p \geq 1} \mathbb{A}_{n,p}^N$ endowed with the projective limit topology of spaces $\mathbb{A}_{n,p}^{N,\star}, p \geq 1$ and $\mathbb{A}_{\infty}^N := \bigcap_{n \in \mathbb{N}_0} \mathbb{A}_{n,\infty}^N$ endowed with the projective limit topology of spaces $\mathbb{A}_{n,\infty}^N, n \in \mathbb{N}_0$, respectively. We also define $\mathbb{A}_\infty := \bigcup_{N=0}^{\infty} \mathbb{A}_{\infty}^N$ endowed with the inductive limit topology of spaces $\mathbb{A}_{\infty}^N, N \in \mathbb{N}_0$.

We endow a base probability space $(\Omega, \mathcal{F}, P)$ with a reference family $\{\mathcal{F}_{n,t};\ t \geq 0\}$ of σ-subalgebras of $\mathcal{F}$, and assume that $\{\mathbf{B}_n(t);\ t \geq 0\}$ is adapted to $(\mathcal{F}_{n,t})$, i.e., for each $t \geq 0$, $\mathbf{B}_n(t)$ is $\mathcal{F}_{n,t}$-measurable.

Let $\{X(t);\ t \geq 0\}$ be an $\mathcal{L}(E_\infty^*; \mathbb{C})$-valued stochastic process. We assume that a function $[0, \infty) \times \Omega \ni (t, \omega) \mapsto X(t)(\omega)(f) \in \mathbb{C}$ is an $(\mathcal{F}_{n,t})$-adapted process for every $f \in E_\infty^*$, and $\sum_{k=-n}^{\infty} \int_0^t E[|X(s)(e_k)|^2] ds < \infty$ for any $t \geq 0$. Then we define the stochastic integral $\int_0^t X(t) d\mathbf{B}_n(t)$ by

$$\int_0^t X(t) d\mathbf{B}_n(t) := \sum_{k=-n}^{\infty} \int_0^t X(t)(e_k) dB_k(t).$$

For any $\varphi \in C^\infty(E_\infty^*)$ with $\sup_{x \in E_\infty^*} |\varphi''(x)|_{\mathcal{L}(E_{n,-p}, \mathcal{L}(E_{n,-p}))} < \infty$ for some $p \geq 1, n \in \mathbb{N}_0$ and any $t \geq 0$ the stochastic integral $\int_0^t \varphi'(\mathbf{B}_n(u))(d\mathbf{B}_n(u))$ is given by

$$\int_0^t \varphi'(\mathbf{B}_n(u))(d\mathbf{B}_n(u)) = \sum_{k=-\infty}^{\infty} \int_0^t \varphi'(\mathbf{B}_n(u))(e_k) dB_k(u).$$

Then we have the following:

Theorem 6.1. [10] *For any $n \in \mathbb{N}_0$, $\varphi \in \mathbb{A}_\infty$ and $m \in \mathbb{N}$, the equality*

$$\begin{aligned} d\varphi(\mathbf{B}_n(t)^m) &= m \sum_{k=-n}^{\infty} \varphi'(\mathbf{B}_n(t)^m)(\mathbf{B}_n(t)^{m-1} e_k) dB_k(t) \\ &\quad + \frac{m(m-1)}{2} \sum_{k=-n}^{\infty} \varphi'(\mathbf{B}_n(t)^m)(\mathbf{B}_n(t)^{m-2} e_{2k}) dt \\ &\quad + \frac{m^2}{2} \sum_{k=-n}^{\infty} \varphi''(\mathbf{B}_n(t)^m)(\mathbf{B}_n(t)^{m-1} e_k, \mathbf{B}_n(t)^{m-1} e_k) dt \end{aligned}$$

holds.

For any $y \in E_\infty^*$, we define the differential operator D_y on $\mathbb{A}_\infty$ as $D_y\varphi(x) := \varphi'(x)(y)$, $\varphi \in \mathbb{A}_\infty$.

Lemma 6.1. [7] *Let $m, j \in \mathbb{N}_0$. Then for any $j \geq -m$, D_{e_j} is a continuous linear operator from $\mathbb{A}_\infty$ into itself. More precisely, the inequality*

$$\|D_{e_j}\varphi\|_{\mathbb{A}_{m,p}^N} \leq \frac{2^{N-1}}{\pi}\ell_j^{-p}\|\varphi\|_{\mathbb{A}_{m,p}^N}$$

holds for any $N \in \mathbb{N}_0$, $p \in \mathbb{R}$ and $\varphi \in \mathbb{A}_{m,p}^N$.

For $x \in E_\infty^*$, we introduce operator $\mathcal{A}_{n,x}$ defined on $\mathbb{A}_\infty$ by $\mathcal{A}_{n,x}\varphi := \sum_{k=-n}^\infty D_{xe_k}^2\varphi$, $\varphi \in \mathbb{A}_\infty$. For $x \in E_\infty^*$, we also introduce operators $\nabla_{n,x}^*$ and $\mathcal{A}_{n,x}^*$ defined on $\mathbb{A}_\infty^*$ by $\nabla_{n,x}^* X(\varphi) := \sum_{k=-n}^\infty X(D_{xe_k}\varphi)e_k$, $\varphi \in \mathbb{A}_\infty$, and $\mathcal{A}_{n,x}^* X(\varphi) := X(\mathcal{A}_{n,x}\varphi)$, $\varphi \in \mathbb{A}_\infty$.

For any $n \in \mathbb{N}_0$, $t \geq 0$ and $m \in \mathbb{N}_0$, we set $X_m(t)(\varphi) := \varphi(\mathbf{B}_n(t)^m)$, $\varphi \in \mathbb{A}_\infty$. Then we have the following:

Lemma 6.2. *Let $N \in \mathbb{N}_0$. Then for any $m \in \mathbb{N}$ and $p \geq 1$, the following estimate holds:*

$$E[|X_m(t)(\varphi)|] \leq \sum_{j=-\infty}^{\infty} \ell_j^{-2} \sum_{k=0}^{N} a_{m,k}\ {}_N C_k\ t^{km}\|\varphi\|_{\mathbb{A}_{mn,p}^N}, \quad t \geq 0, \varphi \in \mathbb{A}_\infty^N$$

where

$$a_{m,k} = \frac{1}{(m-1)!}\rho^{2(mn+m-2)\gamma}\sum_{\nu=1}^{m}\frac{1}{(\nu-1)!}\sum_{\substack{p_1+\cdots+p_\nu=m\\ 1\leq p_\nu\leq m}}(2kp_1-1)!!\cdots(2kp_\nu-1)!!$$

for some $\gamma > 0$.

By Lemma 6.2, $\{X_m(t);\ t \geq 0\}$ is an $\mathbb{A}_\infty^*$-valued stochastic process, then it satisfies the stochastic differential equation by Theorem 6.1:

$$\begin{aligned} dX_m(t) &= m\nabla_{n,\mathbf{B}_n(t)^{m-1}}^* X_m(t)d\mathbf{B}_n(t) \\ &\quad + \left(\frac{m(m-1)}{2}D_{\mathbf{B}_n(t)^{m-2}\delta_n(0)}^* + \frac{m^2}{2}\mathcal{A}_{n,\mathbf{B}_n(t)^{m-1}}^*\right)X_m(t)dt. \quad (1) \end{aligned}$$

We note that the equation (1) includes random differential operators in $\nabla_{n,\mathbf{B}_n(t)^{m-1}}$ and $\mathcal{A}_{n,\mathbf{B}_n(t)^{m-1}}$, for $m \geq 2$.

Acknowledgments

The authors express their sincere appreciation to the organizer Professor Noboru Watanabe at Tokyo University of Science for warm hospitality during 38th International Conference on IDAQP 2017. The authors also pay their respects to Professors Luigi Accardi, Takeyuki Hida and Igor V. Volovich for leading the field of Infinite Dimensional Analysis and Quantum Probability and their great achievements.

References

1. L. Accardi, U. C. Ji and K. Saitô: The exotic Laplacians and associated stochastic processes, *Infinite Dimensional Analysis, Quantum Probability and Related Topics* **12**, no. 1 (2009) 1–19.
2. L. Accardi, U. C. Ji and K. Saitô: The exotic Laplacians and derivatives of white noise, *Infinite Dimensional Analysis, Quantum Probability and Related Topics* **14**, no. 1 (2011) 1–14.
3. L. Accardi, U. C. Ji and K. Saitô: The exotic (higher order Lévy) Laplacians generate the Markov processes given by distribution derivatives of white noise, *Infinite Dimensional Analysis, Quantum Probability and Related Topics* **16**, no. 3 (2013) 1–26.
4. T. Hida: Analysis of Brownian Functionals, *Carleton Math. Lect. Notes* **13**, Carleton University, Ottawa, 1975.
5. K. Itô: Distribution-Valued Processes Arising from Independent Brownian Motions, *Math. Z.* **182** (1983) 17–33.
6. H.-H. Kuo, N. Obata and K. Saitô: Lévy Laplacian of generalized functions on a nuclear space, *J. Funct. Anal.* **94**, no. 1, (1990) 74–92.
7. H.-H. Kuo, K. Saitô and Y. Shibata: The product of distributions and white noise distribution-valued stochastic differential equations, *Communications on Stochastic Analysis* **10**, no. 2 (2016) 163–184.
8. P. Lévy: *Problèmes Concrets d'Analyse Fonctionnelle*, Gauthier-Villars, Paris, 1951.
9. N. Obata: White Noise Calculus and Fock Space, *Lecture Notes in Mathematics* **1577**, Springer-Verlag, 1991.
10. K. Saitô and T. Shimada: Powers of an infinite dimensional Brownian motion associated with the product of distributions, *Communications on Stochastic Analysis* **8**, no. 3 (2014) 343–364.
11. K. Yosida: *Functional Analysis*, Third Edition, Springer 1971.

Infinite Dimensional Analysis, Quantum Probability and Related Topics

pp. 183–188

INTERACTION INDEPENDENT QUANTUM PROBING

GNIEWOMIR SARBICKI

Institute of Physics, Nicolaus Copernicus University, Grudziądzka 5/7, 87–100, Toruń, Poland
E-mail: gniewko@fizyka.umk.pl

MIKKO TUKIAINEN and HENRY LYYRA

Turku Centre for Quantum Physics, Department of Physics and Astronomy, University of Turku, FI-20014 Turun yliopisto, Finland

SABRINA MANISCALCO

Centre for Quantum Engineering, Department of Applied Physics, School of Science, Aalto University, P.O. Box 11000, FIN-00076 Aalto, Finland

The paper presents results from the work I2QP. Assume, that we are able to put the system into the interaction with a given environment prepared in two different temperatures. We fix an evolution duration and for both temperatures we perform a process tomography of the quantum channels related to the evolution. Then we calculate the α-fidelities of the channels. We derive an inequality between the experimental data and the partition function of the environment, which bases on our assumptions about the energy spectrum of the environment. If the inequality is not satisfied, our assumption about the energy spectrum of the environment were wrong. To implement the method, one does not to make any assumptions about the Hamiltonian of the system or the interaction term.

Keywords: Quantum entropies, quantum programming, open quantum systems.

1. Introduction

First, let us remind the definition of a Shannon entropy:

$$S(p) = -\sum_i p_i \log_2 p_i, \tag{1}$$

and the Kulback-Leiber divergence related to it:

$$S(p||q) = \sum_i p_i \log_2(p_i/q_i). \tag{2}$$

The entropy is a measure of the (lack of) information contained in the state (pure states has the entropy equal zero). The divergence can be understood as (non-symmetric) measure of distance of two states (probability distributions). It is a measure of additional uncertainty if a real distribution is different from the assumed one.

The Shannon entropy is a limit case of a one-parameter family of entropies called Rényi entropies defined as follows:

$$H_\alpha(p) = \frac{1}{1-\alpha} \log_2 \sum_i p_i^\alpha, \tag{3}$$

where the paramater α takes its values from the range $(0, \infty) \setminus 1$. One gets the Shannon entropy if $\alpha \to 1$. The corresponding divergences are defined as

$$H_\alpha(p||q) = \frac{1}{\alpha-1} \log_2 \sum_{i:q_i>0} p_i^\alpha q_i^{1-\alpha}, \tag{4}$$

and are called Rényi divergencies of order α.

The Rényi entropies can be easily generalised to the quantum case:

$$H_\alpha(\varrho) = \frac{1}{1-\alpha} \mathrm{Tr}\varrho^\alpha. \tag{5}$$

When $\alpha \to 1$, we reconstruct the von Neumann entropy. There is no unique way to generalise the Rényi divergences to the quantum case, due to non-commutativity of quantum states. The non-symmetric version [2] is:

$$H_\alpha(\varrho||\sigma) = \frac{1}{\alpha-1} \log_2 \mathrm{Tr}\left(\varrho^\alpha \sigma^{1-\alpha}\right) \tag{6}$$

and the symmetrised one [3]:

$$H_\alpha(\varrho||\sigma) = \frac{1}{\alpha-1} \log_2 \mathrm{Tr}\left(\left(\sigma^{\frac{1-\alpha}{2\alpha}} \varrho \sigma^{\frac{1-\alpha}{2\alpha}}\right)^\alpha\right). \tag{7}$$

In our derivation we will use the second one.

The Rényi divergences for $\alpha \in (0, 1)$ have the following properties:

(1) $H_\alpha(\varrho||\sigma) \geq 0$, equality iff $\sigma = \varrho$,
(2) $H_\alpha(\varrho_1 \otimes \xi_1||\varrho_2 \otimes \xi_2) = H_\alpha(\varrho_1||\varrho_2) + H_\alpha(\xi_1||\xi_2)$,
(3) $H_\alpha(\mathcal{U}(\varrho)||\mathcal{U}(\sigma)) = H_\alpha(\varrho||\sigma)$ for any unitary channel $\mathcal{U}$.

For $\alpha \in [\frac{1}{2}, 1)$ the following additional property holds:

4. $H_\alpha(\varepsilon(\varrho)||\varepsilon(\sigma)) \leq H_\alpha(\varrho||\sigma)$ for any quantum channel ε.

It appears to be useful to consider α-fidelities — closely related quantities defined as $F_\alpha(\varrho||\sigma) = \operatorname{Tr}\left(\left(\sigma^{\frac{1-\alpha}{2\alpha}}\varrho\sigma^{\frac{1-\alpha}{2\alpha}}\right)^\alpha\right)$. In the case $\alpha = 1/2$ we reconstruct the usual (Uhlmann) fidelity and this fact stays behind our choice of the symmetrised quantum Rényi divergence.

The α-fidelities has the following properties:

(1) $\mathcal{F}_\alpha(\varepsilon_1, \varepsilon_2) \in [0, 1]$
(2) $\mathcal{F}_\alpha(\varepsilon_1, \varepsilon_2) = 1 \Rightarrow \varepsilon_1 = \varepsilon_2$
(3) $\mathcal{F}_\alpha(\mathcal{U} \circ \varepsilon_1, \mathcal{U} \circ \varepsilon_2) = \mathcal{F}_\alpha(\varepsilon_1, \varepsilon_2)$
for all unitary channels $\mathcal{U}$.

Additionally, if $\alpha \in [\frac{1}{2}, 1)$

5. $\mathcal{F}_\alpha(\varepsilon \circ \varepsilon_1, \varepsilon \circ \varepsilon_2) = \mathcal{F}_\alpha(\varepsilon_1, \varepsilon_2)$
6. $\mathcal{F}_\alpha(\varepsilon_1 \circ \varepsilon, \varepsilon_1 \circ \varepsilon) = \mathcal{F}_\alpha(\varepsilon_1, \varepsilon_2)$.

2. Quantum programming inequalities and fidelities of channels

Any quantum channel acting on a quantum system can be realised as a partial trace of an unitary channel acting on a bigger closed system:

$$\varepsilon(\varrho) = \operatorname{Tr}_B(\mathcal{U}(\varrho \otimes \xi)). \tag{8}$$

We say then that a channel ε is induced by a state ξ. It can be easily shown, that for a finite family of channels the always exists a common bigger closed system, a common unitary operator U and programming states ξ_i such that (8) holds.

Using the properties of Rényi divergences one can prove the following inequality involving channels and their inducing states:

$$H_\alpha(\varepsilon_1(\varrho_1)||\varepsilon_2(\varrho_2)) \leq H_\alpha(\xi_1||\xi_2) + H_\alpha(\varrho_1||\varrho_2) \tag{9}$$

for $\alpha \in (\frac{1}{2}, 1)$.

One can get give an estimation of $H_\alpha(\xi_1||\xi_2)$ independent on the arguments of the related channels:

$$H_\alpha(\xi_1, \xi_2) \geq \inf \varrho_1, \varrho_2 \left(H_\alpha(\varepsilon_1(\varrho_1)||\varepsilon_2(\varrho_2)) - H_\alpha(\varrho_1||\varrho_2)\right), \tag{10}$$

or, using the α-fidelities, as:

$$F_\alpha(\xi_1, \xi_2) \leq \inf \varrho_1, \varrho_2 \left(\frac{F_\alpha(\varepsilon_1(\varrho_1), \varepsilon_2(\varrho_2))}{F_\alpha(\varrho_1, \varrho_2)}\right) \tag{11}$$

the RHS will be the definition of *fidelity of channels* ε_1, ε_2 and the above inequality says, that the α-fidelity of two channels is always greater or equal to the α-fidelity of their inducing states.

A programmable quantum computer is a device performing a unitary operation on two registers — programming register and data register, or — equivalently — a set of channels on data register induced by states from the programming register. The inequalities (11) relate the α-fidelity of channels to the α-fidelity of their inducing (programming) states. That is why we call them the quantum programming inequalities.

In particular for two unitary channels, the α-fidelity is 1 if they differ by a phase factor, otherwise 0. This implies the orthogonality of their programming states, hence a quantum programming register of N qubits one can program only 2^N unitary operations on the data register. That's why an universal quantum computer (implementing all possible unitary gates) is not possible (see [5,4]).

3. Probing the environment

Assume, that we couple our system for a time t to a thermal bath with the Hamiltonian H_E prepared in two different inverse temperatures β_1 and β_2. The states of environment programming the corresponding dynamical maps are Gibbs states. We use the formula (11) which simplifies due to commutativity of Gibbs states:

$$\frac{Z(\alpha\beta_1+(1-\alpha)\beta_2,H_E)}{Z(\beta_1,H_E)^{\alpha}Z(\beta_2,H_E)^{1-\alpha}} \leq \inf_{t\geq 0}\begin{cases}\mathcal{F}_\alpha\big(\varepsilon_2^{(t)},\varepsilon_1^{(t)}\big), & \text{for } \alpha\in(0,\frac{1}{2})\\ \mathcal{F}_\alpha\big(\varepsilon_1^{(t)},\varepsilon_2^{(t)}\big), & \text{for } \alpha\in[\frac{1}{2},1)\end{cases} \tag{12}$$

(the extension to the range $(0,\frac{1}{2})$ possible due to equivalence $\alpha \longleftrightarrow 1-\alpha \Longleftrightarrow \beta_1 \longleftrightarrow \beta_2$).

The RHS is determined doing the **experimental** quantum process tomography for both temperatures to identify $\varepsilon_2^{(t)}$, $\varepsilon_1^{(t)}$ and then the numerical optimisation due to calculate $\mathcal{F}_\alpha\big(\varepsilon_2^{(t)},\varepsilon_1^{(t)}\big)$.

The LHS is calculated from our **theoretical** assumptions about the H_E. We can falsify our assumption if the inequality is broken. Observe, that no knowledge about the mechanism of interaction is involved.

The RHS can be called a *fidelity of dynamical maps.*

4. Example

Assume, that the system of interest is a 2-level atom in a multi-mode vacuum:

$$\begin{aligned} H &= H_S + H_E + H_I \\ &= \frac{\hbar\omega_0}{2}\sigma_z + \sum_k \hbar\omega_k \left(b_k^\dagger b_k + \frac{1}{2} I \right) + \sum_k \sigma_z (g_k b_k^\dagger + g_k^* b_k). \end{aligned} \tag{13}$$

This model [6] is exactly solvable (up to Born, Markov and RW approx.) and one gets:

$$\begin{aligned} &\varepsilon^{(t)}(\varrho) = \begin{bmatrix} \varrho_{00} & \Gamma(t)\,\varrho_{01} \\ \Gamma(t)\,\varrho_{10} & \varrho_{11} \end{bmatrix}, \\ &\text{where} \quad \Gamma(t) = \exp\left[-\sum_k \frac{4|g_k|^2}{\hbar\omega_k^2} \coth\left(\frac{1}{2}\beta\hbar\omega_k\right) [1-\cos(\omega_k t)] \right]. \end{aligned} \tag{14}$$

One has

$$\begin{aligned} \mathcal{F}_\alpha(\varepsilon_1^{(t)}, \varepsilon_2^{(t)}) &= \left(\frac{1+\Gamma_2(t)}{2}\right)^{1-\alpha} \left(\frac{1+\Gamma_1(t)}{2}\right)^{\alpha} \\ &+ \left(\frac{1-\Gamma_2(t)}{2}\right)^{1-\alpha} \left(\frac{1-\Gamma_1(t)}{2}\right)^{\alpha}. \end{aligned} \tag{15}$$

Let us assume now, that the **true** environment is a one-mode vacuum of an unknown frequency ω_1 — this determine the RHS (from experiment). We assume, that environment has a frequency ω (unknown) and we calculate the LHS of (12):

$$\begin{aligned} &-\log_2(1-\exp(-(\alpha\beta_1 + (1-\alpha)\beta_2)\omega)) \\ &\qquad + \alpha\log_2(1-\exp(-\beta_1\omega)) + (1-\alpha)\log_2(1-\exp(-\beta_2\omega)). \end{aligned} \tag{16}$$

We calculate (measure) the RHS of for $\beta_1\hbar = 2$, $\beta_2\hbar = 20$, $g_1/\sqrt{\hbar} = 0.3$ and $\omega_1 = 1$, and draw it in the picture with blue color. For 3 assumed values of ω:

- $\omega = 3$
- $\omega = 3.1$
- $\omega = 3.5$.

We calculate the LHS and draw it on the picture with colors green, red and cyan respectively. One can observe, that the limit value is ≈ 3.1. The value 3.5 (cyan) is forbidden.

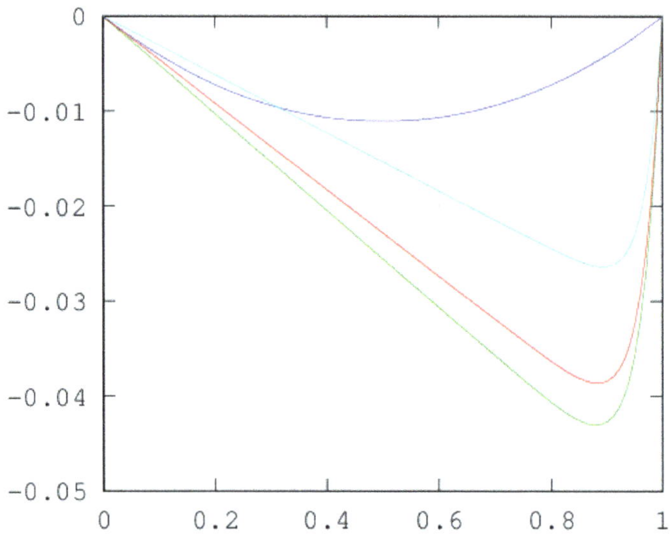

Acknowledgements

The work was supported by National Science Centre (project 2015/17/B/ST2/02026).

References

1. Mikko Tukiainen, Henri Lyyra, Gniewomir Sarbicki, Sabrina Maniscalco, *Fidelity of dynamical maps*, Phys. Rev. A 95, 052102 (2017).
2. Dénes Petz, *Quasi-entropies for finite quantumsystems*, Reports in Mathematical Physics **23**:57–65, 1986.
3. Mark M. Wilde, Andreas Winter, Dong Yang, *Strong converse for the classical capacity of entanglement-breaking and Hadamard channels via a sandwiched Rényi relative entropy*, Comm. Math. Phys. **331**, pp. 593–622 (2014).
4. Teiko Heinosaari, Mikko Tukiainen, *Notes on Deterministic Programming of Quantum Observables and Channels*, Quantum Information Processing, Volume 14, Issue 8, pp. 3097–3114, August 2015.
5. M.A. Nielsen, I.L. Chuang, *Quantum Computation and Quantum Information*, Cambridge University Press 2000.
6. H.P. Breuer, F. Petruccione, *The Theory of Open Quantum Systems*, OUP Oxford, (2007).

Infinite Dimensional Analysis, Quantum Probability and Related Topics

pp. 189–197

ROUGH-PATHS AND NON-COMMUTATIVE PROBABILITY

REN SCHOTT

Institut Élie Cartan and Loria, Nancy, B.P. 239,
54506 Vandoeuvre-lès-Nancy Cedex, France
E-mail: schott@loria.fr

Based on a joint work with Aurèlien Deya

We present, shortly, T. Lyons' formulation of the rough-paths theory. Then we show how to apply the principles of rough-paths theory in a non-commutative probability setting. The free Brownian motion is treated as an example.

1. Introduction

T. Lyons'[28] rough-paths theory goes back to the 90's and is a very active research area with applications in many fields. Among the huge bibliography related to this domain, we recommend Friz-Hairer's book [19], Gubinelli's paper [22] as well as the references therein. Attempts to extend the rough-paths type approach to non-commutative stochastic integration can be found in [14,15,16,17,18]. Our "rough" presentation of results obtained recently is mostly based on [14]. The organization of this paper is as follows: Section 2 contains some hints of the classical rough-paths theory. In Section 3, we study different possibilities to apply the principles of rough-paths theory in a non-commutative probability setting (see [14] for details). Some concluding remarks are given in Section 4.

2. Basics on classical rough-paths theory

The presentation of this section is based on [19].

Consider a *non-differentiable* path

$$x : [0, T] \to \mathbb{R}^n,$$

with *Hölder regularity* $\gamma \in (0, 1)$ (i.e., $\|x_t - x_s\| \le c\,|t - s|^{\gamma}$).

Given a smooth $f : \mathbb{R}^n \to \mathcal{L}(\mathbb{R}^n, \mathbb{R}^n)$, how can we define

$$\int f(x_t)\, dx_t \quad ?$$

We would like this definition to be sufficiently extensible to cover *differential equations*

$$dy_t = f(y_t)\, dx_t.$$

Rough-paths theory tries to give a sense to

$$\int f(x_t)\, dx_t$$

when $x : [0, T] \to \mathbb{R}^n$ is a γ-Hölder path with $\gamma \in (0, 1)$.

The construction of the integral *depends on* γ:

2.1. *If* $\gamma > 1/2$ *("Young case")*

We can use the following result:

Theorem 2.1. ***Young (1936)*** *If x is γ-Hölder and y is κ-Hölder with $\gamma + \kappa > 1$, then the Riemann sum $\sum_{t_i \in \mathcal{P}_{[s,t]}} y_{t_i}(x_{t_{i+1}} - x_{t_i})$ converges as the mesh of the partition $\mathcal{P}_{[s,t]}$ tends to 0. We define*

$$\int_s^t y_u\, dx_u := \lim_{|\mathcal{P}_{[s,t]}| \to 0} \sum_{t_i \in \mathcal{P}_{[s,t]}} y_{t_i}(x_{t_{i+1}} - x_{t_i}).$$

Application:

Let $x : [0, T] \to \mathbb{R}^n$ be γ-Hölder with $\gamma > 1/2$.

If f is smooth then $(t \mapsto f(x_t))$ is γ-Hölder. Therefore $\int f(x_t)\, dx_t$ is a Young integral.

Here is another way to see Young's result:

$$\int_s^t f(x_u)\, dx_u = f(x_s)\,(x_t - x_s) + \int_s^t [f(x_u) - f(x_s)]\, dx_u.$$

The main term is:

$$f(x_s)\,(x_t - x_s), \quad \text{with } |f(x_s)\,(x_t - x_s)| \le c\,|t - s|^{\gamma}.$$

The residual term is:

$$\int_s^t [f(x_u) - f(x_s)]\, dx_u \quad \text{with } \left|\int_s^t [f(x_u) - f(x_s)]\, dx_u\right| \le c\,|t - s|^{2\gamma}.$$

As $2\gamma > 1$, the residual term "disappears" in Riemann sum:

$$\int_s^t f(x_u)\,dx_u = \sum_{t_i\in\mathcal{P}_{[s,t]}} \int_{t_i}^{t_{i+1}} f(x_u)\,dx_u$$

$$= \sum_{t_i\in\mathcal{P}_{[s,t]}} f(x_{t_i})\,(x_{t_{i+1}} - x_{t_i}) + \sum_{t_i\in\mathcal{P}_{[s,t]}} \int_{t_i}^{t_{i+1}} [f(x_u) - f(x_{t_i})]\,dx_u\,.$$

2.2. *If* $\frac{1}{3} < \gamma \leq \frac{1}{2}$

If $x \in \mathcal{C}^\gamma$ with $\gamma \leq 1/2$, we cannot guarantee the convergence of the sum

$$\sum_{t_i\in\mathcal{P}_{[s,t]}} f(x_{t_i})\,(x_{t_{i+1}} - x_{t_i}).$$

We try to "correct" the Riemann sum:

$$\sum_{t_i\in\mathcal{P}_{[s,t]}} \{f(x_{t_i})\,(x_{t_{i+1}} - x_{t_i}) + C_{t_i,t_{i+1}}\}.$$

To find out a proper C, a few heuristic considerations are helpful. Suppose that we can define $\int_s^t f(x_u)\,dx_u$. Then,

$$\int_s^t f(x_u)\,dx_u = f(x_s)\,(x_t - x_s) + \int_s^t [f(x_u) - f(x_s)]\,dx_u$$

$$= f(x_s)\,(x_t - x_s) + \int_s^t \nabla f(x_s)(x_u - x_s)\,dx_u + \int_s^t y_{s,u}\,dx_u\,,$$

where $|y_{s,u}| = |f(x_u) - f(x_s) - \nabla f(x_s)(x_u - x_s)| \leq c\,|u-s|^{2\gamma}$.
$|\int_s^t y_{s,u}\,dx_u| \leq c\,|t-s|^{3\gamma}$. As $3\gamma > 1$, this term "disappears" in Riemann sum.

⇒ *The main term is:*

$$f(x_s)\,(x_t - x_s) + \int_s^t \nabla f(x_s)(x_u - x_s)\,dx_u.$$

This suggests a natural definition when $\gamma \in (\frac{1}{3}, \frac{1}{2}]$,

$$\int_s^t f(x_u)\,dx_u \text{ ``$:=$'' } \lim \sum_{t_k} \left\{ f(x_{t_k})\,(x_{t_{k+1}} - x_{t_k}) + C_{t_k,t_{k+1}} \right\}$$

where

$$C_{s,t} := \int_s^t \nabla f(x_s)(x_u - x_s)\,dx_u$$

$$= \partial_i f_j(x_s) \int_s^t (x_u^{(i)} - x_s^{(i)})\,dx_u^{(j)} = \partial_i f_j(x_s) \int_s^t \int_s^u dx_v^{(i)}\,dx_u^{(j)}$$

where ∇ stands for the gradient.

To define the integral when $\gamma \in (\frac{1}{3}, \frac{1}{2}]$, we need to *assume the a priori existence of the Lévy area*

$$\mathbf{x}^{\mathbf{2},(i,j)}_{s,t} := \int_s^t \int_s^u dx_v^{(i)} dx_u^{(j)}$$

above x.

Assume that we can define the Lévy area $\mathbf{x^2} = \iint dx dx$. Then:

(i) We can indeed *define the integral* $\int f(x_u)\, dx_u$ as

$$\int f(x_u)\, dx_u := \lim \sum_{t_k} \Big\{ f(x_{t_k})\,(x_{t_{k+1}} - x_{t_k}) + \nabla f(x_{t_k}) \cdot \mathbf{x}^{\mathbf{2}}_{t_k,t_{k+1}} \Big\}.$$

(ii) We can extend this definition to $\int f(y)\, dx$ for a large class of paths y, and then *solve the equation*

$$dy_t = f(y_t)\, dx_t.$$

(iii) We can show that the solution y is a *continuous function of the pair* $(x, \mathbf{x^2})$, i.e., $y = \Phi(x, \mathbf{x^2})$ with Φ continuous with respect to Hölder topology.

The procedure can be extended to any Hölder coefficient $\gamma \in (0, \frac{1}{3}]$ provided one can define the iterated integrals of x:

$$\int dx, \quad \iint dx dx, \quad \iiint dx dx dx, \ldots$$

It applies to stochastic processes in a pathwise way.

$$Y(\omega) = \Phi\big(B(\omega), \mathbf{B^2}(\omega)\big),$$

where Φ is continuous and deterministic.

3. Non-commutative probability theory and rough paths

In the rough-paths machinery, what happens if we replace the γ-Hölder path

$$x : [0, T] \to \mathbb{R}^n$$

with a γ-Hölder path

$$X : [0, T] \to \mathcal{A},$$

where $\mathcal{A}$ is a *non-commutative probability space*?

More precisely, what happens if we replace the γ-Hölder path

$$x : [0,T] \to \mathbb{R}^n \quad (\|x_t - x_s\| \le |t-s|^\gamma)$$

with a γ-Hölder path

$$X : [0,T] \to \mathcal{A} \quad (\|X_t - X_s\|_{\mathcal{A}} \le |t-s|^\gamma),$$

where $\mathcal{A}$ is a *non-commutative probability space*? For instance, how can we define

$$\int P(X_t) \cdot dX_t$$

when P is a polynomial and $\cdot$ refers to the product in $\mathcal{A}$?

If $\gamma > \frac{1}{2}$, we can use Young's theorem:

$$\int P(X_t) \cdot dX_t := \lim \sum_{t_i} P(X_{t_i}) \cdot (X_{t_{i+1}} - X_{t_i}).$$

If $\gamma \in (\frac{1}{3}, \frac{1}{2}]$, we use corrected Riemann sums

$$\int P(X_t) \cdot dX_t := \lim \sum_{t_i} \{P(X_{t_i}) \cdot (X_{t_{i+1}} - X_{t_i}) + C_{t_i,t_{i+1}}\}.$$

Let us find out C in this context. As in the finite-dimensional case, one has (morally)

$$\int_s^t P(X_u) \cdot dX_u$$

$$= P(X_s) \cdot (X_t - X_s) + \int_s^t \nabla P(X_s)(X_u - X_s) \cdot dX_u + \int_s^t Y_{s,u} \cdot dX_u \ ,$$

with $\|\int_s^t Y_{s,u} \cdot dX_u\| \le c\,|t-s|^{3\gamma}$. Since $3\gamma > 1$, main term:

$$P(X_s) \cdot (X_t - X_s) + \int_s^t \nabla P(X_s)(X_u - X_s) \cdot dX_u.$$

Again, we have a natural definition:

$$\int P(X_t) \cdot dX_t \text{ ``: ="}$$

$$\lim \sum_{t_k} \Big\{P(X_{t_k}) \cdot (X_{t_{k+1}} - X_{t_k}) + \int_{t_k}^{t_{k+1}} \nabla P(X_{t_k})(X_u - X_{t_k}) \cdot dX_u\Big\}.$$

Remember that in finite dimension,

$$\int_s^t \nabla f(x_s)(x_u - x_s) \cdot dx_u = \partial_i f_j(x_s) \int_s^t \int_s^u dx_v^{(i)} dx_u^{(j)} \ .$$

This is no longer possible for $\int_s^t \nabla P(X_s)(X_u - X_s) \cdot dX_u$.

The question is: *What can play the role of the Lévy area?*

For instance, when $P(x) = x^2$,

$$\begin{aligned}&\int_s^t \nabla P(X_s)(X_u - X_s) \cdot dX_u \\ &= X_s \cdot \int_s^t (X_u - X_s) \cdot dX_u + \int_s^t (X_u - X_s) \cdot X_s \cdot dX_u \\ &= X_s \cdot \int_s^t \int_s^u dX_v \cdot dX_u + \int_s^t \int_s^u dX_v \cdot X_s \cdot dX_u \ .\end{aligned}$$

For $s < t$, we "define" the Lévy area $\mathbf{X}^{\mathbf{2}}_{s,t}$ as the *operator on* $\mathcal{A}$

$$\mathbf{X}^{\mathbf{2}}_{s,t}[Y] = \int_s^t \int_s^u dX_v \cdot Y \cdot dX_u \quad \text{for } Y \in \mathcal{A}.$$

For instance, when $P(x) = x^2$,

$$\int_s^t \nabla P(X_s)(X_u - X_s) \cdot dX_u = X_s \cdot \mathbf{X}^{\mathbf{2}}_{s,t}[1] + \mathbf{X}^{\mathbf{2}}_{s,t}[X_s] \ .$$

Assume that we can define the Lévy area $\mathbf{X}^{\mathbf{2}}$ in the previous sense. Then, for every polynomial P, we can indeed define the integral

$$\int P(X_t) \cdot dX_t$$

as the limit of corrected Riemann sums involving X and $\mathbf{X}^{\mathbf{2}}$.

Given $\mathbf{X}^{\mathbf{2}}$, we can then successively define:

(i) The integral

$$\int P(X_t) \cdot dX_t \cdot Q(X_t)$$

for all polynomials P, Q.

(ii) The integral

$$\int f(X_t) \cdot dX_t \cdot g(X_t)$$

for a large class of functions $f, g : \mathbb{C} \to \mathbb{C}$ (where $f(X_t), g(X_t)$ are understood in the functional calculus sense).

(iii) The integral

$$\int f(Y_t) \cdot dX_t \cdot g(Y_t)$$

for a large class of processes $Y : [0, T] \to \mathcal{A}$.

With this definition in hand, we can solve the equation

$$dY_t = f(Y_t) \cdot dX_t \cdot g(Y_t).$$

More precisely: $Y = \Phi(X, \mathbf{X}^2)$, where Φ is a continuous function.

3.1. *The free Brownian motion case*

Remember that the free Brownian motion has been introduced and investigated by Biane and Speicher [2].

Let $X : \mathbb{R}_+ \to (\mathcal{A}, \varphi)$ be a *free Brownian motion.* In this case: $\|X_t - X_s\|_{\mathcal{A}} \leq c\,|t-s|^{1/2}$.

To define a stochastic calculus with respect to X, it suffices to be able to give a sense to

$$\mathbf{X}^2_{s,t}[Y] = \int_s^t \int_s^u dX_v \cdot Y \cdot dX_u \quad \text{for } Y \in \mathcal{A}. \tag{1}$$

We can use the results of Biane-Speicher [2] to define (1) as an "Itô" integral, denoted by $\mathbf{X}^{2;\text{Itô}}$. The "rough-paths" calculus based on $\mathbf{X}^{2;\text{Itô}}$ then coincides with Itô's stochastic calculus with respect to X.

Here is another way to define (1): consider the linear interpolation X^n of X along $t_i^n = \frac{i}{n}$ $(i = 1, \ldots, n)$, and

$$\mathbf{X}^{2,n}_{s,t}[Y] = \int_s^t \int_s^u dX^n_v \cdot Y \cdot dX^n_u \quad \text{(Lebesgue integral)}.$$

Then $X^n \to X$ and $\mathbf{X}^{2,n} \to \mathbf{X}^{2,\text{Strato}}$. Moreover,

$$\mathbf{X}^{2,\text{Strato}}_{s,t}[Y] = \mathbf{X}^{2,\text{Itô}}_{s,t}[Y] + \frac{1}{2}\varphi(Y)\,(t-s).$$

See [14] for details.

4. Concluding remarks

We have presented elements of the classical and of the non-commutative rough-paths theory at a comprehensive level. Recently, extensions have been provided by Aurélien Deya and the author for the stochastic integration with respect to q-Brownian motion [15,16] and for the non-commutative fractional Brownian motion [17]. Both studies involve aspects of the rough-paths theory.

Acknowledgements

I want to thank Noboru Watanabe and the organizing committee for the warm hospitality during the conference QP38.

This short presentation is based on joint work with Aurélien Deya. I'm very grateful to him.

References

1. Biane, Philippe: Free probability for probabilists. Quantum probability communications, Vol. XI (Grenoble, 1998), 55–71, QP-PQ, XI, World Sci. Publ., River Edge, NJ, 2003.
2. Biane, Philippe; Speicher, Roland: Stochastic calculus with respect to free Brownian motion and analysis on Wigner space. *Probab. Theory Related Fields* **112** (1998), no. 3, 373–409.
3. Biane, Philippe; Speicher, Roland: Free diffusions, free entropy and free Fisher information. *Ann. Inst. H. Poincaré Probab. Statist.* **37** (2001), no. 5, 581–606.
4. Bożejko, Marek: Ultracontractivity and strong Sobolev inequality for q-Ornstein-Uhlenbeck semigroup $(-1 < q < 1)$. *Infin. Dimens. Anal. Quantum Probab. Relat. Top.* **2** (1999), no. 2, 203–220.
5. Bożejko, Marek; Kémerer, Burkhard; Speicher, Roland: q-Gaussian processes: non-commutative and classical aspects. *Comm. Math. Phys.* **185** (1997), no. 1, 129–154.
6. Bożejko, Marek; Speicher, Roland: An example of a generalized Brownian motion. *Comm. Math. Phys.* **137** (1991), no. 3, 519–531.
7. Capitaine, Mireille; Donati-Martin, Catherine: The Lévy area process for the free Brownian motion. *J. Funct. Anal.* **179**, 153–169 (2001).
8. Capitaine, Mireille; Donati-Martin, Catherine: Free Wishart processes. *J. Theoret. Probab.* **18** (2005), 413–438.
9. Coutin, Laure; Qian, Zhongmin: Stochastic analysis, rough path analysis and fractional Brownian motions. *Probab. Theory Related Fields* **122** (2002), no. 1, 108–140.
10. Demni, Nizar: Free Jacobi process. *J. Theoret. Probab.* **21** (2008), 118–143.
11. Deya, Aurélien; Gubinelli, Massimiliano; Tindel, Samy: Non-linear rough heat equations. *Probab. Theory Related Fields* **153** (2012), no. 1-2, 97–147.
12. Deya, Aurélien; Neuenkirch, Andreas; Tindel, Samy: A Milstein-type scheme without Lévy area terms for SDEs driven by fractional Brownian motion. *Ann. Inst. H. Poincaré Probab. Statist.* **48**, 2 (2012), 518–550.
13. Deya, Aurélien; Tindel, Samy: Rough Volterra equations 2: Convolutional generalized integrals. *Stochastic Process. Appl.* **121** (2011), no. 8, 1864–1899.
14. Deya, Aurélien; Schott, René: On the rough-paths approach to non-commutative stochastic calculus. *Journal of Functional Analysis* **264** (2013), no. 4, 594–628.

15. Deya, Aurélien; Schott, René: On stochastic calculus with respect to q-Brownian motion. *Journal of Functional Analysis* **274** (2018), no. 4, 1047–1075.
16. Deya, Aurélien; Schott, René: On multilpication of q-chaoses. *Electronic Communications in Probability* **121** (2018), no. 8, 1864–1899.
17. Deya, Aurélien; Schott, René: Integration with respect to the non-commutative fractional Brownian motion. *ArXiv*.
18. Donati-Martin, Catherine: Stochastic integration with respect to q Brownian motion. *Probab. Theory Related Fields* **125** (2003), no. 1, 77–95.
19. Friz, Peter; Hairer, Martin: A course on rough-paths: With an introduction to regularity structures. Springer, 2014.
20. Friz, Peter; Victoir, Nicolas: Differential equations driven by Gaussian signals. *Ann. Inst. Henri Poincaré Probab. Stat.* **46** (2010), no. 2, 369–413.
21. Friz, Peter; Victoir, Nicolas: Multidimensional stochastic processes as rough paths. Theory and applications. Cambridge Studies in Advanced Mathematics, 120. Cambridge University Press, Cambridge, 2010.
22. Gubinelli, Massimiliano: Controlling rough paths. *J. Funct. Anal.* **216** (2004), no. 1, 86–140.
23. Gubinelli, Massimiliano; Tindel, Samy: Rough evolution equations. *Ann. Probab.* **38** (2010), no. 1, 1–75.
24. Hairer, Martin: Rough stochastic PDEs. *Comm. Pure Appl. Math.* **64** (2011), no. 11, 1547–1585.
25. Kargin, Vladislav: On Free Stochastic Differential Equations. *J. Theoret. Probab.* **24** (2011), no. 3, 821–848.
26. Kemp, Todd; Nourdin, Ivan; Peccati, Giovanni; Speicher, Roland: Wigner chaos and the fourth moment. *Ann. Probab.* **40** (2012), no. 4, 1577–1635.
27. Kümmerer, Burkhard; Speicher, Roland: Stochastic integration on the Cuntz algebra, *J. Funct. An.* **103** (1992), 372–408.
28. Lyons, Terry; Qian, Zhongmin: System control and rough paths. Oxford Mathematical Monographs. Oxford Science Publications. Oxford University Press, Oxford, 2002. x+216 pp.
29. Neuenkirch, Andreas; Nourdin, Ivan; Tindel, Samy: Delay equations driven by rough paths. *Electron. J. Probab.* **13** (2008), no. 67, 2031–2068.
30. Nica, Alexandru; Speicher, Roland: Lectures on the Combinatorics of Free Probability. Cambridge University Press, 2006.
31. Speicher, Roland: Stochastic integration on the full Fock space with respect to a kernel calculus. *Publ. R.I.M.S.* **27** (1991), 1149–184.
32. Victoir, Nicolas: Levy area for the free Brownian motion: existence and non-existence. *J. Funct. Anal.* **208** (2004), no. 1, 107–121.
33. Voiculescu, Dan: Addition of certain non-commuting random variables. *J. Funct. Anal.* **66** (1986), 323–346.
34. Voiculescu, Dan: Limit laws for random matrices and free products. *Invent. Math.* **104** (1991), no. 1, 201–220.

Infinite Dimensional Analysis, Quantum Probability and Related Topics

pp. 199–218

A QUICK VIEW ON PROFESSOR TAKEYUKI HIDA'S WORK

SI SI

Emeritus Professor
Aichi Prefectural University, Japan
E-mail: ohnbin@gmail.com

1. Introduction

This paper is based on my talk, dedicated to Professor Takeyuki Hida ninetieth birthday, at the international conference "QP38" held at Tokyo University of Science. We would like to write the summary of Professor Takeyuki Hida's contribution in his mathematical life. However, it is impossible to cover the whole of his work, so some important parts of his work will be discussed as much as we can.

In 1975, T. Hida introduced White noise Theory in his celebrated monograph "Analysis of Brownian Functionals". In reality, a few years before 1975 he had presented his idea of white noise theory in various conferences. His idea was to introduce a mathematical theory so that white noise $\dot{B}(t)$ is meaningful for each t and the collection $\{\dot{B}(t), t \in R\}$ of white noise can be used as a continuum coordinate system. In 2000, American Mathematical Society recognized his work and gave the Mathematical Subject Classification 2000:

60H40 white noise theory.

He was inspired by Paul Lévy's work on Brownian motion and Functional analysis. But his mathematical theory of white noise is based on his own earlier work, for instance we can see at his monograph "Canonical representation of Gaussian Processes and their Applications", refer [3].

White noise analysis setup motivated to study Poisson noise which is a time derivative of Poisson process done by himself and his colleagues. Poisson process $P(t, \lambda)$ depends on time t and also on intensity λ so he considered the another Poisson noise which is called a space parameter Poisson noise that is the derivative with respect to λ and it has been done to some extent together with the author. In addition he introduced Complex white noise analysis and Infinite dimensional rotation group.

He was also interested in other fields such as Physics, Biology, Social science, and others. So he contributed in those fields, but we will not discuss them in this paper.

We would like to discuss his work under the topics as follows:

1. Gaussian processes and others
2. White noise analysis, the so called Hida distributions
3. Infinite dimensional Harmonic analysis and rotation group
4. Variational calculus and random fields
5. Feynman integrals and Dirichlet forms
6. Application to biology

2. Gaussian processes

T. Hida has contributed in many aspects of Gaussian process, Poisson process and others. The most important contribution can be seen in his paper [2], "Canonical representations of Gaussian processes" which will be discussed in the following.

Canonical representations of Gaussian processes

Let T be a parameter set which is taken to be a closed interval, $(-\infty, \infty)$ or $[0, \infty)$. If a real Gaussian process $X(t), t \in T$, has mean zero and it can be expressed as

$$X(t) = \int^{t} F(t, u) dB(u) \tag{2.1}$$

with a real Borel measurable function $F(t, u)$ of u vanishing for $u > t$ and is belonging to $L^2(\nu)$ where $d\nu = E(dB(u)^2$ then (2.1) is called a representation of $X(t)$ and F is called a **kernel of the representation.**

Definition 2.1. The representation (2.1) is called **canonical** if

$$E(X(t)|\mathcal{B}_s) = \int^{s} F(t, u) dB(u) \tag{2.2}$$

for every $s \leq t$, where $\mathcal{B}_s$ is the smallest Borel field of measurable ω-sets with respect to which all the $X(\tau)$'s for $(\tau \leq s)$ are measurable.

P. Lévy had proved that a canonical representation of any process is uniquely determined if it exists, but not for the existence. T. Hida proved the existence of canonical representation using Helinger Hahn Theorem in the theory of Hilbert space.

Existence of canonical representation

In order to prove the existence of canonical representation he had used some theorems in the theory of Hilbert space. By using the reproducing kernel and Hellinger Hahn's theorem he proved the following theorem.

Theorem 2.1. *Let $\mathcal{M}_t$ be a closed linear manifold generated by $\{X(\tau), \tau \leq t\}$. The process $X(t)$ has a canonical representation if and only if the following three conditions hold:*

1) $\mathcal{M} = \bigcup_{t\in T} \mathcal{M}_t$ is separable (as a subspace of $L^2(\Omega)$.)
2) $\bigcap_{t\in T} \mathcal{M}_t = \{0\}$.
3) The multiplicity of $X(t)$ is one.

Theorem 2.2. *A canonical representation (2.1) is called proper if*

$$\mathcal{M}_t = \mathcal{M}_t(B)$$

for every $t \in T$, where $\mathcal{M}_t$ is the closed linear manifolds generated by $\{X(\tau), \tau \leq t\}$ and $\mathcal{M}_t(B) =$ the closed linear manifold generated by

$$\left\{ \int_{(-\infty,t)} \cdot \quad dB(u); \mathcal{M} \in \mathcal{B}_T \right\}.$$

Canonical Criterion

T. Hida gave the canonical criterion by the following theorem.

Theorem 2.3. *The representation (2.1) is proper canonical if for any fixed $t_0 \in T$ and $f \in L^2(\nu, t_0)$*

$$\int^t F(t,u) f(u) d\nu(u) = 0 \text{ for every } t \leq t_0$$

implies

$$f(u) = 0 \quad \text{a.e}, (\nu) \text{ on } (-\infty, t_0].$$

Multiple Markov property

T. Hida gave the definition of Multiple Markov property different with the definition given by J. L. Doob (for stationary processes) and P. Lévy who used the derivatives up to the $(N-1)$-th order of the processes.

Definition 2.2. For any choice of $\{t_i\}$ with $t_0 \leq t_1 < ... < t_N < t_{N+1}$,

1) $E[X(t_i)|\mathcal{B}_{t_0}(X)], i = 1, 2, ..., N$, are linearly independent and
2) $E[X(t_i)|\mathcal{B}_{t_0}(X)], i = 1, 2, ..., N+1$ are linearly dependent

then, $X(t)$ is said to be ***N**-ple Markov* Gaussian process.

And there is a beautiful result for $\boldsymbol{N}$-ple Markov process by the expression with Goursat kernel.

Theorem 2.4. *The process $X(t)$ is $\boldsymbol{N}$-ple Markov if and only if the kernel F, expressed in (2.1), is a Goursat kernel of order N. Namely,*

$$F(t, u) = \sum_{1}^{N} f_i(t) g_i(u) \tag{2.3}$$

where $det(f_i(t_j)) \neq 0$ for different t_j's and $\{g_i(u), i = 1, 2, \cdots\}$ are linearly independent on $L^2([0, t])$, for any t.

A Gaussian process $X(t)$ is said to be *$\boldsymbol{N}$-ple Markov in the restricted sense*, if $X(t)$ is differentiable in strong topology up to $(\boldsymbol{N}-\mathbf{1})$ times and satisfy

$$L_t X(t) = \dot{B}(t)$$

for an $\boldsymbol{N}$-th order differential operator L_t. The $\boldsymbol{N}$-th derivative is taken in the topology of Hida distribution.

Applying Markov property in the sense of Hida, we can discuss the duality as follows (Ref. [7]). In our setup, the differential equation is well defined in $H_1^{(-1)}$, Hida distribution space, which is isomorphic to Sobolev space $K^{(-1)}(R^1)$ of order -1.

We use the notation $K^N(R^n)$ to express the Sobolev space of order N over R^n. (N can be any rational number.)

Theorem 2.5. *There is a solution to the differential equation*

$$L_t X(t) = \dot{B}(t)$$

expressed in the form

$$X(t) = \int_0^t R(t,u)\dot{B}(u)du, \tag{2.4}$$

where $R(t,u)$ is the Riemann function for an ordinary differential operator L_t. The representation of $X(t)$ is canonical.

We can now give an interpretation to the triple $(X(t), L_t, R)$ determined above from the view point of duality.

Theorem 2.6. *If a triple $(X(t), L_t, R), t \in [0,1]$, is given, then $\mathbf{N}$-ple Markov process $X^*(t)$ can be constructed in such a way that $(X^*(t),\ L_t^*,\ R^*)$, $t \in [0,1]$ is a dual system of $(X(t), L_t, R)$ in the sense that*

1) *L_t^* is the formal adjoint of L_t,*
2) *R^* is the Riemann function for L_t^*,*
3) *$X^*(t) = \int_t^1 R^*(t,u)\dot{B}(u)du$ is the solution of*

$$L_t^* X^*(t) = \dot{B}(t),$$

where $\dot{B}(t)$ is the same $\dot{B}$ given in (2.4) up to sign.

In addition,

$$(X^*)^*(t) = X(t).$$

Proposition 2.1. *If $X(t), t \in (-\infty, \infty)$ is a stationary $\mathbf{N}$-ple Markov in a restricted sense such that*

$$X(t) = \int_{-\infty}^t R(t-u)\dot{B}(u)du, \tag{2.5}$$

then there exists an ordinary differential operator L_t of order N with constant coefficients such that

$$L_t X(t) = \dot{B}(t)$$

and the dual process $X^*(t)$ is

$$X^*(t) = \int_t^\infty R^*(t-u)\dot{B}(u)du, \tag{2.6}$$

where $R^*(u) = R(-u)$.

Note. Comparing the expressions of $X(t)$ and $X^*(t)$, one can see duality.

3. White noise analysis, the so called Hida distributions

Professor Hida initiated White noise analysis as is described in the introduction. We describe white noise $\dot{B}(t)$ as the time derivative of Brownian motion $B(t)$ with respect to time parameter t. However, almost all sample functions $B(t,\omega)$ are no where differentiable although they are continuous. Thus the white noise $\{\dot{B}(t), t \in R^1\}$ is defined as a generalized stochastic process. Let E be a nuclear space which is dense in $L^2(R^1)$. It is taken to be the space of test functions.

We have a Gel'fand triple

$$E \subset L^2(R) \subset E^*.$$

Take $\xi \in E$. Then we can define the derivative $\dot{B}(t)$ of a Brownian motion $B(t)(= B(t,\omega))$ with respect to parameter t in the distribution sense with test function ξ, namely, $\langle \dot{B}, \xi \rangle$.

With this choice of the test function space E, a white noise $\dot{B}(t)$ through the bilinear form $\langle \dot{B}, \xi \rangle$ is understood in such a way that

1) for almost all $\omega, \dot{B}(\xi)(\omega), \xi \in E$, is a continuous linear functional of ξ,
2) $\dot{B}(\xi)$ has mean 0 and variance $||\xi||^2$, $\|\cdot\|$ is the $L^2(R^1)$-norm,
3) $\{\dot{B}(\xi), \xi \in E\}$ is a Gaussian system.

Thus, we can see that the characteristic functional $C(\xi)$, $\xi \in E$, is given by

$$C(\xi) = \exp[-\frac{1}{2}||\xi||^2], \ \xi \in E. \tag{3.1}$$

So far, we have proceeded in line with the classical theory of stochastic analysis.
Since (3.1) is a characteristic functional, by Bochner Minlos theorem, there exists a unique probability measure μ on the measurable space $(E^*, \mathcal{B})$ such that

$$C(\xi) = \int_{E^*} \exp[i\langle x, \xi \rangle] d\mu(x). \tag{3.2}$$

The probability measure μ determined by the characteristic functional (3.1) is the white noise.

Definition 3.1. The measure space $(E^*, \mathcal{B}, \mu)$ is called white noise.

The sigma-field generated by $\{\dot{B}(t), t \in R^1\}$ will be denoted by $\mathcal{B} = \mathcal{B}(\dot{B})$.

White noise functionals

Take the measure space $(E^*, \mathcal{B}, \mu)$ being established above. Since μ on E^* is the probability distribution of $\{\dot{B}(t); t \in R^1\}$, a member $x \in E^*$ is viewed as a sample function of white noise $\dot{B}(t,\omega) \in E^*$.

Definition 3.2. A complex valued $\mathcal{B}$-measurable function $\varphi(x), x \in E^*$, is called a *white noise functional.*

We can form a complex Hilbert space $(L^2) = L^2(E^*, \mu)$ involving μ-square integrable white noise functionals.

Proposition 3.1. *The algebra $\mathcal{A}$ generated by the polynomials $e^{a\langle x,\xi\rangle}$ is dense in (L^2).*

Proposition 3.2. *The algebra generated by $\{P(\langle x, \xi_1\rangle, \cdots, \langle x, \xi_n\rangle)\}$ is dense in (L^2).*

Corollary A white noise functional $\varphi(x)$ may be expressed in the form

$$\varphi(x) = \varphi(\langle x, \xi_1\rangle, \langle x, \xi_2\rangle, \cdots)$$

where $\{\xi_n\}$ is a complete orthonormal system in $L^2(R^1)$ such that $\xi_n \in E$ for every n.
Here we note that

$$(\langle x, \xi_1\rangle, \langle x, \xi_2\rangle, \cdots)$$

looks like a coordinate representation of x, while $\{\langle x, \xi_n\rangle\}$ is a sequence of independent identically distributed (i.i.d.) standard Gaussian random variables on the probability space $(E^*, \mathcal{B}, \mu)$.

The complex Hilbert space $(L^2) = L^2(E^*, \mu)$ has a representation as the Fock space in the sense that it admits a direct sum decomposition into the spaces of homogeneous chaos $\mathcal{H}_n$, $n = 0, 1, 2, ...$, in terms of N. Wiener,

$$(L^2) = \bigoplus_n \mathcal{H}_n. \tag{3.3}$$

More precisely, each subspace $\mathcal{H}_n$ *is spanned by the Fourier-Hermite polynomials of the form*

$$c \prod_k H_{n_k}\left(\frac{\langle x, \xi_k\rangle}{\sqrt{2}}\right), \quad \Sigma n_k = n,$$

where $\{\xi_k\}$ is a system defined above. Those polynomials are orthogonal to each other in (L^2) and span the entire space. The direct sum decomposition of the form (3.3) of the space (L^2) is called the *Fock space.*

Infinite dimensional spaces spanned by generalized linear functionals of white noise

According to his idea of **Reduction, Synthesis and Analysis**, Hida wished to use $\dot{B}(t)$ itself, without smearing by test function. We know that $\dot{B}(t)$ is elemental, atomic and mutually independent Gaussian random variable although it is infinitesimal. In addition, the parameter t stands for the time, so that it can be used to express evolution phenomena that propagate as t changes. Namely, he liked to keep the time variable t not letting it be smeared.

1) Space $\mathcal{H}_1^{(-1)}$ of generalized linear functionals of white noise

He first introduced the space $\mathcal{H}_1^{(-1)}$, *a space of generalized white noise linear functionals.* The collection of linear functionals of $\dot{B}(t)$'s of the form

$$\dot{B}(f) = \int f(u)\dot{B}(u)du, \quad f \in L^2(R^1), \tag{3.4}$$

spans a Hilbert space $\mathcal{H}_1$ under the $L^2(\Omega, P)$-topology. The above formula proves that the space $\mathcal{H}_1$ is isomorphic to $L^2(R^1)$:

$$\mathcal{H}_1 \cong L^2(R^1). \tag{3.5}$$

Thus, we can see that the above correspondence let the isometric correspondence

$$f \mapsto \dot{B}(f) \in \mathcal{H}_1, \quad f \in L^2(R^1),$$

extend to that between the Sobolev space $K^{-1}(R^1)$ of order minus one (which is an extension of $L^2(R^1)$) We make use of a correspondence

$$\dot{B}(t) \;\leftrightarrow\; \delta_t(\cdot).$$

The delta-functions naturally generate the Sobolev space of order -1 over R^1, denoted by $K^{(-1)}(R^1)$.
Based on the above basic correspondence, we have a completion: $\mathcal{H}_1^{(-1)}$ so as to have an isomorphism

$$\mathcal{H}_1^{(-1)} \cong K^{(-1)}(R^1). \tag{3.6}$$

Namely, we have a generalization of (3.5):

The Hilbert space $\mathcal{H}_1^{(-1)}$ *is taken to be the domain of the white noise functions (functionals),* where the system $\{\dot{B}(t), t \in R^1\}$ is total and each member is viewed as a variable of white noise functionals. In reality, $\mathcal{H}_1^{(-1)}$ is defined by (3.6).

First take the class of polynomials of degree 1, that is a linear form. As a generalization of $\sum a_j \dot{B}(t_j)$, we may take

$$\int f(u)\dot{B}(u)du,$$

where f should be in $K^{(-1)}(R^1)$. The collection of those linear functionals will span the space $\mathcal{H}_1^{(-1)}$. Such a consideration extends to the nonlinear functionals of white noise.

2) Space $\mathcal{H}_1^{(-n)}$ of generalized linear functionals of white noise

Let E_n be the Sobolev space $K^n(R^1)$ of order n over R^1 and let its dual space be denoted by E_{-n}. Namely, there is a Gel'fand triple:

$$E_n \subset L^2(R^1) \subset E_{-n}.$$

We know that $\mathcal{H}_1^{(-1)} \cong K^{(-1)}(R^1)$ which can be generalized in the following line.

Let $K^n(R^1), n > 0$ be the Sobolev space to be over R^1 of order n. The space $K^{(-\frac{n+1}{2}}(R^1)$ is the dual space of $K^{\frac{n+1}{2}}(R^1)$ with respect to the $L^2(R^1)$-topology. Hida usually take the order of Sobolev space $\frac{n+1}{2}$, since it is convenient to carry out the analysis on $\mathcal{H}_\backslash$.

Corresponding to Gel'fand triple

$$K^{\frac{n+1}{2}}(R^1) \ \subset \ L^2(R^1) \ \subset \ K^{-\frac{n+1}{2}}(R^1),$$

we have a triple

$$\mathcal{H}_1^{(n)} \ \subset \ \mathcal{H}_n \ \subset \ \mathcal{H}_1^{(-n)}.$$

Next we wish to define the infinite dimensional space $\mathcal{H}_n^{(-n)}$, for $n > 1$, generated by Hermite polynomials in $\dot{B}(t)$'s of higher degree. So it is necessary to define the renormalization. For that we first have to define the $\mathcal{T}$-transform and S-transform.

$\mathcal{T}$-transform and S-transform

The $\mathcal{T}$ and S-transform plays an essential role in White Nose Analysis.

The $\mathcal{T}$-transform and S-transform of a white noise functional $\varphi(x)$ is defined as follows:

$$(\mathcal{T}\varphi)(\xi) = \int e^{\langle x,\xi\rangle}\varphi(x)d\mu(x), \ \xi \in E.$$

$$(S\varphi)(\xi) = \int e^{-\frac{1}{2}\|\xi\|^2} e^{\langle x,\xi\rangle}\varphi(x)d\mu(x),\ \xi \in E.$$

Note: Here we can see that the S-transform is defined by using the generating functional. The following propositions can be seen as an example of the S-transform and which will be also important fact for the renormalization.

Proposition 3.3. *i) Let $\varphi = (\dot{B}(t))$ of polynomials in $\dot{B}(t)$). Then we have*

$$(S\varphi)(\xi) = p(\xi(t)) + O\left(\frac{1}{dt}\right),$$

where p is a polynomial. For a product $\Pi\varphi_j(\dot{B}(t_j)$ of polynomials (using the same notation as above) we have the S-transform

$$(S(\Pi\varphi_j(\dot{B}(t_j)))(\xi) = \Pi p_j(\xi(t_j)) + O\left(\Pi\frac{1}{dt_j}\right).$$

In fact the S-transform gives us a favorable representation to carry out the white noise calculus.

Renormalization

Polynomials (given algebraically) in $x(t)$'s (i.e. in $\dot{B}(t)$'s) always need the so-called renormalizations except linear case. The technique to exclude ∞ from polynomials in an algebraic form can automatically done through the S-transform, where we are given enough interpretation taking the term $\frac{1}{dt^n}$ to be removed.

Example 3.1. Consider $\dot{B}(t)^2$. Then

$$S(\dot{B}(t)^2) = \xi(t)^2 + \frac{1}{dt}$$

Renormalization of $\dot{B}(t)^2$ is obtained by removing $\frac{1}{dt}$. That is

$$:\dot{B}(t)^2 := \dot{B}(t)^2 - \frac{1}{dt}.$$

It is necessary to modify polynomials in $\dot{B}(t)$'s to have generalized functions. We see, a simplest case,

$$:\dot{B}(t)^p := p!H_n\left(\dot{B}(t);\frac{1}{dt}\right), \tag{3.7}$$

where $H_n(x, \sigma^2)$ is the Hermite polynomial with parameter σ^2 of degree n.

In reality : $\dot{B}(t)^p$: is the result of additive renormalization of $\dot{B}(t)^p$ which has only formal meaning before renormalized. Even renormalized, it is not an (L^2)-functional, but a generalized functional.

Such a renormalization technique can be applied to all the polynomials in $\dot{B}(t)$'s of degree n to have the space $\mathcal{H}_n^{(-n)}$, which is the extension of $\mathcal{H}_n$ for $n > 1$, generated by Hermite polynomials in $\dot{B}(t)$'s of higher degree.

$\mathcal{H}_n^{(-n)}$ of generalized white noise functionals of higher degree

First we wish to use a result obtained by S-transform. It is

$$\mathcal{H}_n \cong \sqrt{n!}\widehat{L^2(R^n)},$$

where $\mathcal{H}_n$ is spanned by Hermite polynomials of degree n. Let $\widehat{K^m(R^n)}, n \geq 1$ be Sobolev space of order m, involving symmetric functions, over R^n. Take the dual space $\widehat{K^{-m}(R^n)}$ of $\widehat{K^m(R^n)}$ with respect to the topology being equipped with $L^2(R^n)$. Then, we have th Gel'fand triple

$$\widehat{K^{\frac{n+1}{2}}} \subset \widehat{L^2(R^n)} \subset \widehat{K^{-\frac{n+1}{2}}},$$

where inclusions are continuous injection and we form the triple

$$\mathcal{H}_n^{(n)} \subset \mathcal{H}_n \subset \mathcal{H}_n^{(-n)},$$

where each space of the triple is isomorphic to the corresponding symmetric Sobolev space.

Note. T. Hida took $K^{-\frac{n+1}{2}}(R^n)$ instead of $K^n(R^n)$. Since it is convenient to do so since the restrictions of members of $K^{-\frac{n+1}{2}}(R^1)$ to an $(n-1)$ dimensional hyper plane, they belong to $K^{-\frac{n}{2}}(R^{n-1})$, that is the decreased by $\frac{1}{2}$ when the restriction is done to a space of one dimensional lower.

The Hilbert space $\mathcal{H}_n^{(-n)}, n \geq 0$ is called **a space of generalized functionals of degree n, on the Hida distributions of degree n.**

We have

$$\mathcal{H}_n^{(-n)} \cong \widehat{K^{-\frac{n+1}{2}}(R^n)}.$$

Denote the $\mathcal{H}_n^{(n)}$-norm by $\|\cdot\|_n$ which is Sobolev norm defined on $K^{\frac{n+1}{2}}(R^n)$. Define

$$\|\cdot\|_+ = \left(\sum_n c_n^2 \|\cdot\|_n^2\right)^{\frac{1}{2}},$$

where c_n is a non-decreasing sequence of positive numbers which are suitably chosen according to the problem to be discussed.

The completion of the algebraic sum $(L^2)^+$ of $\mathcal{H}_n^{(n)}$ is topologized by the norm $\|\cdot\|_+$. That is

$$(L^2)^+ = \bigoplus_n c_n \mathcal{H}_n^{(n)} \tag{3.8}$$

is the Hilbert space which plays the role of *test functionals.*

Definition 3.3. The dual space of $(L^2)^+$ with respect to the (L^2)-norm is a space of generalized white noise functionals or Hida distributions and is denoted by $(L^2)^-$.

By definition, the space $(L^2)^-$ is expressed as

$$(L^2)^- = \bigoplus_{n=0}^{\infty} c_n^{-1} \mathcal{H}_n^{(-n)} \tag{3.9}$$

and it is a Hilbert space too.
The norm introduced to $(L^2)^-$ is $(\sum_n c_n^{-2}\|\cdot\|_{-n)}^2)^{\frac{1}{2}}$, where $\|\cdot\|_{-n}$ is the $\mathcal{H}_n^{-n}$-norm which is defined by the Sobolev norm defined on $K^{-\frac{n+1}{2}}$.
Hence, we obtain the triple

$$(L^2)^+ \subset (L^2) \subset (L^2)^-. \tag{3.10}$$

The canonical bilinear form connecting $(L^2)^+$ and $(L^2)^-$ is denoted by $\langle \cdot, \cdot \rangle_\mu$.

White noise differentiation

White noise derivative can neither be an analogue of non-random case, nor discrete random case. However, with the help of S-transform and Fréchet derivative we can give a correct definition of the white noise derivative:

$$\partial_t \varphi = S^{-1}\left(\frac{\delta}{\delta\xi(t)}(S\varphi)(\xi)\right)(x),$$

where ∂_t denotes the derivative with respect to the white noise $\dot{B}(t)$.

Example 3.2.

$$\varphi = \int f(u) : \dot{B}(u)^2 : du$$

where f is bounded.

Take the S-transform of φ, denoted by $U\xi)$ then we have

$$U(\xi) = (S\varphi)(\xi) = \int f(u)\xi(u)^2 du.$$

Then the functional derivative is

$$\frac{\delta}{\delta\xi(t)} U(\xi) = 2f(t)\xi(t).$$

Thus, we have

$$\partial_t \varphi = 2f(t)\dot{B}(t).$$

Laplacian

Using the white noise, he discussed the number operator given by

$$\Delta_\infty = \int \partial_t^* \partial_t dt.$$

For $\varphi \in \mathcal{H}_n$, we have

$$\Delta_\infty \varphi = -n\varphi.$$

This can be extended to an operator on $(L^2)^-$.

Lévy Laplacian

An infinite dimensional Laplacian

$$\lim_{n\to\infty} \frac{1}{n} \sum_1^\infty \frac{\partial^2}{\partial \xi_i^2}$$

is rephrased in terms of the ∂_t

$$\Delta_L = \int \partial_t^2 (dt)^2,$$

by H.-H. Kuo.

Gauss kernel is an eigen-function of Δ_L.

Rotation group and harmonic analysis

A transformation g on $L^2(R)$ is a rotation of E if it satisfies

$$\|g\xi\| = \|\xi\|, \text{ for every } \xi \in E,$$

and g is a homeomorphism of E.

The collection of all rotations of E forms a group since a rotation g is isomorphism of E, and so g^{-1} is defined. The product of rotations is defined in the usual manner such that

$$(g_1g_2)\xi = g_1(g_2\xi), \xi \in E.$$

It is topologized, say by the compact open topology, to have a topological group $O(E)$. It is called an *infinite dimensional rotation group.*

For any $g \in E$, the adjoint g^* is a linear transformation acting on E^*:

$$\langle x, g\xi\rangle = \langle g^*x, \xi\rangle.$$

Theorem 3.1. *The white noise measure μ is invariant under g^*, i.e.*

$$g^* \cdot \mu = \mu.$$

The collection

$$O^*(E^*) = \{g^*; g \in O(E)\}$$

forms a group isomorphic to $O(E)$ under the mapping

$$g \to (g^*)^{-1} \in O^*(E^*), \quad g \in O(E).$$

Note that

$$(g^*)^{-1} = (g^{-1})^*.$$

Define the product $(g^* \cdot \mu)(B) = \mu(g^*B), B \in \mathcal{B}$.

Theorem 3.2. *Each g^* is a μ measure preserving transformation acting on (E^*, μ). That is, μ is $O^*(E^*)$-invariant.*

Subgroups of $O(E)$

T. Hida is interested in subgroups of $O^*(E^*)$ which is isomorphic to $O(E)$, so he first discussed subgroups of $O(E)$ which is somewhat easier to be dealt with.

I. Lévy group and normal subgroup

Let $\{\xi_n\}$ in $L^2(R)$ such that $\xi_n \in E$, for every n. Let E_n be the n-dimensional subspace of E spanned by $\xi_k, k = 1, 2, \cdots, n$. Let

$$\tilde{G_n} = \{g \in O(E); g|_{E_n^\perp} = \text{identity}\},$$

where $E_n^\perp$ is orthogonal complement of E_n and $g|_{E_n^\perp}$ is the restriction of g to $E_n^\perp$. Obviously we have

$$\tilde{G_n} \cong O(n).$$

There is a subgroup G_n of $\tilde{G_n}$ such that

$$G_n \cong SO(n), \quad n = 1, 2, \cdots .$$

Their inductive limit is also defined:

$$G_\infty = \vee_n G_n.$$

The Lévy group $\mathcal{G}$

Set

$$\mathcal{G} = \{g \in O(E),\ g = g_\pi\},$$

where $g_\pi : \xi = \sum_n a_n \xi_n \to \sum_n a_n \xi_{\pi(n)}$, π is the permutation of $(1, 2, 3, \cdots)$ satisfying $\forall \epsilon > 0, \exists N$ such that $\forall n > N$,

$$\#\{i; i \leq n \ \&\ \pi(i) > n\} < \varepsilon.$$

Obviously $\mathcal{G}$ forms a group and it is called Lévy group.

Proposition 3.4. *G_∞ is a normal subgroup of G, where $G = \mathcal{G} \vee G_\infty$ generated by $\mathcal{G}$ and G_∞.*

II. Whiskers

A whisker means a one-parameter subgroup of $O(E)$ that comes from diffeomorpism of $\overline{R_1}$. It is expressed in the form

$$\{g_t\} \subset O(E)$$

where

$$(g_t\xi)(u) = \xi(\psi_t(u))\sqrt{|\psi_t'(u)|} \tag{3.11}$$

$$\psi_t \cdot \psi_s = \psi_{t+s} \quad (\Rightarrow g_t g_s = g_{t+s}) \tag{3.12}$$

where ψ is some monotone mapping of R onto itself.

Examples of whiskers

1) Shift $S_t^j : (S_t^j)(u) = \xi(u - te_j), j = 1, 2, \cdots, d$,
where $\{e_1, e_2, \cdots, e_d\}$ is a coordinate system in R^d.
2) (Isotropic) Dilation $\tau_t : (\tau_t\xi)(u) = \xi(ue^t)e^{\frac{t}{2}}$.
3) Rotations of $R^d : (g_\psi\xi)(u) = \xi(\psi(u)), \psi \in SO(d)$.
4) Special conformal transformation $\kappa_t^j : \kappa_t^j = wS_t^j w$,
where w is the reflection with respect to the unit sphere
$(w\xi)(u) = \xi(\frac{u}{|u|^2})|u|^{-d}$.

Note. He also introduced Complex White Noise Analysis and Infinite dimensional unitary group.

4. Variational calculus and random fields

T. Hida also considered the Gaussian random field $X(C)$ parametrized by C and is assumed to be a linear function of the white noise x.
The class $\mathbf{C}$ of the family of C is taken as

$$\mathbf{C} = \{C; C : \text{closed convex surface in } \mathrm{R}^{\mathrm{d}}, \text{diffeomorphic to } \mathrm{S}^{\mathrm{d}-1}\}$$

and is topologised by the Euclidean metric.
First variational formula for a random field is given. Before discussing the Gaussian random fields depending on a contour, it is essential to consider a non-random function $G(C)$ of C in $\mathbf{C}$.

I. Non-random function

Consider a non-random function $G(C)$ defined on $\mathbf{C}$, where $G(C)$ is in R^1 and $\mathbf{C}$ is defined as in the above. Take $C + \delta C$ ($\in \mathbf{C}$) which is a slight deformation of C. We write δC as only a symbolic expression of a contour sitting outside of C determined by

$$\delta C = \{\delta n(s); s \in C\}, \tag{4.1}$$

where s is the arc length representing the parameter of C, $\delta n(s)$ denotes the normal vector to C to the outward direction at the point s and $|\delta n(s)|$ denotes the distance from s to $C + \delta C$.

Definition 4.1. If $\|\delta n\| = \sup_s |\delta n(s)| \to 0$ then we say that $C+\delta C$ tends to C.

We can now assume that $\delta n(s)$ is continuous. Let us assume that G satisfies the following.

$$G(C + \delta C) - G(C) = \delta G(C) + g(C, \delta C) \tag{4.2}$$

such that

1) $\delta G(C)$ is continuous and linear in $\delta n(s)$ and
2) $g(C, \delta C)$ is $o(\|\delta n\|)$.

According to the above fact (1), there is φ such that $\delta G(C)$ can be expressed as

$$\delta G(C) = \int_C \varphi(s)\delta n(s)ds. \tag{4.3}$$

Denote $\varphi(s)$ by $\frac{\partial G(C)}{\partial n}(s)$. Thus we have

$$\delta G(C) = \int_C \frac{\partial G(C)}{\partial n}(s)\delta n(s)ds. \tag{4.4}$$

Note. It is to note that the normal vector $\delta n(s)$ is taken to the outward direction from C, since the interior of C is tacitly understood to be the past in a sense so that δC is taken towards the future.

II. Random fields

For the random field $X(C)$, the variation formula can be given. Let $X(C)$ be an (L^2)-functional of white noise $x(u)$. Then we can apply the S-transform to have a functional $U(C;\xi)$.

Fix ξ for a moment and use the notation simply $U(C)$. Then the variation of $X(C)$ is defined by

$$\delta X(C) = S^{-1}\delta(S(X(C))(\xi)) = S^{-1}(\delta U(C)).$$

Set

$$Y(C) = \int_{(C)} g(u)x(u)du, \tag{4.5}$$

where g is a continuous function. Using the above formula, the variation of $Y(C)$ is given by the following proposition.

Proposition 4.1. *The variation of $Y(C)$, expressed in the form (4.5) is*

$$\delta Y(C) = \int_C g(s)x(s)\delta n(s)ds, \tag{4.6}$$

where $g(s)$ is the restriction of g on C and $\delta n(s)$ determines the variation δC of C.

Take the S-transform of (4.5) and obtain the variation. And by taking the S^{-1}-transform the assertion is obtained.

We are now interested in a random field which can be expressed by a homogeneous chaos.

Let $Y(C)$ be a Gaussian martingale with respect to $\mathbf{B}_C(Y)$ such that $Y(C) \neq 0$ for every C. Then $Y(C)$ can be expressed in the form

$$Y(C) = \int_{(C)} g(u)x(u)du \tag{4.7}$$

and its variance is

$$\sigma^2 = \int_{(C)} g(u)^2 du.$$

Define

$$Z(C) = H_n(Y(C), \sigma^2) \tag{4.8}$$

where $H_n(x;\sigma^2)$ is a Hermite polynomial with parameter.

Proposition 4.2. *The variation equation for* $Z(C)$ *is given by*

$$\delta Z(C) = H_{n-1}(Y(C),\sigma^2)\int_C g(s)x(s)\delta n(s)ds. \tag{4.9}$$

For the proof we take the S-transform to obtain the U-functional and take its variation and take back the S^{-1} transform.

He also considered a variation of a functional $F(X(C))$ of a Markov random field $\{X(C); C \in \mathbf{C}\}$.

Variational equation

Let C be represented by a C^∞-function η. Thus the variation of η defines that of C. Thus U-functional is $U(\eta,\xi)$, however ξ is arbitrary and is fixed so we write $U(\eta)$ omitting ξ. Consider the variation equation

$$\delta U(\eta) = \int_C f(\eta, U, s)\delta n(s)ds,$$

where $f(\eta, U, s)$ is continuous in U and has Fréchet derivative in η in the $C^\infty(S^{n-1})$-topology and has partial derivative in U. Assume that the Fréchet derivative and the partial derivative are continuous in all variables.

Theorem 4.1. *The integrability condition for the variational equation is given by*

$$\frac{\delta f(\eta,U,s)}{\delta\eta(t)} + \frac{\delta f(\eta,U,s)}{\partial U}f(\eta,U,t) = \frac{\delta f(\eta,U,t)}{\delta\eta(s)} + \frac{\delta f(\eta,U,t)}{\partial U}f(\eta,U,s).$$

Existence theorem for a variational equation

Consider the particular case where C is diffeomorphic to a unit circle S^1, so that $U(C)$ may be expressed as a function of ξ, U and t:

$$U(C) = U(\xi, U, t),\ \xi \in E,\ t \in I = [0,1],$$

where E being a nuclear subspace of $L^2(I)$.

Hence the variation $\delta U(C)$ may be expressed in the form

$$\delta U(C) = \int_C f(\xi, U, s)\delta n(s)ds.$$

Theorem 4.2. *The solution of the above variational equation exists and is unique under the following assumptions:*

1) $f(\xi, U, s)$ *is continuous in the three variables* (ξ, U, s) *and satisfies the integrability condition,*
2) *In some neighbourhood of* U_0, *e.g.* $|U - U_0| < |V - U_0| < C$, *there exist constants* M *and* K *such that*
   (i) $\int_I f(\xi, U, s)^2 ds < M^2$,
   (ii) $\int_I [f(\xi, U, s) - f(\xi, V, s)]^2 ds < K^2(U - V)^2$,
3) $\xi = \xi_0 + \lambda\eta, \|\eta\| = 1, \lambda \geq 0$, *where* $\| \ \|$ *is the* $L^2(I)$*-norm.*

5. Feynman Integrals and Dirichlet forms

In 1981 T. Hida and L. Streit proposed a white noise approach to path integrals to have quantum mechanical propagators. The key point is the use of a Brownian bridge to express fluctuation around the classical paths.

Following the Lagrangian dynamics, the possible trajectories are sample paths $y(s), s \in [0, 1]$, expressed in the form

$$y(s) = x(s) + \sqrt{\frac{\hbar}{m}} B(s),$$

where $B(t)$ is an ordinary Brownian motion. Hence the action S is expressed in the form

$$S = \int_0^t L(y(s), \dot{y}(s)) ds.$$

Note that the bridged effect is done by putting the delta-function $\delta_0(y(t) - y_2)$, where $y_2 = x(t)$.

Now we set

$$S(t_0, t_1) = \int_{t_0}^{t_1} L(t) dt,$$

and set

$$\exp\left[\frac{i}{\hbar}\int_{t_0}^{t_1} L(t)dt\right] = \exp\left[\frac{i}{\hbar} S(t_0, t_1)\right] = B(t_0, t_1).$$

Then, we have (see Dirac's monograph 1930), for $0 < t_1 < t_2 < \cdots < t_n < t$,

$$B(0, t) = B(0, t_1) \cdot B(t_1, t_2) \cdots B(t_n, t).$$

Theorem 5.1. *The quantum mechanical propagator* $G(0,t;y_1,y_2)$ *is given by the following average*

$$G(0,t;y_1,y_2)$$

$$= \left\langle N \exp\left[\frac{i}{\hbar}\int_0^t L(y,\dot{y})ds + \frac{1}{2}\int_0^t \dot{B}(s)^2 ds\right] \delta_o(y(t)-y_2)\right\rangle .$$

There are many other fields that he has contributed but we now stopped here.

Acknowledgement: The author wishes to express her deep thanks to Professor Watanabe for his great effort in organizing the conference and for his kind invitation.

References

1. T. Hida, Analysis of Brownian Funcionals, Carleton Mathematical Lecture Notes N0. 13.1975.
2. T. Hida, Canonical representation of Gaussian Processes and their Applications, Memoirs Coll. Sci. Univ., Kyoto, A33, 1960, 109–155.
3. T. Hida, Stationary Stochastic Processes, Math. Notes, Princeton Univ. Press 1970.
4. T. Hida, Brownian Motion, Springer-Verlag 1980.
5. T. Hida and M. Hitsuda, Gaussian Processes, American Math. Soc. Translations of Mathematical Monographs vol. 12, 1993.
6. T. Hida and Si Si, Innovation Approach to Random Fields: An Application of White Noise Theory, World Scientific Pub. Co. Ltd., 2004.
7. Si Si and Win Win Htay, Duality arising from Multiple Markov Gaussian Processes. Quantum Bio-Informatics III. Proceedings of the International Conference on From Quantum Information to Bio-Informatics, QP-PQ Vol. 26, World Scientific Pub. 2010, 317–326.
8. Si Si, Introduction to Hida Distributions, World Scientific Pub. Co. Ltd., 2012.

Infinite Dimensional Analysis, Quantum Probability and Related Topics

pp. 219–229

REPRESENTING THE QUANTUM OPERATORS IN TERMS OF MULTIPLICATION AND DIFFERENTIATION OPERATORS

GABRIELA POPA

Department of Physics,
Ohio University,
College of Arts and Science,
240 Elton Hall,
Zanesville, OH 43701, USA
E-mail: popag@ohio.edu

AUREL I. STAN

Department of Mathematics,
The Ohio State University at Marion,
1465 Mount Vernon Avenue,
Marion, OH 43302, USA
E-mail: stan.7@osu.edu

The quantum operators generated by a random variable X, having finite moments of all orders, are expressed as a sum of compositions of powers of the multiplication operator by X and differentiation operator $D = d/dX$. The multiplication by X operator is located to the left of the differentiation operator.

Keywords: Szegö–Jacobi parameters, annihilation, preservation, creation, differentiation, and multiplication operators.

1. Introduction

The joint moments of a finite family of random variables, having finite moments of all orders, can be recovered from the commutators of their joint annihilation and creation operators, and their preservation operators, see Ref. 2. In fact, the complete knowledge of the joint preservation operators is not needed, being enough to know the commutators of the joint annihilation and preservation operators, provided that the first order moments are given. A method of computing the joint moments from these commutators was introduced and applied successfully, mainly on the Meixner class, in Ref. 12.

This method was improved furthermore in Ref. 8, where concrete computations were carried out in the two-dimensional and non-commutative case. Finally, the method was simplified more, by introducing the semi-quantum operators, which were obtained by splitting each preservation operator in two halves and adding one half to the corresponding annihilation operator, and the other half to the creation operator, in Refs. 9, 10, and 11. It must also be mentioned, that a method of recovering the joint moments of a finite family of not-necessary commuting random variables, from the q-commutators of the joint quantum operators was developed in Ref. 4.

The main idea of the method of recovering the moments from the commutators of the quantum operators, developed in the papers mentioned above is to repeatedly commute (permute) an annihilation operator with the multiplication operator generated by each of the involved random variables, until that annihilation operator, a^-, reaches the vacuum vector (constant polynomial equal to 1), in which moment $a^-1 = 0$. These repeated computations can take a long time and require a lot of patience and combinatorics to keep track of the new terms that appear during the computation.

In this paper we propose a different approach, which comes from the following intuition. Since the annihilation operators are mapping a polynomial of degree n into a polynomial of degree $n-1$, that means they reduce the degree of a polynomial by 1 unit, and the classic differentiation operator from Calculus does the same thing, it must be a connection between these operators. This paper being a beginning work in this direction, we will consider only the one-dimensional case, working with a single random variable X, which is assumed to have finite moments of all orders.

The paper is structured as follows. In Section 2 we provide a brief background about the quantum operators generated by a random variable. In Section 3 we show how the quantum operators, generated by a random variable X, having finite moments of all orders, can be uniquely expressed in terms of the multiplication by X and the differentiation operator D. We end the paper with a simple application related to the Gaussian random variables.

2. Background

Let $(\Omega,\ \mathcal{F},\ P)$ be a probability space and $X : \Omega \to \mathbb{R}$ a random variable (measurable function). We assume the X has finite moments of all orders, that means:

$$E\left[|X|^r\right] := \int_\Omega |X(\omega)|^r dP(\omega) \tag{1}$$

$$< \infty, \tag{2}$$

for all $r > 0$, where E denotes the expectation.
For all non-negative integers n, we define the space F_n of all random variables of the form $f(X)$, where $f(x) = a_0 + a_1x + \cdots + a_nx^n$ is a polynomial function of degree at most n, with complex coefficients. Since X has finite moments of all orders, we have:

$$\mathbb{C} \equiv F_0 \subseteq F_1 \subseteq F_2 \subseteq \cdots \subseteq L^2(\Omega, \mathcal{F}, P). \tag{3}$$

Since, for all $n \geq 0$, F_n is finitely generated by 1, X, ..., X^n, F_n is a closed subspace of $L^2(\Omega,\, \mathcal{F},\, P)$. We define $G_0 := F_0 \equiv \mathbb{C}$, and for all $n \geq 1$,

$$G_n := F_n \ominus F_{n-1}, \tag{4}$$

that means G_n is the orthogonal complement of F_{n-1} into F_n. Because the codimension of the vector space F_{n-1} into F_n is at most 1, we conclude that the dimension of the vector space G_n is at most 1, for all $n \geq 0$. In fact, if the probability distribution of X has an infinite support, then for all $n \geq 0$, the dimension of G_n is 1. If the probability distribution of X has a finite support of cardinality k, then the dimension of G_n is:

$$dim(G_n) = \begin{cases} 1 \text{ if } n \leq k-1 \\ 0 \text{ if } \quad n \geq k \end{cases}. \tag{5}$$

If we view X not as a random variable, but as a multiplication operator that maps a polynomial random variable $f(X)$ into the polynomial random variable $Xf(X)$, $f(X) \mapsto Xf(X)$, then we have the following lemma (see Ref. 1):

Lemma 2.1. *For all non-negative integers n, we have:*

$$XG_n \perp G_k, \tag{6}$$

for all $k \neq n-1$, n, and $n+1$.

Since, for all $n \geq 1$, G_n has dimension at most 1, if G_n is not the null space, then there exists a unique polynomial random variables $f_n(X)$, with a leading coefficient equal to 1, in G_n. We call f_n the *n-th orthogonal monic polynomial generated by X*. Since $Xf_n(X) \in G_0 \oplus G_1 \oplus G_2 \oplus \cdots$, where

"$\oplus$" means "orthogonal sum", the above lemma implies that $Xf_n(X) \in G_{n-1} \oplus G_n \oplus G_{n+1}$. Thus, there exist α_n and ω_n real numbers, such that:

$$Xf_n(X) = f_{n+1}(X) + \alpha_n f_n(X) + \omega_n f_{n-1}(X). \tag{7}$$

In the above equality, the coefficient of $f_{n+1}(x)$ in the right-hand side is 1, since the coefficient of X^{n+1} in the left-hand side is 1, and the only polynomial in the right that contains the term X^{n+1} is $f_{n+1}(X)$, which also has the leading coefficient (the coefficient of X^{n+1}) equal to 1.
Also, in the above equality, if $n = 0$, then $f_{n-1} = f_{-1} := 0$ is the null polynomial, and we can take ω_0 however we want.
The numbers $\{\alpha_n\}_{n\geq 0}$ and $\{\omega_n\}_{n\geq 1}$ are called the *Szegö-Jacobi parameters* of X.
We consider the following operators:

$$D_n^+ : G_n \to G_{n+1}, \tag{8}$$

defined by:

$$D_n^+ (cf_n(X)) = cf_{n+1}(X), \tag{9}$$

and call it a *creation operator* since it increases the degree of an orthogonal polynomial by one unit,

$$D_n^0 : G_n \to G_n, \tag{10}$$

defined by:

$$D_n^0 (cf_n(X)) = \alpha_n cf_n(X), \tag{11}$$

and call it a *preservation operator* since it preserves the degree of an orthogonal polynomial, and

$$D_n^- : G_n \to G_{n-1}, \tag{12}$$

defined by:

$$D_n^- (cf_n(X)) = \omega_n cf_{n-1}(X), \tag{13}$$

and call it a *creation operator* since it decreases the degree of an orthogonal polynomial by one unit.
Lemma 2.1 can be written now as:

Lemma 2.2. *For each non-negative integer n, the restriction of the multiplication operator by X to the space G_n can be written as:*

$$X|_{G_n} = D_n^+ + D_n^0 + D_n^-. \tag{14}$$

We extend now the definition of the creation, preservation, and annihilation operators, called generically the *quantum operators* generated by X, to the space F of all polynomial random variables in X, where:

$$F = \cup_{n=0}^{\infty} F_n. \tag{15}$$

If f is a polynomial, then there exists a sequence of complex numbers $\{c_n\}_{n\geq 0}$, with only finitely many non-zero terms, such that:

$$f(X) = \sum_{n=0}^{\infty} c_n f_n(X). \tag{16}$$

We define:

$$a^+ f(X) := \sum_{n=0}^{\infty} c_n D_n^+ f_n(X) \tag{17}$$

$$= \sum_{n=0}^{\infty} c_n f_{n+1}(X), \tag{18}$$

$$a^0 f(X) := \sum_{n=0}^{\infty} c_n D_n^0 f_n(X) \tag{19}$$

$$= \sum_{n=0}^{\infty} c_n \alpha_n f_n(X), \tag{20}$$

and

$$a^- f(X) := \sum_{n=0}^{\infty} c_n D_n^- f_n(X) \tag{21}$$

$$= \sum_{n=0}^{\infty} c_n \omega_n f_{n-1}(X). \tag{22}$$

Thus, the last lemma can be written as:

Lemma 2.3. *The multiplication operators by X, can be written as:*

$$X = a^+ + a^0 + a^-. \tag{23}$$

In this equality the domain of the operators X, a^+, a^0, and a^- is considered to be the space F of all polynomial random variables in X.

3. Expressing the quantum operators in terms of the position and momentum operators

Let us assume that X is a random variable, having finite moments of all orders, whose probability distribution has an infinite support, so that, for all $n \geq 0$, the space G_n has dimension 1. That means, for each non-zero polynomial function $f(x) = a_0 + a_1x + \cdots + a_nx^n$, the random variable $f(X) = a_0 + a_1X + \cdots + a_nX^n$ is not almost surely equal to zero.
We call the multiplication operator by X, the *position* operator. We call the differentiation operator $D = d/dX$, the *momentum* operator, where

$$D\left(a_0 + a_1X + \cdots + a_nX^n\right) := a_1 + 2a_2X + \cdots + na_nX^{n-1}, \tag{24}$$

for all $n \geq 0$, and a_0, a_1, ..., a_n complex numbers. This operator is well defined on F, viewed as a subspace of $L^2(\Omega, \mathcal{F}, P)$ (in which two random variables that are almost surely equal are considered to be equal), since no two different polynomial function $f(x)$ and $g(x)$ can generate two polynomial random variables in X, $f(X)$ and $g(X)$ that are equal almost surely. Here, we can see the importance of the fact that the probability distribution of X has an infinite support.
We have the following result.

Theorem 3.1. *If X is a random variable, having finite moments of all orders, whose distribution has an infinite support, then its annihilation and preservation operators can be written uniquely in the form:*

$$a^- = \sum_{n=1}^{\infty} A_{n-1}(X)D^n \tag{25}$$

and

$$a^0 = \sum_{n=0}^{\infty} B_n(X)D^n, \tag{26}$$

where for each $k \geq 0$, A_k and B_k are polynomial functions of degree at most k.

Before, we present the proof of this theorem, we would like to make the remark that no convergence is needed for the sums from the right-hand side of (25) and (26), since when these sums are applied to a polynomial random variable in X, $f(X)$, all the terms containing $D^nf(X)$, for n greater than the degree of f, are equal to zero.

Proof. We will construct inductively the polynomials A_{n-1}, for $n \geq 1$, using the monic orthogonal polynomials f_n and the Szegö-Jacobi parameters.

Indeed, for $n = 1$, $f_1(X) = X + c_1$, where c_1 is a real number, and the equality:

$$a^- f_1(X) = \sum_{n=1}^{\infty} A_{n-1}(X) D^n f_1(X), \tag{27}$$

becomes:

$$\omega_1 f_0(X) = A_0(X) \cdot 1. \tag{28}$$

Since $f_0(X) = 1$, we conclude that $A_0 = \omega_1$ is a polynomial of degree zero. Let us suppose now that $A_0, A_1, \cdots, A_{k-1}$ have been constructed, such that for all $i \in \{0, 1, \ldots, k\}$, we have:

$$a^- f_i(X) = \sum_{n=1}^{k} A_{n-1}(X) D^n f_i(X). \tag{29}$$

Assuming that formula (25) works when applied to $f(X) := f_{k+1}(X)$, we should have:

$$\begin{aligned} a^- f_{k+1}(X) &= \sum_{n=1}^{\infty} A_{n-1}(X) D^n f_{k+1}(X) \\ &= A_k(X) D^{k+1} \left(X^{k+1} + c_k X^k + \cdots + c_0\right) \\ &\quad + \sum_{n=1}^{k} A_{n-1}(X) D^n f_{k+1}(X) \\ &= A_k(X) \left((k+1)!\right) + \sum_{n=1}^{k} A_{n-1}(X) D^n f_{k+1}(X). \end{aligned} \tag{30}$$

Since $a^- f_{k+1}(X) = \omega_{k+1} f_k(X)$, solving for $A_k(X)$, we obtain:

$$A_k(X) = \frac{1}{(k+1)!} \left[\omega_{k+1} f_k(X) - \sum_{n=1}^{k} A_{n-1}(X) D^n f_{k+1}(X) \right], \tag{31}$$

and so, A_k is a polynomial of degree at most k.
We can see now, that for all $i \in \{0, 1, \ldots, k+1\}$, we have:

$$\begin{aligned} a^- f_i(X) &= \sum_{n=1}^{i} A_{n-1}(X) D^n f_i(X) \\ &= \sum_{n=1}^{k+1} A_{n-1}(X) D^n f_i(X), \end{aligned}$$

since $D^{i+1} f_i(X) = D^{i+2} f_i(X) = \cdots = D^{k+1} f_i(X) = 0$.
Thus the existence in (25) has been proved.

To prove the uniqueness of the terms, $A_0(X)$, $A_1(X)$, ..., we have to prove that if $C_0(x)$, $C_1(x)$, ... are polynomial functions of degree at most 0, 1, ..., such that:

$$\sum_{n=1}^{\infty} C_{n-1}(X)D^n = 0, \tag{32}$$

then $C_0(x) = C_1(x) = \cdots = 0$.
Indeed, assuming that (32) holds, but not all C_n are the null polynomials, for $n \geq 0$, then let $n_0 := \min\{n \geq 0 \mid C_n(x) \neq 0\}$. Let us apply formula (32) to $f(X) := X^{n_0+1}$. We have:

$$\begin{aligned} 0 &= \sum_{n=1}^{\infty} C_{n-1}(X)D^n X^{n_0+1} \\ &= \sum_{n=n_0+1}^{\infty} C_{n-1}(X)D^n X^{n_0+1} \\ &= (n_0+1)!C_{n_0}(X). \end{aligned} \tag{33}$$

Thus, we conclude that $C_{n_0}(X) = 0$, which is a contradiction!
In the same way, we can prove the existence and uniqueness of $\{B_n(x)\}_{n\geq 0}$ in (26). □

We close the paper with an application.

Example 3.1. Find the symmetric random variable X, having finite moments of all orders, that satisfies the following commutator problem:

$$\left[a^-, a^+\right] = I. \tag{34}$$

where I denotes the identity operator.

Solution. Since X is symmetric, its preservation operator $a^0 = 0$ (see Ref. 1). Writing

$$a^- = \sum_{n=1}^{\infty} A_{n-1}(X)D^n \tag{35}$$

as in the previous theorem, we have:

$$\begin{aligned} I &= \left[a^-, a^+\right] \\ &= \left[a^-, a^+ + a^-\right] \\ &= \left[a^-, X\right] \\ &= \left[\sum_{n=1}^{\infty} A_{n-1}(X)D^n, X\right] \\ &= \sum_{n=1}^{\infty} \left[A_{n-1}(X)D^n, X\right]. \end{aligned} \tag{36}$$

Using now Leibniz rules for commutators:

$$[P, Q_1Q_2 \cdots Q_n] = \sum_{i=1}^{n} Q_1 \cdots Q_{i-1} [P, Q_i] Q_{i+1} \cdots Q_n, \tag{37}$$

for any operators P, Q_1, Q_2, $\cdots$, Q_n, we obtain:

$$\begin{aligned} I &= \sum_{n=1}^{\infty} \left[A_{n-1}(X)D^n, X\right] \\ &= \sum_{n=1}^{\infty} \left\{ \left[A_{n-1}(X), X\right] D^n + \sum_{i=1}^{n} A_{n-1}(X)D^{i-1} [D, X] D^{n-i} \right\} \\ &= \sum_{n=1}^{\infty} \left\{ 0 + \sum_{i=1}^{n} A_{n-1}(X)D^{i-1} I D^{n-i} \right\} \\ &= \sum_{n=1}^{\infty} n A_{n-1}(X)D^{n-1}. \end{aligned} \tag{38}$$

Due to the uniqueness of position-momentum decomposition from our main theorem, we conclude that:

$$A_{n-1}(X) = \begin{cases} 1 \text{ if } n = 1 \\ 0 \text{ if } n \geq 2 \end{cases}. \tag{39}$$

Thus, we have:

$$a^- = D. \tag{40}$$

Therefore, if $\langle\ ,\ \rangle$ denotes the inner product of $L^2(\Omega,\ \mathcal{F},\ P)$, since a^+ is the polynomial dual of a^-, then for all natural numbers n, we have:

$$\begin{aligned} E\left[X^{2n}\right] &= \langle X^{2n}1, 1\rangle \\ &= \langle \left(a^+ + a^-\right) X^{2n-1}1, 1\rangle \\ &= \langle X^{2n-1}1, a^-1\rangle + \langle a^- X^{2n-1}1, 1\rangle \\ &= 0 + \langle DX^{2n-1}1, 1\rangle \\ &= (2n-1)\langle X^{2n-2}1, 1\rangle \\ &= (2n-1)E\left[X^{2n-2}\right] \qquad (41) \\ &= (2n-1)(2n-3)E\left[X^{2n-4}\right] \\ &\vdots\ \vdots \\ &= (2n-1)(2n-3)\cdots 1E\left[X^0\right] \\ &= (2n-1)!!. \qquad (42) \end{aligned}$$

Since X is symmetric, $E\left[X^{2n-1}\right] = 0$, for all natural numbers n.
We recognize that these are the moments of a standard normal distribution. Since the moment problem can be uniquely solved for the moments of a standard normal distribution, we conclude that X is Gaussian with mean 0 and variance 1.

References

1. Accardi, L., Kuo, H.-H. and Stan, A. I. 2004, Characterization of probability measures through the canonically associated interacting Fock spaces, *Infin. Dimens. Anal. Quantum Probab. Relat. Top.* **7** No. 4, 485–505.
2. Accardi, L., Kuo, H.-H. and Stan, A. I. 2007, Moments and commutators of probability measures, *Infin. Dimens. Anal. Quantum Probab. Relat. Top.* **10** No. 4, 591–612.
3. Chihara, T. S. 1978, *An Introduction to Orthogonal Polynomials* (New York: Gordon & Breach).
4. Drożdżewicz, K. and Matysiak, W. 2011, Moments and q-commutators of noncommutative random vectors, *Infin. Dimens. Anal. Quantum Probab. Relat. Top.* **14** No. 4, 629–45.
5. Kuo, H.-H. 1996, *White Noise Distribution Theory* (Boca Raton, FL: CRC Press).
6. Meixner, J. Orthogonale 1934, Polynomsysteme mit einer besonderen Gestalt der erzeugenden Funktion, *J. London Math. Soc.* **9** 6–13.
7. Obata, N. 1994, *White Noise Calculus and Fock Space* (Berlin Heidelberg: Springer-Verlag).
8. Popa, G. and Stan, A. I. 2013, Moments of two noncommutative random

variables in terms of their joint quantum operators, *J. Phys.: Conf. Ser.* 474 012027 doi:10.1088/1742-6596/474/1/012027.
9. Popa, G. and Stan, A. I. 2014, Gamma distributed random variables and their semi-quantum operators, Journal of Physics: Conference Series, Volume 563, Number 1, 2014, pp. 12029–12038.
10. Popa, G. and Stan, A. I. 2015, Two-dimensional 1-Meixner random vectors and their semi-quantum operators, Communications on Stochastic Analysis, Vol. 9, No. 4, 425–455.
11. Popa, G. and Stan, A. I. 2017, 2-Meixner random variables and their semi-quantum operators, J. Phys.: Conf. Ser. 819 012002.
12. Stan, A. I. and Whitaker, J. J. 2009, A study of probability measures through commutators, J. Theoret. Probab. **22** No. 1 123–45.
13. Stan, A. I., Whitaker, J. J. 2009, A method of recovering the moments of a probability measure from commutators. In: Quantum Probability and Related Topics: Proceedings of the 28th Conference (Quantum Probability and White Noise Analysis). Edited by Garcia, J.C., Quezada, R., Sontz, S.B.: World Scientific Publishing Company: 242–253.
14. Stan, A. I. 2011, Two-dimensional Meixner random vectors of class $\mathcal{M}_L$, J. Theoret. Probab. **24** No. 1 39–65.
15. Szegö, M. 1975, *Orthogonal Polynomials*, Coll. Publ. **23** Amer. Math. Soc.

Infinite Dimensional Analysis, Quantum Probability and Related Topics

pp. 231–250

HOLOGRAPHIC PHOTOSYNTHESIS AND ENTANGLEMENT ENTROPY*

IRINA AREF'EVA* and IGOR VOLOVICH†

Steklov Mathematical Institute,
Russian Academy of Sciences,
Gubkina str. 8, 119991,
Moscow, Russia
E-mail: *arefeva@mi.ras.ru, †volovich@mi.ras.ru

We apply the holographic approach to photosynthesis that is an important example of nontrivial quantum phenomena relevant for life which is being studied in the emerging field of quantum biology. Light harvesting complexes of photosynthetic organisms are many-body quantum systems, in which quantum coherence has recently been experimentally shown to survive for relatively long time scales even at the physiological temperature despite the decohering effects of their environments. We use the holographic approach to evaluate the time dependence of entanglement entropy and quantum mutual information in the Fenna-Matthews-Olson (FMO) protein-pigment complex in green sulfur bacteria during the transfer of an excitation from a chlorosome antenna to a reaction center. It is demonstrated that the time evolution of the mutual information simulating the master equation in some cases can be obtained by means of a dual gravity describing black hole formation in the AdS-Vaidya spacetime. The wake up and scrambling times for various partitions of the FMO complex are investigated.

1. Introduction

In last years, there has been a growing interest to study the entanglement entropy and quantum mutual information [1,2,3,4,5,6,7] for various quantum systems by using the holographic approach [8,9,10,11,12,13,14,15,16] and refs therein. The entanglement entropy of a boundary region is determined by the area of the minimal bulk surface that coincides with the entangling surface at the boundary [8,9].

In this paper we apply the holographic approach to photosynthesis. Light harvesting complexes (LHC) in bacteria and plants are important

*Irina Aref'eva's work is supported by the Russian Science Foundation, project 17-71-20154.

examples of nontrivial quantum phenomena relevant for life which are being studied in the emerging field of quantum biology. Recent investigations of quantum effects in biology include also the process of vision, the olfactory sense, the magnetic orientation of migrant birds as well as photon antibunching in proteins, the quantum delocalization of biodyes in matter-wave interferometry and quantum tunneling in biomolecules [1,2,17,18,19,20,21,22,23,24,25,26].

Photosynthesis is vital for life on Earth [27,1]. Photosynthesis changes the energy from the sun into chemical energy and splits water to liberate oxygen and convert carbon dioxide into organic compounds, especially sugars. Nearly all life either depends on it directly as a source of energy, or indirectly as the ultimate source of the energy in their food.

Light-harvesting complexes in plants and photosynthetic bacteria include protein scaffolds into which pigment molecules are embedded, e.g. chlorophyll or bacteriochlorophyll molecules. The pigment molecules absorb light and the resulting electronic excitation, exciton, is transported between the pigment molecules until it reaches a reaction center complex, where its energy is converted into separated charges. The process whereby the light energy is transported through the cell is extremely efficient — higher than any artificial energy transport process.

One models many photosynthetic light harvesting complexes by a general three-part structure comprising the antenna, the transfer network, and the reaction center. The antenna captures photons from sunlight and subsequently excites the electrons of the pigment from their ground state. The excited electrons, which combine with holes to make excitons, travel from the antenna through an intermediate protein exciton transfer complex to the reaction center where they participate in the chemical reaction that generates oxygen.

Quantum coherences have been observed in two-dimensional spectroscopic studies of energy transfer within several different light harvesting complexes. The simplest light harvesting complex is the Fenna-Matthews-Olson (FMO) protein-pigment complex in green sulfur bacteria. We demonstrate that some numerical results on the time evolution of the mutual information for the FMO complex [28] can be obtained by using the holographic approach.

Experiments with the FMO complex have shown the presence of quantum beats between excitonic levels at both cryogenic ($77°K$) and ambient ($300°K$) temperatures. Quantum coherence has also been seen in light harvesting antenna complexes of green plants and marine algae. There are also studies of quantum coherence within the reaction center.

Theoretical studies of excitation dynamics in the FMO complex in the single excitation subspace have demonstrated the presence of long-lived, multipartite entanglement. A study of the temporal duration of entanglement in the FMO complex using a simulation of excitation energy transfer dynamics under conditions that approximate the real environment was performed. Entanglement within the Markovian description of the FMO complex has been considered. The time evolution of the mutual information in the FMO complex was studied in [21] by using simulation of the Gorini-Kossakowski-Sudarshan-Lindblad master equation. Quantum nonlocality as a function of time is also studied.

The anti-de Sitter/conformal field theory correspondence [29], AdS/CFT, and more general holographic gravity/gauge duality play an important role in modern theoretical physics. They have been used for description of strong interacting equilibrium and non-equilibrium systems in high energy physics, in particular, to describe heavy-ion collisions and formation of quark-gluon plasma [30,31], as well as in condensed matter physics [32,33].

According to the holographic correspondence the quantum gravity (string theory) on anti-de Sitter spacetime (AdS) is equivalent to a certain quantum field theory on the AdS boundary. The holographic approach provides a powerful method for studying strongly coupled quantum field theories by means of the dual classical gravitational theory in the AdS space which is more tractable.

The paper is organized as follows.

In Sec. 2 we remind the main objects and tools of our study of the FMO complex. In Sec. 2.1 we briefly describe the FMO complex and write down the corresponding master equation. In Sec. 2.1.3 the standard consideration of LHC as quantum system and entropy of entanglement and mutual information for the FMO complex are sketched. We list different reductions of the FMO complex used in modern theoretical studies. In Sec. 2.2 definitions of holographic entanglement entropy and mutual information, as well as the basic formula that we use in the main text are presented. In Sec. 2.3 we present the phase structure of the holographic mutual information $I(A, B)$ for two belts in the static backgrounds, empty AdS_{d+1} and AdS_{d+1} black brane.

Section 3 is devoted to the study of the time evolution of holographic mutual information for the simplest reduction of the FMO complex. We start, Sec. 3.1, by calculation of holographic entanglement entropy for two site system during a quench at nonzero temperature. Then in Sec. 3.2 we study holographic mutual information for two site system at nonzero temperature. In Sec. 3.3 we compare the results of our calculations with the mutual information $I(A,B)$ calculated for the reduced FMO system in [21]. In Sec. 3.4 we discuss the dependencies of wake up time and scrambling times on the geometrical parameters and the initial temperature.

In Sec. 4 we consider the holographic mutual information for the reduction of the FMO complex with one composite part. For this purpose we study the phase structure of holographic entanglement entropy $S(A\cup B\cup C)$ for the Vaidya shell in the AdS_d black brane background. In Sec. 4.1 the mutual information for 3 segments and scrambling time for this background is calculated and in Sec. 4.2 we make a fit of time dependence of the mutual information at the physiological temperature ($300^\circ K$) calculated in [21] for one mixed state by the time dependence of the holographic mutual information $I(A_1\cup A_6, A_3)$ under the global quench by the Vaidya shell in AdS_4. One can find more details in the preprint [25].

2. Setup

2.1. *FMO complex*

2.1.1. *Seven bacteriochlorophills*

The FMO protein complex [28] is the main light-harvesting component of the green sulfur bacteria *Prosthecochloris aestuarii.* It is a trimer, consisting of three identical molecular sub-units, Fig. 1A. Each of the sub-units is a network of seven interconnected bacteriochlorophylls $\{1,2,...,7\}$ arranged in two weakly connected branches that are separately connected to the antenna (bacteriochlorophyll sites one and six) and jointly connected to the reaction center via site three, Fig. 1B.

The theoretical models in the literature study the dynamics of one sub-unit of the trimer. The total Hamiltonian of the system includes the non-relativistic QED Hamiltonian in the dipole approximation, phonon Hamiltonian and other environmental fields, for a derivation of the master equation in the weak coupling stochastic limit see [34] and also [4].

We consider one sub-unit of the trimer which consists of seven bacteriochlorophyll sites $\{1,2,...,7\}$ transferring energetic excitations from a photon-receiving antenna to a reaction center, see Fig. 1.

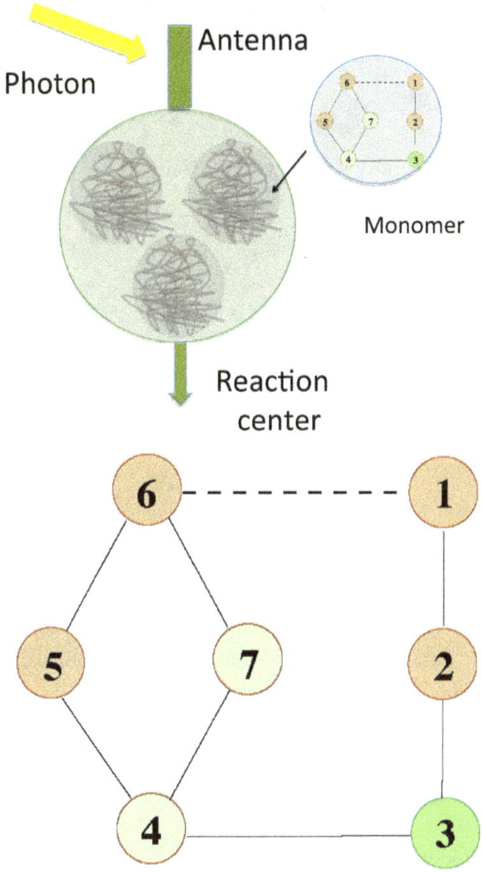

Fig. 1. The schematic picture of the FMO complex. A. The FMO complex trimer. B. The schematic picture for one monomer.

2.1.2. *Master equation*

We consider the nine-state model [20] for excitation transfer in the single excitation approximation which is described by a 9-dimensional subspace in the Hilbert space $(\mathbb{C}^2)^{\otimes 8}$. The possible states for the exciton will be expressed in the site basis $\{|m\rangle\}_{m=1}^{7}$ where the state $|m\rangle$ indicates that the excitation is present at site m. There are also a ground state $|G\rangle$ corresponding to the loss or recombination of the excitation and a sink state $|S\rangle$ corresponding to the trapping of the exciton at the reaction center (there is a discussion of the "local" and "global" bases in theory of open quantum systems in [5]). The density operator for this quantum system has the following representation in the site basis:

$$\rho = \sum_{m,n\in\{G,1,\ldots,7,S\}} \rho_{m,n}|m\rangle\langle n|. \tag{1}$$

It is assumed that the density matrix satisfies the GKS-Lindblad master equation of the following form

$$\frac{d}{dt}\rho = -i[H,\rho] + \mathcal{L}(\rho). \tag{2}$$

Here the Hamiltonian H describes the coupling between the seven sites states $|1\rangle, ..., |7\rangle$:

$$H = \sum_m E_m|m\rangle\langle m| + \sum_{m<n} V_{mn}(|m\rangle\langle n| + |n\rangle\langle m|), \tag{3}$$

where E_m is the energy of the site m and V_{mn} describes the coupling between sites m an n. A Lindblad superoperator $\mathcal{L}(\rho)$ has the general form

$$\mathcal{L}(\rho) = \sum_m \gamma_m(2L_m\rho L_m^\dagger - \{L_m^\dagger L_m, \rho\}), \tag{4}$$

where L_m are arbitrary operators and γ_m are positive constants, see for example [1]. In our case for the FMO complex the Lindblad superoperator is taken in the form [20,21]

$$\mathcal{L}(\rho) = \mathcal{L}_{\text{diss}}(\rho) + \mathcal{L}_{\text{deph}}(\rho) + \mathcal{L}_{\text{sink}}(\rho). \tag{5}$$

Here the first term $\mathcal{L}_{\text{diss}}(\rho)$ describes the dissipative recombination of the exciton,

$$\mathcal{L}_{\text{diss}}(\rho) = \sum_m \Gamma_m(2|G\rangle\langle m|\rho|m\rangle\langle G| - \{|m\rangle\langle m|, \rho\}) \tag{6}$$

where Γ_m is the rate of the recombination at site m.

The second Lindblad superoperator $\mathcal{L}_{\text{deph}}$ accounts for the dephasing interaction with the environment,

$$\mathcal{L}_{\text{deph}}(\rho) = \sum_m \lambda_m(2|m\rangle\langle m|\rho|m\rangle\langle m| - \{|m\rangle\langle m|, \rho\}) \tag{7}$$

where λ_m is the rate of dephasing at site m.

The final term $\mathcal{L}_{\text{sink}}(\rho)$ accounts for the trapping of the exciton in the reaction center:

$$\mathcal{L}_{\text{sink}}(\rho) = \Gamma_{\text{sink}}(2|S\rangle\langle 3|\rho|3\rangle\langle S| - \{|3\rangle\langle 3|, \rho\}). \tag{8}$$

It is supposed that the initial state of the FMO complex is a pure excitation at site one or site six, or a mixture of these two states.

2.1.3. *Entropy of entanglement and mutual information*

Let two parties A and B share a quantum state ρ_{AB} in a Hilbert space $\mathcal{H}_A \otimes \mathcal{H}_B$. The von Neumann entropy of this state is

$$S(AB) = -\mathrm{tr}(\rho_{AB} \log \rho_{AB}). \tag{9}$$

Similarly, one can define the entanglement entropies

$$S(A) = -\mathrm{tr}(\rho_A \log \rho_A), \;\; S(B) = -\mathrm{tr}(\rho_B \log \rho_B), \tag{10}$$

where the reduced density matrices $\rho_A = \mathrm{tr}_B \rho_{AB}$ and $\rho_B = \mathrm{tr}_A \rho_{AB}$.

The quantum mutual information $I(A;B)$ measures the correlations shared between the two parties,

$$I(A;B) = S(A) + S(B) - S(AB). \tag{11}$$

This can be written as a relative entropy and is therefore non-negative:

$$I(A;B) = S(\rho_{AB} || \rho_A \otimes \rho_B) \geq 0 \tag{12}$$

where

$$S(\rho||\sigma) = \mathrm{Tr}(\rho \log \rho) - \mathrm{Tr}(\rho \log \sigma),$$

see for example [1].

Several simulations of the quantum mutual information as a function of time at cryogenic ($77°K$) and physiological ($300°K$) temperatures were conducted in [21]. In particular the simulations calculate the quantum mutual information with respect to several "bipartite cuts" of the sites in the FMO complex. The different cases are as follows, (Figs. 1–6 in [21]):

1. The first cut was picked up where system A consists of site three and system B consists of sites one and six, (Figs. 1 and 2 in [21]). Remind that the initial state of the complex is at site one, six, or the mixture, and the FMO complex transfers the excitation from these initial sites to site three.

2. The next bipartite cut is with the A system consisting of sites one and two and the B system consisting of site three, (Figs. 3 and 4 in [21]).

3. Finally, the third cut was the cut where system A consists of site three and system B consists of all other sites, (Figs. 5 and 6 in [21]). The quantum mutual information for this case should be larger than for the case of the other cuts.

2.2. *Holographic entanglement entropy*

2.2.1. *Static AdS background*

Consider a quantum field theory on a d-dimensional manifold $\mathbb{R} \times \mathbb{R}^{d-1}$, where $\mathbb{R}$ and $\mathbb{R}^{d-1}$ denote the time axis and the $(d-1)$-dimensional space-like manifold, respectively. Let be given a $(d-1)$-dimensional submanifold $A \subset \mathbb{R}^{d-1}$ at fixed time $t = t_0$ and let ∂A be its boundary. Then the formula for the holographic entanglement entropy S_A in a CFT on $\mathbb{R} \times \mathbb{R}^{d-1}$ reads [8]

$$S_A = \frac{A(\gamma_A)}{4G_N^{(d+1)}}, \tag{13}$$

where $A(\gamma_A)$ is the area of γ_A, that is the $d-1$-dimensional static minimal surface in AdS_{d+1} with metric

$$ds^2 = \frac{R^2}{z^2}\left(dz^2 - dt^2 + \sum_{i=1}^{d-1} dx_i^2\right), \tag{14}$$

whose boundary is given by ∂A, and $G_N^{(d+1)}$ is the $d+1$-dimensional Newton constant, R is the radius of AdS_{d+1}.

The simplest example for the shape of A is a straight belt at the boundary $z = 0$,

$$A = \{x_i | x_1 \in [-l, l], x_{2,3,\dots,d-1} \in (-\infty, \infty)\}. \tag{15}$$

In this case the area of the minimal surface, divided on the suitable constants, is [8]

$$\frac{A_{vac,reg}}{R^{d-1}D^{d-2}} = \left(\frac{1}{(d-2)\epsilon^{d-2}} - \frac{c_0}{\ell^{d-2}}\right), \quad d > 2, \tag{16}$$

where D is the length of A in the traversal $x_{2,3,\cdots,d-1}$-direction, $\epsilon > 0$ is the UV regularization, c_0 is a positive constant depending on d. The first term is divergent when $\epsilon \to 0$, but for our purpose it will not play a role, since we will consider the differences of entanglement entropies.

For an excited state whose gravitational dual is provided by the black brane solution with mass m and the Hawking temperature $T_H = dm^{1/d}/4\pi$ (here we assume $R = 1$)

$$ds^2 = \frac{1}{z^2}\left(-f(z)dt^2 + \frac{dz^2}{f(z)} + \sum_{i=1}^{d-1} dx_i^2\right), \quad f(z) = 1 - mz^d, \tag{17}$$

in general, it is not possible to find an explicit expression for the entanglement entropy for $d > 3$. It can be found from the integral equation relating the entropy to an auxiliary parameter z_*

$$\frac{\mathcal{A}_{exc,ren}}{2R^{d-1}D^{d-2}} = \frac{1}{z_*^{d-2}}\int_0^1 \frac{dw}{w^{d-1}}\left[\frac{1}{\sqrt{f(z_*w)\left(1-w^{2(d-1)}\right)}} - 1\right] - \frac{1}{(d-2)z_*^{d-2}}, \tag{18}$$

meanwhile the parameter z_* is related to the width of the belt

$$\frac{\ell}{2} = z_* \int_0^1 \frac{w^{2(d-1)}dw}{\sqrt{f(z_*w)(1-w^{2(d-1)})}}.$$

2.2.2. *Time dependent AdS Vaidya background*

The proposal (13) has been generalized to time dependent geometries in [9]. In the Vaidya AdS_{d+1} spacetime the corresponding minimal surface describes the thermalization process in the d-dimensional boundary theory [11], see also [10,12]. The Vaidya AdS_{d+1} metric has the form

$$ds^2 = \frac{R^2}{z^2}\left(-f(v,z)dv^2 - dvdz + \sum_{i=1}^{d-1} dx_i^2\right), \tag{19}$$

where the function

$$f(v,z) = 1 - m(v)z^d. \tag{20}$$

The form of $m(v)$ is usually chosen to be

$$m(v) = \frac{m}{2}\left(1 + \tanh\frac{v}{\alpha}\right). \tag{21}$$

The metric (19) with this function describes a spacetime model which evolves from pure AdS at early times to the Schwarzschild black brane at late times because of the shell of null dust infalling along $v = 0$. The parameter α determines the thickness of the shell. The case $\alpha \to 0$ corresponds to a step function and one deals with a shock wave.

We will consider also the shell in the AdS_{d+1} black brane background (the Vaidya AdS black brane metric), in this case

$$m(v) = m_0 + \frac{m}{2}\left(1 + \tanh\frac{v}{\alpha}\right). \tag{22}$$

The entanglement entropy is given by the extremum of the functional [11,12]

$$\frac{\mathcal{A}}{2R^{d-1}D^{d-2}} = \int_0^\ell \frac{dx}{z^{d-1}}\sqrt{1 - \left[1 - m(v)z^d\right](v'^2 - 2v'z'}, \tag{23}$$

where $z = z(x), v = v(x)$ and $v^{'} = dv/dx$.

The Euler-Lagrange equations corresponding to the action (23) read

$$\big[1 - m(v)z^d\big]v'' + z'' - \frac{\partial_v m(v)}{2}\, z^d(v'^2 - d\, m(v)z^{d-1}z'v' = 0\,, \quad (24)$$

$$z\, v'' - \frac{d-2}{2}\, m(v)z^d(v'^2 + (d-1)\big[\,(v'^2 + 2v'z' - 1\,\big] = 0\,. \quad (25)$$

For the system of equations (24), (25) we will solve numerically the Cauchy problem with initial data (26), (27)

$$z(0) = z_*, \quad v(0) = v_*, \quad (26)$$

$$z'(0) = v'(0) = 0. \quad (27)$$

We are interested in finding solutions that reach the boundary at some point, that we identify with ℓ in (23) at the boundary time t:

$$z(\ell) = 0, \quad v(\ell) = t. \quad (28)$$

Equations (24), (25) have an integral of motion

$$\left(\frac{z_*}{z}\right)^{2(d-1)} = 1 - \big[\,1 - m(v)z^d\,\big](v'^2 - 2v'z'\,. \quad (29)$$

Due to this identity the integral in (23) can be substantially simplified to give

$$\frac{\mathcal{A}_{reg}}{2R^{d-1}D^{d-2}} = \int_0^{\ell-\epsilon} \frac{z_*^{d-1}}{z^{2(d-1)}}\, dx\ , \quad (30)$$

where we introduce regularization $\epsilon > 0$. This integral contains the UV divergence when $\epsilon \to 0$. The renormalized version of (30) can be written as [11,12]

$$\frac{\mathcal{A}_{ren}}{2R^{d-1}D^{d-2}} = \int_0^{\ell-\epsilon} \frac{z_*^{d-1}}{z^{2(d-1)}}\, dx - \frac{1}{(d-2)\,(z(\ell-\epsilon))^{d-2}}, \quad d > 2, \quad (31)$$

and regularization can be removed. Note, that in the case of $d = 2$ [11,12]

$$\frac{\mathcal{A}_{ren}}{2R} = \int_0^{\ell-\epsilon} \frac{z_*}{z^2}\, dx + z_* \log(z(\ell-\epsilon)). \quad (32)$$

It is quite straightforward to find the dependence of the renormalized entanglement entropy on z_* and v_*. To find the dependence of the entanglement entropy on ℓ and t one has to parametrize the solutions to (24)–(27), not by (z_*, v_*), but by ℓ and t.

It happens that the values of ℓ and t are very sensitive to the initial data (26), and one needs a special numerical procedure to find the pair (z_*, v_*)

corresponding to the given pair ℓ, t with large values of ℓ, t, see [12,16] and refs therein for more details. It is interesting to note that the case when $m_0 > 0$ in (22) is more stable than the case $m_0 = 0$.

2.3. *Holographic mutual information*

The mutual information $I(A; B)$ of two entangling regions A and B is defined by (11):

$$I(A; B) = S(A) + S(B) - S(A \cup B), \tag{33}$$

where $S(A \cup B)$ is the entropy of the union of two regions. Holographic definition of the mutual information $I(A; B)$ just means that we calculate all terms in the RHS of (34) holographically.

To find $S(A \cup B)$ one has to find the minimal surface between two competing ones, "joint" and "disjoint" surfaces, as shown in Fig. 17 in [25]. The winner, in the case of vacuum background, depends only on the ratios of segments to the distance between them. It is obvious that $I(A; B)$ is equal to zero for the case when the minimal surface is realized on the "disjoint" configuration. Generally speaking, $I(A, B)$ undergoes a first order phase transition as one increases the distance between two strips [9,14]

$$I(A; B) = \begin{cases} \mathcal{I}(A; B), & \text{if } \mathcal{I}(A; B) \geq 0 \\ 0 & \text{if } \mathcal{I}(A; B) \leq 0 \end{cases} \tag{34}$$

$$\mathcal{I}(A; B) \equiv S(A) + S(B) - S(A + x + B) - S(x), \tag{35}$$

where $S(A)$ are areas of surfaces entangling the belt (stript) with segment A. In the case of the AdS$_d$ background $\mathcal{I}(A; B)$ is given by

$$\frac{\mathcal{I}(A; B)_{vac}}{2L^{d-2}c_0} = -\frac{1}{\ell_1^{d-2}} - \frac{1}{l_1^{d-2}} + \frac{1}{(\ell_1 + \ell_2 + x)^{d-2}} + \frac{1}{x^{d-2}}, \tag{36}$$

here $d > 2$. The region where the mutual information is nonzero depends on the ratios of parameters $\ell_1/x, \ell_2/x$ in the case of AdS_d, and on the ratios of parameters $\ell_1/x, \ell_2/x$ and z_h/x, $z_h = m^{-1/d}$, in the case of AdS$_d$ black brane.

3. Holographic studies of simplest FMO complex

3.1. *Holographic entanglement entropy for two site system during a quench at nonzero temperature*

The simplest model of LHC is the two systems model. We can consider two parallel infinite strips with non-equal widths $2\ell_1$ and $2\ell_2$ separated by a distance $2x$ in a d-dimensional field theory as a simplest model of LHC and find the mutual information induced by the coming photon. This coming photon we model by the Vaidya shell.

The holographic mutual information for two parallel strips in static backgrounds and in the Vadya AdS has been studied in several papers [12,15]. In particular for equal widths of segment in certain limits the analytical results have been obtained [15]. The specificity of our consideration here is that we want to consider the time dependence of the mutual information during a global quench of a medium that already has high temperature. The corresponding holographical model is given by the Vaidya shell propagating in the AdS_{d+1} black brane background. We also compare the obtained time dependence with the time dependence of the thermalization process started from the zero temperature.

If we are given the two disconnected regions A and B in the boundary, then there are three configurations of hypersurfaces extending in the bulk whose boundaries coincide with $\partial(A \cup B) = \partial A \cup \partial B$: the "disjoint" configuration, the "connected" configuration, given by a bridge connecting A and B through the bulk and the "crossing" configuration, that looks as a configuration with cross-sections in the projection on the (x, z)-plane. The "crossing" configurations do not contribute to the holographic entropy calculations [15,16], in the static AdS and AdS black brane backgrounds. They also do not contribute to the same calculations during the thermalization process described by the Vaidya AdS metric [12]. We have checked numerically that they do not contribute in the case of the Vaidya AdS black brane metric during the transition from the black brane with mass m_0 to the black brane with mass $m_0 + m$.

3.2. *Holographic mutual information for two site system at nonzero temperature*

Let us first consider the holographic mutual information $S(A \cup B)$ for the Vaidya shell in the three dimensional AdS black brane background with $f = f(z, v)$ given by (20) and (22), as function of the boundary time t at fixed l_1, l_2, x. This quantity depends on many parameters, l_1, l_2, x, m, m_0

and α and we performed our analysis numerically.

At the scrambling time any sort of correlations, in particular the exchange of information, is erased. There is also the regime when the mutual information starts from zero at wake up time, t_{wu}, then it increases up to the maximum and when becomes zero at $t = t_{sc}$. In this regime the time dependence has the bell form and it is more interesting for us, see Sec. 3.3. There are also geometrical configurations, when I is zero at all times. These configurations correspond to the cases with large distances x.

3.3. *Matching holographic calculations to the simulation results*

It is obvious that we find the scaling behavior of the holographic mutual information only up to normalization factors. To fit the mutual information for a given physical system there are 3 numerical parameters in our disposal: the $(d+1)$-dimensional gravity constant, the radius of the AdS and the scale, that in our case can be identified with the black brane mass. We compare the results of our calculations with the mutual information $I\,(A;B)$ calculated for the reduced FMO system in [21], where system A consists of site three and system B consists of site one, see Fig. 3a in [21]. We reproduce this curve in Fig. 1 by the purple dashed line. In Fig. 4a of [21] this quantity at physiological temperature (300°K) is presented. We reproduce this curve in Fig. 1 by the red dashed line. We see that in this particular case the purple curve has a maximum approximately twice higher when the red one. As it is shown in Fig. 1 these two curves fit two curves that we take from [21] for particular choices of the geometry of two strips.

It is interesting to note, that the mutual information calculated in [21] for the reduced FMO system, where system A consists of site three and system B consists of site six, or two, see Figs. 1b and 3b in [21], has the two-humped form. We can reproduce this form of the time dependence of the mutual information taking two thin shells Vaidya AdS$_{d+1}$ metric with

$$m(v) = \frac{m_1}{2}\left(1 + \tanh\frac{v}{\alpha}\right) + \frac{m_2}{2}\left(1 + \tanh\frac{v - v_0}{\alpha}\right), \qquad (37)$$

cf.[36].

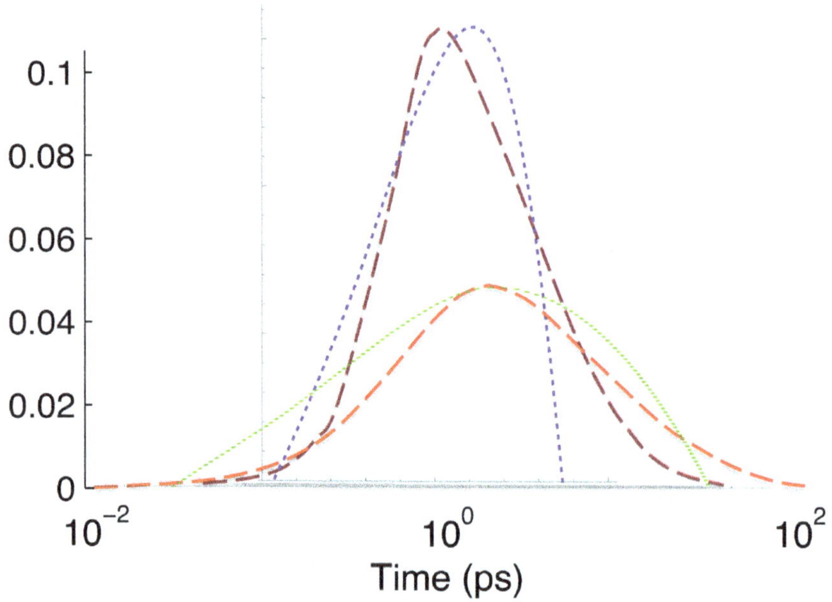

Fig. 2. The purple and red dashed lines show the time dependence of the mutual information of the reduced FMO system calculated in [21] at 70°K and 300°K temperature, respectively. The green and purple solid lines show the rescaled time dependence of the holographical mutual information for the cases of the global quench in AdS_4 and AdS_4 black brane, respectively.

3.4. *Wake up and scrambling times*

As we have seen in Sec. 3.2 there are configurations for which the time dependence of the mutual information has the bell form, see also plots in [25]. We see that for some time the mutual information is equal to zero (or very small) then at the wake up time t_{wu} it starts to increase, reaches the maximum at time t_{max} and then decreases and vanishes at the scrambling time t_{scr}. This behaviour can be explained by a causality argument [11]. It is supposed that before the quench there is a finite correlation length in the system. Therefore after the quench only excitations created at nearest points will be entangled. At later times, such entangled pairs will only contribute to the mutual information between two intervals, if each interval contains one of the two particles.

The scrambling time is the time scale at which all correlations in a system have been destroyed after introducing a perturbation [35]. One defines the scrambling time as the time t_{scr} at which the mutual information of any two subsystems will be vanishing or small. It was conjectured that black holes are fastest scramblers in nature.

Numerical results show that all these time scales depend on the geometry as well as on the form of the quench. By this reason our fitting with results [21] have been performed for different geometries.

4. Holographic studies of the FMO complex with one composite part

In this section we consider holographic description of the quantum mutual information for the reduction of the FMO complex with one composite part.

4.1. *Mutual information for* $(1,6|3)$*-reduction of the FMO complex and scrambling time*

In the section we will consider also the mutual information for a system one part of which consists on two disjoint parts. Just this system corresponds to one of simplest reduced FMO complexes. Applying to this system the general definition (33) we get

$$I(A \cup B; C) = S(A \cup B) + S(C) - S(A \cup B \cup C), \tag{38}$$

where $S(A \cup B \cup C)$ is the entanglement entropy for the union of three subsystems. $I(A \cup B, C)$ is related with the tripartite information which is defined for a system consisting of three disjoint parts as follows

$$\begin{aligned} I_3(A;B;C) = S(A) + S(B) + S(C) - S(A \cup B) - S(A \cup C) \\ - S(B \cup C) + S(A \cup B \cup C). \end{aligned} \tag{39}$$

The tripartite information can be positive, negative or zero, however the holographic tripartite information is always negative, i.e. $I_3(A;B;C) < 0$. The validity of this inequality means that the holographic mutual information is monogamous. The tripartite information can be written in terms of mutual information as follows

$$I_3(A,B,C) = I(A;B) + I(A;C) - I(A;B \cup C). \tag{40}$$

We take $A = A_6$, $B = A_1$, $C = A_3$, where A_i, $i = 1,3,6$ are 1, 3 and 6 sites of the $(1,6|3)$-reduction of the FMO complex. We assume that A_6, A_3, A_1 are three strips of length l_1, l_2 and l_3, and separated by distances x and y.

The time dependence of the holographic mutual information $I(A_6 \cup A_1; A_3)$ for the Vaidya metric in the four dimensional black brane background with $f = f(z, v)$ given by (20) and (22) was studied numerically (for $m_0 = 0.25$, $m = 1$).

We obtain that the evolution of the holographic mutual information for the case when one part of the system is consisting from two disjoints subsystems, globally resembles the case of two simple parts, but locally there are some changes.

More specifically, as in the case of two simple parts, see Sec. 3.2, globally there are 4 different behaviors: the mutual information is 0 at all times; the mutual information starts from a positive value and ends at another positive value; the mutual information starts from a positive value and ends at 0; the mutual information starts from 0, becomes positive for some time and ends at 0 (the bell form of the time dependence).

In the context of the FMO complex study, the last case presents a special interest. Locally, we see phase transitions for some particular configurations.

4.2. *Matching holographic calculations to the simulation results*

In this subsection we fit some results of numerical simulations for he FMO complex presented in [21] by the holographic description. The mutual information at the physiological temperature ($300^\circ K$) calculated in [21], Fig. 3c, when the first system is the site three and the second system is a mixed state of sites one and six is presented in Fig. 3 by the red dashed line.

The time dependence of the holographic mutual information $I(A_6 \cup A_1; A_3)$ under the global quench by the Vaidya shell in AdS_4 is shown by the dark cyan line also in Fig. 3. Note that we make rescaling for this line. We see a rather good fit of these two curves.

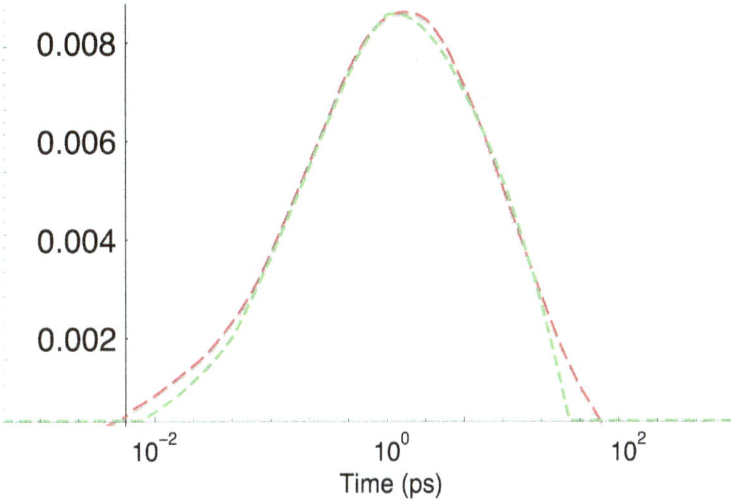

Fig. 3. The dashed red curve shows the time dependence (in the logarithmic scale) of the mutual information at the physiological temperature ($300^{\circ} K$) calculated in [21] Fig. 3c, when the first system is the site ones and the second system is a mixed state of sites one and six. The time dependence of the holographic mutual information $I(A_6 \cup A_1; A_3)$ under the global quench by the Vaidya shell in AdS_4 is shown by the dark cyan line.

5. Conclusions

We have applied the holographic approach to evaluate the time dependence of entanglement entropy and quantum mutual information in the Fenna-Matthews-Olson protein-pigment complex in green sulfur bacteria during the transfer of an excitation from a chlorosome antenna to a reaction center. It is shown that the time evolution of the mutual information simulating the master equation in some cases can be obtained by means of a dual gravity describing black hole formation in the Vaidya AdS spacetime or the Vaidya AdS black brane spacetime. The wake up and scrambling times for various partitions of the FMO complex are discussed. We have demonstrated that some results of numerical simulations [21] for the FMO complex can be fitted by reduced holographic models containing just 2 or 3 segments, Fig. 2 and Fig. 3.

To describe another results known for the FMO complex it would be interesting to study more complex holographic models containing 7 or 8 segments. It would be also suitable in this context to change the geometry of disjoint regions and consider, for example, the disk regions. Considering the global AdS_{d+1} one can also study the compact systems. In the case of the global AdS_3 one can consider the time dependence of the mutual information not only during the global quench, described by the Vaidya metric in BH background [37], but also during the local quench provided by ultra-relativistic particle in AdS_3 [38]. Different configurations have been studied in [39]. Studying Lifshitz backgrounds could be useful for describing not only the FMO complex but also other light-harvesting complexes. We suspect that the holographic approach could provide a useful description of certain appropriate quantities not only for quantum photosynthesis but also for other life science phenomena studied in quantum biology such as the process of vision, the olfactory sense and quantum tunneling in biomolecules.

References

1. Masanori Ohya, Igor Volovich, *Mathematical Foundations of Quantum Information and Computation and Its Applications to Nano- and Bio-systems*, Springer, 2011.
2. *Quantum Bio-Informatics IV*, eds. L. Accardi, W. Freudenberg, and M. Ohya, World scientific, 2011.
3. Noboru Watanabe, "Note on Entropies of Quantum Dynamical Systems", Foundations of Physics 41 (3):549–563 (2011).
4. I. Ya. Aref'eva, I. V. Volovich, S. V. Kozyrev, "Stochastic limit method and interference in quantum many-particle systems", Theoretical and Mathematical Physics, 2015, 183:3, 782–799.
5. A. S. Trushechkin, I. V. Volovich, "Perturbative treatment of inter-site couplings in the local description of open quantum networks", EPL, 113:3 (2016), 30005.
6. Igor V. Volovich, "Cauchy-Schwarz inequality-based criteria for the non-classicality of sub-Poisson and antibunched light", Phys. Lett. A, 380:1-2 (2016), 56–58.
7. Al. N. Pechen, N. B. Il'in, "On the problem of maximizing the transition probability in an n-level quantum system using nonselective measurements", Proc. Steklov Inst. Math., 294 (2016), 233–240.
8. S. Ryu and T. Takayanagi, "Holographic derivation of entanglement entropy from AdS/CFT," Phys. Rev. Lett. **96**, 181602 (2006).
9. V. E. Hubeny, M. Rangamani and T. Takayanagi, "A Covariant holographic entanglement entropy proposal," JHEP **0707**, 062 (2007).
10. T. Albash and C. V. Johnson, "Evolution of Holographic Entanglement Entropy after Thermal and Electromagnetic Quenches," New J. Phys. **13**, 045017 (2011).
11. V. Balasubramanian *et al.*, "Holographic Thermalization," Phys. Rev. D **84**, 026010 (2011).

12. A. Allais and E. Tonni, "Holographic evolution of the mutual information," JHEP **1201**, 102 (2012).
13. R. Callan, J. -Y. He and M. Headrick, "Strong subadditivity and the covariant holographic entanglement entropy formula," JHEP **1206**, 081 (2012).
14. W. Fischler, A. Kundu and S. Kundu, Holographic Mutual Information at Finite Temperature, Phys. Rev. D **87**, no. 12, 126012 (2013).
15. M. Alishahiha, M. R. M. Mozaffar and M. R. Tanhayi, "On the Time Evolution of Holographic n-partite Information," JHEP **1509**, 165 (2015).
16. O. Ben-Ami, D. Carmi and J. Sonnenschein, "Holographic Entanglement Entropy of Multiple Strips," JHEP **1411**, 144 (2014).
17. Graham R. Fleming, Gregory D. Scholes, Yuan-Chung Cheng, "Quantum effects in biology," Procedia Chemistry 3 (2011) 3857.
18. Quantum Effects in Biology, eds. Masoud Mohseni, Yasser Omar, Gregory S. Engel, Martin B. Plenio, Cambridge University Press, 2014.
19. Collini, E.; Wong, C. Y.; Wilk, K. E.; Curmi, P. M. G.; Brumer, P.; Scholes, G. D. "Coherently Wired LightHarvesting in Photosynthetic Marine Algae at Ambient Temperature", Nature, 2010, 463, 644.
20. M. Sarovar, A. Ishizaki, G.R. Fleming, K.B. Whaley, "Quantum entanglement in photosynthetic light harvesting complexes", Nature Physics, 6:462-467, April 2010.
21. Kamil Bradler, Mark M. Wilde, Sai Vinjanampathy, Dmitry B. Uskov, "Identifying the quantum correlations in light-harvesting complexes", Phys. Rev. A 82, 062310 (2010), [arXiv:0912.5112 [quant-ph]].
22. S.V. Kozyrev and I.V. Volovich, "Dark states in quantum photosynthesis," Trends in Biomathematics: Modeling, Optimization and Computational Problems, Proceedings of BIOMAT 2017 Conference, eds. R.P. Mondaini, Springer, Berlin, arXiv:1603.07182 [physics.bio-ph].
23. I. V. Volovich, S. V. Kozyrev, "Manipulation of states of a degenerate quantum system", Proc. Steklov Inst. Math., 294 (2016), 241–251.
24. S. V. Kozyrev, A. A. Mironov, A. E. Teretenkov, I. V. Volovich, "Flows in nonequilibrium quantum systems and quantum photosynthesis", Infin. Dimens. Anal. Quantum Probab. Relat. Top., 20:4 (2017), 1750021, 19 pp.
25. I. Ya. Aref'eva and I. V. Volovich, Holographic Photosynthesis, arXiv:1603.09107.
26. Yuta Mitome, Satoshi Iriyama, Keiko Sato, Igor V. Volovich, Efficient Energy Transfer in Network Model of Photosynthesis, In: Khrennikov A., Toni B. (eds.) Quantum Foundations, Probability and Information. STEAM-H: Science, Technology, Engineering, Agriculture, Mathematics and Health. Springer (2018).
27. R.E. Blankenship, *Molecular Mechanism of Photosynthesis*, (Blackwell Science, London, 2002).
28. R. E. Fenna and B. W. Matthews, "Chlorophyll arrangement in a bacteriochlorophyll protein from chlorobium limicola", Nature, 258, 573–577, December 1975.
29. O. Aharony, S. S. Gubser, J. M. Maldacena, H. Ooguri and Y. Oz, "Large N field theories, string theory and gravity," Phys. Rept. **323**, 183 (2000).

30. I. Ya. Aref'eva, "Holographic approach to quark-gluon plasma in heavy ion collisions," Phys. Usp. **57**, 527 (2014).
31. I. Ya. Aref'eva, "Formation time of quark-gluon plasma in heavy-ion collisions in the holographic shock wave model," Theoret. and Math. Phys., 184:3 (2015), 1239–1255, arXiv:1503.02185 [hep-th].
32. S. A. Hartnoll, C. P. Herzog and G. T. Horowitz, "Holographic Superconductors," JHEP **0812**, 015 (2008) [arXiv:0810.1563 [hep-th]].
33. S. Sachdev, "Condensed Matter and AdS/CFT," Lect. Notes Phys. **828**, 273 (2011).
34. L. Accardi, Yu.G. Lu, and I. Volovich, *Quantum theory and its stochastic limit*, Springer, 2002.
35. N. Sircar, J. Sonnenschein, W. Tangarife, "Extending the scope of holographic mutual information and chaotic behavior", [arXiv:1602.07307 [hep-th]].
36. I. Ya. Aref'eva and I. V. Volovich, On holographic thermalization and dethermalization of quark-gluon plasma; Theor. Math. Phys. 174, 186 (2013), arXiv:1211.6041 [hep-th].
37. D. Ageev, I. Ya. Aref'eva, "Holographic non-equilibrium heating" FF, JHEP, 2018:3 (2018), 103, 19 pp., arXiv: 1704.07747.
38. Irina Ya. Aref'eva, Mikhail A. Khramtsov, Maria D. Tikhanovskaya, "Thermalization after holographic bilocal quench", JHEP, 9 (2017), 115 , 66 pp., arXiv: 1706.07390.
39. I. Y. Aref'eva, I. V. Volovich and O. V. Inozemcev, "Holographic control of information and dynamical topology change for composite open quantum systems," Theor. Math. Phys. **193**, no. 3, 1834 (2017); I. Y. Aref'eva, I. V. Volovich and O. V. Inozemcev, Evolution of holographic entropy quantities for composite quantum systems, Theor. Math. Phys. (2018), to appear.

Infinite Dimensional Analysis, Quantum Probability and Related Topics

pp. 251–271

NOTE ON COMPLEXITY OF COMMUNICATION PROCESSES

NOBORU WATANABE

Department of Information Sciences,
Tokyo University of Science,
Noda City, Chiba 278-8510, Japan
E-mail: watanabe@is.noda.tus.ac.jp

Dedicated to Professor Luigi Accardi, Professor Igor V. Volovich and late Professor Takeyuki Hida.

In order to discuss the efficiency of information transmission of the quantum communication processes consistently, we consider the entropy type functional and the mutual entropy type functional with respect to the initial state and the quantum communication channel. In this study, the mutual entropy type measures are constructed by the compound state between the initial and final systems.

In this paper, we modify the compound state and examine the entropy functional and the mutual entropy functional defined by the modified compound states by means of the initial state and the completely positive channel to study the efficiency of information transmission of the quantum communication processes.

Keywords: Quantum entropy, quantum compound states, quantum communication theory.

1. Introduction

In 1932, von Neumann [4] defined quantum entropy and Umegaki [5] introduced the quantum relative entropy, and Araki [6,7], Uhlmann [8] and Donald [9] extended this relative entropy to general quantum system. In information theory, the efficiency of the communication processes with respect to the initial state ρ and the channel Λ^* is discussed by means of the entropy and the mutual entropy (information), which fulfill the fundamental inequalities such as

$$0 \leq I(\rho; \Lambda^*) \leq \min \{S(\rho), S(\Lambda^* \rho)\}.$$

In the quantum information theory, it is necessary to formulate the mutual entropy of the quantum system by improving joint states (compound states) [10] in quantum compound systems. Ohya introduced the compound state defined by using the von Neumann-Schatten decomposition of the input state and the quantum channel. By applying the Ohya compound state σ_E, the quantum mutual entropy $I(\rho; \Lambda^*)$ [29,11,12,14,15,31,36] was defined by taking the supremum of the quantum relative entropy between the Ohya compound state σ_E and the trivial compound state σ_0 for every von Neumann-Schatten decomposition of the input state ρ, which holds the fundamental inequalities $0 \leq I(\rho; \Lambda^*) \leq S(\rho)$ [11,12]. The entropy and the mutual entropy of C*-systems were introduced by Ohya in [17] and [14], and these properties are discussed in [18,29] and in a book [15].

The entangled state is an important topic for investigating quantum information theory. One of the notable results to discuss the entanglement is the Jamiołkowski's isomorphism [2].

In this paper, we briefly review the construction of the Ohya compound state based on the Jamiołkowski isomorphism, and we introduce the mutual entropy type complexities consisted of the quantum relative entropy between the modified entangled compound states and trivial compound states. The purpose of this paper is to consider the efficiency of information transmission measured by the quantum mutual entropy type complexities formulated by the modified entangled compound states.

2. Quantum Entropy for Density Operators

The entropy of a quantum state was introduced by von Neumann. This entropy of a state ρ is defined by

$$S(\rho) = -\mathrm{tr}\rho \log \rho.$$

For a state ρ, there exists a unique spectral decomposition

$$\rho = \Sigma_k \lambda_k P_k,$$

where λ_k is an eigenvalue of ρ and P_k is the associated projection for each λ_k. The projection P_k is not one-dimensional when λ_k is degenerated, so that the spectral decomposition can be further decomposed into one-dimensional projections. Such a decomposition is called a Schatten decomposition, namely,

$$\rho = \Sigma_{kj} \lambda_{kj} E_{kj},$$

where E_{kj} is the one-dimensional projection associated with λ_k and the degenerated eigenvalue λ_k repeats $\dim P_k$ times; for instance, if the eigenvalue λ_1 has the degeneracy 3, then $\lambda_{11} = \lambda_{12} = \lambda_{13} < \lambda_2$. To simplify notations we shall write the Schatten decomposition as

$$\rho = \Sigma_k \lambda_k E_k,$$

where the numbers $\{\lambda_k\}$ form a probability distribution $\{\lambda_k\}$:

$$\sum_k \lambda_k = 1, \quad \lambda_k \geq 0.$$

This Schatten decomposition is not unique unless every eigenvalue is non-degenerated. Then the entropy (von Neumann entropy) $S(\rho)$ of a state ρ equals to the Shannon entropy of the probability distribution $\{\lambda_k\}$:

$$S(\rho) = -\sum_k \lambda_k \log \lambda_k.$$

Therefore the von Neumann entropy contains the Shannon entropy as a special case.

3. Quantum Channels

Let $(\mathbf{B}(\mathcal{H}_1), \mathfrak{S}(\mathcal{H}_1))$ and $(\mathbf{B}(\mathcal{H}_2), \mathfrak{S}(\mathcal{H}_2))$ be the input and output quantum systems, respectively, where $\mathbf{B}(\mathcal{H}_1)$ (resp. $\mathbf{B}(\mathcal{H}_2)$) is a the set of all bounded operators on a separable complex Hilbert space $\mathcal{H}_1$ (resp. $\mathcal{H}_2$) and $\mathfrak{S}(\mathcal{H}_1)$ (resp. $\mathfrak{S}(\mathcal{H}_2)$ is the set of all density operators on $\mathcal{H}_1$ (resp. $\mathcal{H}_2$). A quantum channel is defined by a mapping Λ^* from the input quantum state space $\mathfrak{S}(\mathcal{H}_1)$ to the output quantum state space $\mathfrak{S}(\mathcal{H}_2)$. A quantum channel Λ^* satisfying an affine property such as

$$\Lambda^*\left(\sum_k \lambda_k \rho_k\right) = \sum_k \lambda_k \Lambda^*(\rho_k)$$

for any $\rho_k \in \mathfrak{S}(\mathcal{H}_1)$ and any $\sum_k \lambda_k = 1$ with $\lambda_k \geq 0 \quad (\forall k)$ is said to be a linear. Λ^* is a completely positive (C.P.) channel [3,29,15,31,36] if its dual map Λ from $\mathbf{B}(\mathcal{H}_2)$ to $\mathbf{B}(\mathcal{H}_1)$ satisfies

$$\sum_{i,j=1}^{n} A_i^* \Lambda(B_i^* B_j) A_j \geq 0$$

for any $n \in \mathbf{N}$, any $B_j \in \mathbf{B}(\mathcal{H}_2)$ and any $A_j \in \mathbf{B}(\mathcal{H}_1)$, where the dual map Λ of Λ^* is given by $tr\Lambda^*(\rho)B = tr\rho\Lambda(B)$ for any $\rho \in \mathfrak{S}(\mathcal{H}_1)$ and any $B \in \mathbf{B}(\mathcal{H}_2)$.

3.1. *Quantum communication processes*

Let $\mathcal{K}_1$ and $\mathcal{K}_2$ be two Hilbert spaces describing noise and loss systems, respectively. Put $\mathbf{B}(\mathcal{K}_1)$ (resp. $\mathbf{B}(\mathcal{K}_2)$) and $\mathfrak{S}(\mathcal{K}_1)$ (resp. $\mathfrak{S}(\mathcal{K}_2)$), where $\mathfrak{S}(\mathcal{K}_1)$ is the set of all density operators on a separable complex Hilbert space $\mathcal{K}_1$ (resp. $\mathcal{K}_2$). A mathematical scheme of quantum communication process including the influence of noise and loss is discussed in [11] as follows: Let ρ and ξ be input and noise states in $\mathfrak{S}(\mathcal{H}_1)$ and $\mathfrak{S}(\mathcal{K}_1)$. γ^* is a CP channel from $\mathfrak{S}(\mathcal{H}_1)$ to $\mathfrak{S}(\mathcal{H}_1 \otimes \mathcal{K}_1)$ given by

$$\gamma^*(\varphi) = \rho \otimes \xi, \quad \rho \in \mathfrak{S}(\mathcal{H}_1)$$

and a^* is a CP channel from $\mathfrak{S}(\mathcal{H}_2 \otimes \mathcal{K}_2)$ to $\mathfrak{S}(\mathcal{H}_2)$ defined by

$$a^*(\sigma) = tr_{\mathcal{K}_2}\sigma, \quad \sigma \in \mathfrak{S}(\mathcal{H}_2 \otimes \mathcal{K}_2).$$

The map π^* is a CP channel from $\mathfrak{S}(\mathcal{H}_1 \otimes \mathcal{K}_1)$ to $\mathfrak{S}(\mathcal{H}_2 \otimes \mathcal{K}_2)$ determined by the physical property of the communication processes. The quantum channel for the communication process is given by

$$\begin{aligned}\Lambda^*(\rho) &\equiv tr_{\mathcal{K}_2}\pi^*(\rho \otimes \xi) \\ &= a^* \circ \pi^* \circ \gamma^*(\rho)\end{aligned}$$

for any $\rho \in \mathfrak{S}(\mathcal{H}_1)$. Some examples of the quantum communication channels are the noisy optical channel, the attenuation channel are given as follows:

3.2. *Noisy quantum channel*

Noisy optical channel Λ^* with a noise state ξ was introduced in [19] such as

$$\begin{aligned}\Lambda^*(\rho) &\equiv tr_{\mathcal{K}_2}\pi^*(\rho \otimes \xi) \\ &= a^* \circ \pi^* \circ \gamma^*(\rho)\end{aligned}$$

where $\xi = |m_1\rangle\langle m_1|$ is the m_1 photon number state in $\mathfrak{S}(\mathcal{K}_1)$ and V is a linear mapping from $\mathcal{H}_1 \otimes \mathcal{K}_1$ to $\mathcal{H}_2 \otimes \mathcal{K}_2$ defined by

$$V\left(|n_1\rangle \otimes |m_1\rangle\right) = \sum_{j=0}^{n_1+m_1} C_j^{n_1,m_1} |j\rangle \otimes |n_1+m_1-j\rangle ,$$

$$C_j^{n_1,m_1} = \sum_{r=L}^{K} (-1)^{n_1+j-r} \frac{\sqrt{n_1!m_1!j!(n_1+m_1-j)!}}{r!(n_1-j)!(j-r)!(m_1-j+r)!} \times \alpha^{m_1-j+2r} \left(-\bar{\beta}\right)^{n_1+j-2r},$$

and $|n_1\rangle$ is the n_1 photon number state vector in $\mathcal{H}_1$, and α, β are complex numbers holding $|\alpha|^2 + |\beta|^2 = 1$, $K = \min\{n_1, j\}$ and $L = \max\{m_1 - j, 0\}$. The following theorem is proved [41].

Theorem 3.1. *The noisy optical channel Λ^* with noise state*

$$\xi = |m_1\rangle \langle m_1| \in \mathfrak{S}(\mathcal{K}_1)$$

is described by

$$\Lambda^*(\rho) = \sum_{i=0}^{\infty} O_i V Q^{(m)} \rho Q^{(m)*} V^* O_i^*, \tag{1}$$

where $Q^{(m)} \equiv \sum_{l=0}^{\infty} \left(|y_l\rangle \otimes |m\rangle\right) \langle y_l|$, $O_i \equiv \sum_{k=0}^{\infty} |z_k\rangle \left(\langle z_k| \otimes \langle i|\right)$, $\{|y_l\rangle\}$ is a CONS in $\mathcal{H}_1$, $\{|z_k\rangle\}$ is a CONS in $\mathcal{H}_2$ and $\{|i\rangle\}$ is the set of number states in $\mathcal{K}_2$.

π^* defined in [19] is called a generalized beam splitting. The noisy optical channel with a vacuum noise is called an attenuation channel defined in [11].

3.3. *Attenuation channel*

The noisy optical channel Λ^* with a vacuum noise state ξ_0 is called the attenuation channel defined in [11] by

$$\Lambda_0^*(\rho) \equiv tr_{\mathcal{K}_2} \pi_0^* \left(\rho \otimes \xi_0\right)$$

where $\xi_0 = |0\rangle\langle 0|$ is the vacuum noise state in $\mathfrak{S}(\mathcal{K}_1)$ and π_0^* is a CP channel from $\mathfrak{S}\left(\mathcal{H}_1 \otimes \mathcal{K}_1\right)$ to $\mathfrak{S}\left(\mathcal{H}_2 \otimes \mathcal{K}_2\right)$. On generalized Fock spaces, the mathematical representation of beam splitting [25] is given by

$$\mathcal{E}_0^*\left(|\xi\rangle\langle\xi|\right) = |\alpha\xi\rangle\langle\alpha\xi| \otimes |\beta\xi\rangle\langle\beta\xi|$$

where $\mathcal{E}_0^*$ is a lifting from $\mathfrak{S}(\mathcal{H})$ to $\mathfrak{S}(\mathcal{H} \otimes \mathcal{K})$ in the sense of Accardi and Ohya [38].

4. Quantum Mutual Entropy

The mutual entropy of a pure quantum system represented by $I(\rho;\Lambda^*)$ depends on the input quantum state ρ, and the quantum channel Λ^* should satisfy the following three conditions.

(1) The quantum mutual entropy of the identity channel $\Lambda^* = id$ is equal to the von Neumann entropy: $I(\rho; id) = S(\rho)$.

(2) Quantum mutual entropy w.r.t. the classical system is reduced to the classical system.

(3) Shannon's basic inequality $0 \leq I(\rho;\Lambda^*) \leq S(\rho)$ is satisfied.

Ohya defined the compound state σ_E of the input state ρ and the output state $\Lambda^*\rho$ instead of the joint state of the classical system as follows.

$$\sigma_E = \sum_k \lambda_k E_k \otimes \Lambda^* E_k,$$

where E denotes a one dimensional orthogonal base $\{E_k\}$ of the Schatten decomposition [13] $\rho = \sum_k \lambda_k E_k$ of ρ, which is not always unique unless every eigenvalue of ρ is not degenerated. The compound state depends on how we distinguish the state ρ into basic states. By using two compound states σ_E and $\sigma_0 = \rho \otimes \Lambda^* \rho$, Ohya defined [11,12] the quantum mutual entropy (information) by

$$I(\rho;\Lambda^*) = \sup\left\{S(\sigma_E, \sigma_0); E = \{E_k\}\right\},$$

where the supremum is taken over all Schatten decompositions of ρ, and $S(\sigma_E, \sigma_0)$ is the Umegaki's relative entropy [5] given by

$$S(\sigma_E, \sigma_0) = \begin{cases} tr\sigma_E (\log \sigma_E - \log \sigma_0) & (s(\sigma_E) \ll s(\sigma_0)), \\ \infty & (\text{else}), \end{cases}$$

$s(\sigma_E) \ll s(\sigma_0)$ means the support projection $s(\sigma_0)$ of σ_0 is greater than the support projection $s(\sigma_E)$ of σ_E. The above conditions (1)–(2) are held for the quantum mutual entropy. Moreover the condition (3) follows from the monotonicity of relative entropy [11]:

Theorem 4.1.

$$0 \leq I(\rho;\Lambda^*) \leq \min\{S(\rho), S(\Lambda^*\rho)\}.$$

For a linear channel, one has the following form [11]:

Theorem 4.2. *The quantum mutual entropy is described by*

$$I(\rho;\Lambda^*) = \sup\left\{\sum_k \lambda_k S(\Lambda^* E_k, \Lambda^* \rho); E = \{E_k\}\right\}.$$

If the input system is classical, Ohya's quantum mutual entropy includes the semiclassical mutual entropies by Levitin [39] and Holevo [16] as a special case.

5. Compound States

Based on [40], we will briefly explain some of the consequences of entangled compound states.

Consider some structures of compound states. Given the initial state ρ and the quantum channel Λ^* (the signal is sent from the initial system to the final system), the compound state Φ should hold the following marginal conditions:

$$
\begin{aligned}
tr_2\Phi &= \rho \quad \text{(marginal condition 1)} \ , \\
tr_1\Phi &= \Lambda^*\rho \quad \text{(marginal condition 2)} \ .
\end{aligned}
$$

The Schatten-von Neumann decomposition of ρ is given by

$$
\rho = \sum_n \lambda_n E_n .
$$

It is not unique unless every eigenvalue are not degenerate. One can consider several compound states with respect to the input state ρ and the quantum channel Λ^* as follows:

$$
\begin{aligned}
\sigma_0 &= \rho \otimes \Lambda^*\rho \quad \text{(trivial compound state)} , \\
\sigma_E &= \sum_n \lambda_n E_n \otimes \Lambda^* E_n \quad \text{(Ohya compound state)} .
\end{aligned}
$$

They are separable compound states. σ_0 does not depend on the correlation between the two marginal systems. σ_E represents a specific correlation between the two compound systems. When the Stinespring-Sudarshan-Kraus form of the quantum channel Λ^* is given by the following equation

$$
\Lambda^*(\rho) = \sum_k V_k \rho V_k^* \quad \text{with} \ \sum_k V_k^* V_k = I,
$$

where V_k is a linear mapping from $\mathcal{H}_1$ to $\mathcal{H}_2$, then we have the following results. For the separable compound states such as

$$
\begin{aligned}
\omega &= \rho \otimes \rho, \\
\omega_E &= \sum_n \lambda_n E_n \otimes E_n,
\end{aligned}
$$

one can construct the compound states σ_0 and σ_E by using the Jamiołkowski isomorphism channel [2]

$$\sigma_0 = \sum_k (I \otimes V_k)(\omega)(I \otimes V_k^*),$$
$$\sigma_E = \sum_k (I \otimes V_k)(\omega_E)(I \otimes V_k^*).$$

The relative entropy between σ_E and σ_0 satisfies

$$S(\sigma_E, \sigma_0) = S(\Lambda^*\rho) - \sum_n \lambda_n S(\Lambda^* E_n, \Lambda^*\rho) \leq S(\rho).$$

For the attenuation channel, we have the following theorems [40].

Theorem 5.1. *For the attenuation channel Λ_0^* and the input state*

$$\rho = \lambda|0\rangle\langle 0| + (1-\lambda)|\theta\rangle\langle\theta|,$$

if $\lambda = \frac{1}{2}$ and α=1, then there exists a compound state Φ satisfying

$$S(\Phi, \rho \otimes \Lambda_0^*\rho) = S(\rho).$$

Let Φ_E be a compound state w.r.t. the initial state ρ, the quantum channel Λ^* and a Schatten decomposition of $\rho = \sum_k \lambda_k E_k$ defined by

$$\Phi_E = \sum_n (id \otimes V_n)\left[\sum_k \sqrt{\lambda_k}(|x_k\rangle \otimes |x_k\rangle)\right]\left[\sum_{k'} \sqrt{\lambda_{k'}}(\langle x_{k'}| \otimes \langle x_{k'}|)\right](id \otimes V_n^*)$$

under the condition

$$\sum_n (id \otimes V_n^*)(id \otimes V_n) = id \otimes id.$$

One has the following theorem [41]

Theorem 5.2. *It holds two marginal conditions*

$$tr_{\mathcal{H}_2}\Phi_E = \rho \quad and \quad tr_{\mathcal{H}_1}\Phi_E = \Lambda^*(\rho).$$

One has

$$S(\Phi_E, \rho \otimes \Lambda^*\rho) \leq 2S(\rho).$$

In [41], we introduced the construction of the hybrid entangled compound state $\Phi_E^{(\Delta)}$. However, they were not sufficient as the representation of the compound states. Therefore we here modify the compound states by means of the initial and final systems.

Let $\mathcal{A}$ be a partition of the total index set Q with respect to a Schatten-von Neumann decomposition of the initial state ρ given by

$$\begin{aligned}
\mathcal{A} &= \left\{ A_j \subset Q;\quad \#(A_j) \neq \varnothing \quad (j \in J), \right. \\
&\qquad \left. A_i \cap A_j = \varnothing \ (i \neq j), \quad Q = \bigcup_{j \in J} A_j \right\} \\
\mathcal{A}_\Delta &= \left\{ A_j \in \mathcal{A};\quad \#(A_j) \geq 2 \quad (j \in J) \right\}, \\
\mathcal{A}_1 &= \left\{ A_j \in \mathcal{A};\quad \#(A_j) = 1 \quad (j \in J) \right\}, \\
\mathcal{A} &= \mathcal{A}_1 \cup \mathcal{A}_\Delta, \\
J_\ell &= \left\{ n_i \in J;\quad \sharp(A_{n_i}) = \ell\, (\geq 2) \right\} \\
J_1 &= \left\{ n_i \in J;\quad \sharp(A_{n_i}) = 1 \right\} \\
\Delta_\ell &= \bigcup_{n_i \in J_\ell} A_{n_i}, \quad \Delta = \bigcup_{\ell=2} \Delta_\ell, \quad Q \diagdown \Delta = \bigcup_{n_i \in J_1} A_{n_i} \\
\mathcal{A}_{\Delta_\ell} &= \left\{ A_j \in \mathcal{A};\quad \#(A_j) = \ell \quad (j \in J) \right\} \subset \mathcal{A}_\Delta \\
\mathcal{A}_\Delta &= \bigcup_{\ell=2} \mathcal{A}_{\Delta_\ell}.
\end{aligned}$$

We denote the set of all partitions of the total index set Q by $\mathcal{P}(Q)$ such as

$$\begin{aligned}
\mathcal{P}(Q) = \Big\{ &\mathcal{A}; \mathcal{A} = \mathcal{A}_1 \cup \mathcal{A}_\Delta, \forall A_i, A_j\, (\neq \varnothing) \in \mathcal{A} \ (i \neq j), \\
&A_i \cap A_j = \varnothing,\ Q = \bigcup_{j \in J} A_j = \bigcup_{\ell=2} \left(\bigcup_{n_i \in J_\ell} A_{n_i} \right) \cup \left(\bigcup_{n_i \in J_1} A_{n_i} \right) \Big\}.
\end{aligned}$$

Now we introduce modified compound states $\Phi_E^{(\mathcal{A})}$ by means of the partition $\mathcal{A}$ of the total index set Q defined by

$$\begin{aligned}
&\Phi_E^{(\mathcal{A})} \\
&= \sum_{\ell=2} \sum_{n_\ell \in J_\ell} \left(\sum_{n_i \in A_{n_\ell}} \sqrt{\lambda_{n_i}} \left| x_{n_i} \right\rangle \otimes \left| x_{n_i} \right\rangle \right) \left(\sum_{n_j \in A_{n_\ell}} \sqrt{\lambda_{n_j}} \left\langle x_{n_j} \right| \otimes \left\langle x_{n_j} \right| \right) \\
&\quad + \sum_{n_k \in Q \diagdown \Delta} \lambda_{n_k} E_{n_k} \otimes E_{n_k}.
\end{aligned}$$

Let us consider the completely positive channel Λ_t^* given by $\Lambda_t^*(\rho) = V_t\rho V_t^*$ for any $\rho \in \mathfrak{S}(\mathcal{H}_1)$, $t \geq 0$ with $V_t^* V_t = I$ and $\lim_{t\to 0} V_t = \lim_{t\to 0} V_t^* = I$.

By using the Jamiołkowski isomorphic channel, we defined (1)–(3) compound states in [41], and we here introduce a modified compound state (4):

(4) The modified compound state $\Psi_{E,\Lambda_t^*}^{(\mathcal{A})}$ concerned with the partitions $\mathcal{A}$ of the total index set Q with respect to the Schatten decomposition of the initial state ρ and the completely positive channel Λ_t^* is defined by

$$
\begin{aligned}
&\Psi_{E,\Lambda_t^*}^{(\mathcal{A})} \\
&= (id \otimes V_t)\,\Phi_E^{(\mathcal{A})}\,(id \otimes V_t^*) \\
&= \sum_{\ell=2} \sum_{n_\ell \in J_\ell} \left(\sum_{n_i \in A_{n_\ell}} \sqrt{\lambda_{n_i}}\, |x_{n_i}\rangle \otimes V_t\, |x_{n_i}\rangle \right) \\
&\quad \times \left(\sum_{n_j \in A_{n_\ell}} \sqrt{\lambda_{n_j}}\, \langle x_{n_j}| \otimes \langle x_{n_j}|\, V_t^* \right) \\
&\quad + \sum_{n_k \in Q \setminus \Delta} \lambda_{n_k} E_{n_k} \otimes \Lambda_t^*(E_{n_k}).
\end{aligned}
$$

Since the modified compound state $\Psi_{E,\Lambda_t^*}^{(\mathcal{A})}$ with the partition $\mathcal{A} = \mathcal{A}_1 \cup \mathcal{A}_\Delta$ satisfying $\Delta = Q \setminus \bigcup_{n_i \in J_1} A_{n_i}$ is the hybrid compound state $\Psi_{E,\Lambda_t^*}^{(\Delta)}$, it means that the hybrid compound state $\Psi_{E,\Lambda_t^*}^{(\Delta)}$ is a special case of the modified compound state $\Psi_{E,\Lambda_t^*}^{(\mathcal{A})}$ We define the mutual entropy type measure as follows: For a Schatten decomposition $\sum_{n\in Q} \lambda_n E_n$ of the initial state ρ, let $\Psi_{E,\Lambda_t^*}^{(\mathcal{A})}$ be a modified compound state with respect to a partition $\mathcal{A}$ of the total index set Q and the CP channel $\Lambda_t^*(\rho) = V_t\rho V_t^*$ for any $\rho \in \mathfrak{S}(\mathcal{H}_1)$, $t \geq 0$ with $V_t^* V_t = I$ and $\lim_{t\to 0} V_t = \lim_{t\to 0} V_t^* = I$. The mutual entropy type measure $I^{(\mathcal{A})}(\rho; \Lambda_t^*)$ with respect to a partition $\mathcal{A}$ of Q and the CP channel Λ_t^* is defined by taking the supremum of the relative entropy between $\Psi_{E,\Lambda_t^*}^{(\mathcal{A})}$ and $\rho \otimes \Lambda_t^* \rho$ for all Schatten decomposition $E = \{E_n\}$ of the initial state ρ

$$
I^{(\mathcal{A})}(\rho; \Lambda_t^*) = \sup \left\{ S\left(\Psi_{E,\Lambda_t^*}^{(\mathcal{A})}, \rho \otimes \Lambda_t^* \rho \right); \quad E = \{E_n\} \right\}.
$$

Theorem 5.3. *For a Schatten decomposition $\sum_{n\in Q} \lambda_n E_n$ of the initial state ρ, let $\Psi_{E,\Lambda_t^*}^{(\mathcal{A})}$ be an entangled compound state with respect to a partition $\mathcal{A}$ of the total index set Q and the CP channel $\Lambda_t^*(\rho) = V_t\rho V_t^*$ for any*

$\rho \in \mathfrak{S}(\mathcal{H}_1)$, $t \geq 0$ with $V_t^* V_t = I$ and $\lim_{t\to 0} V_t = \lim_{t\to 0} V_t^* = I$. It holds two marginal conditions

$$tr_{\mathcal{H}_2} \Psi^{(\mathcal{A})}_{E,\Lambda_t^*} = \rho \quad \text{and} \quad tr_{\mathcal{H}_1} \Psi^{(\mathcal{A})}_{E,\Lambda_t^*} = \Lambda_t^*(\rho)$$

and the relative entropy between $\Psi^{(\mathcal{A})}_{E,\Lambda_t^*}$ and $\sigma_0 = \rho \otimes \Lambda_t^* \rho$ satisfies the following inequality:

$$I^{(\mathcal{A})}(\rho; \Lambda_t^*) \geq S\left(\Psi^{(\mathcal{A})}_{E,\Lambda_t^*}, \rho \otimes \Lambda_t^*(\rho)\right) \geq S(\Lambda_t^* \rho).$$

Proof. Since

$$\Psi^{(\mathcal{A})}_{E,\Lambda_t^*} = \sum_{\ell=2} \sum_{n_\ell \in J_\ell} \left\| \phi_{A_{n_\ell}} \right\|^2 \frac{\left|\phi_{A_{n_\ell}}\right\rangle}{\left\|\phi_{A_{n_\ell}}\right\|} \frac{\left\langle\phi_{A_{n_\ell}}\right|}{\left\|\phi_{A_{n_\ell}}\right\|} + \sum_{n_k \in Q \backslash \Delta} \lambda_{n_k} E_{n_k} \otimes V_t E_{n_k} V_t^*$$

is held, then one has

$$tr \Psi^{(\mathcal{A})}_{E,\Lambda_t^*} \log \Psi^{(\mathcal{A})}_{E,\Lambda_t^*} = \sum_{\ell=2} \sum_{n_\ell \in J_\ell} \left(\sum_{n_i \in A_{n_\ell}} \lambda_{n_i} \right) \log \left(\sum_{n_i \in A_{n_\ell}} \lambda_{n_i} \right) + \sum_{n_k \in Q \backslash \Delta} \lambda_{n_k} \log \lambda_{n_k}$$

under the condition

$$(I \otimes V_t^*)(I \otimes V_t) = I \otimes I \quad \text{and} \quad \Lambda_t^*(\rho) = V_t \rho V_t^*,$$

where $\left|\phi_{A_{n_\ell}}\right\rangle$ is

$$\left|\phi_{A_{n_\ell}}\right\rangle = \sum_{n_i \in A_{n_\ell}} \sqrt{\lambda_{n_i}} \left|x_{n_i}\right\rangle \otimes \left|x_{n_i}\right\rangle.$$

For any $x, x' \in \mathcal{H}_1$, one has

$$\begin{aligned} &\left\langle x, tr_{\mathcal{H}_2} \Psi^{(\mathcal{A})}_{E,\Lambda_t^*} x' \right\rangle \\ &= \left\langle x, \left(\sum_{\ell=2} \sum_{n_\ell \in J_\ell} \sum_{n_i \in A_{n_\ell}} \lambda_{n_i} E_{n_i} + \sum_{n_k \in Q \backslash \Delta} \lambda_{n_k} E_{n_k} \right) x' \right\rangle \\ &= \langle x, \rho x' \rangle \end{aligned}$$

and for any $y, y' \in \mathcal{H}_2$, one obtains

$$\begin{aligned}
&\left\langle y, tr_{\mathcal{H}_1} \Psi_{E,\Lambda_t^*}^{(\mathcal{A})} y' \right\rangle \\
&= \left\langle y, V_t \left(\sum_{\ell=2} \sum_{n_\ell \in J_\ell} \sum_{n_i \in A_{n_\ell}} \lambda_{n_i} E_{n_i} + \sum_{n_k \in Q \backslash \Delta} \lambda_{n_k} E_{n_k} \right) V_t^* y' \right\rangle \\
&= \langle y, V_t \rho V_t^* y' \rangle \\
&= \langle y, \Lambda_t^* (\rho) y' \rangle .
\end{aligned}$$

Then $\Psi_{E,\Lambda_t^*}^{(\mathcal{A})}$ holds two marginal conditions

$$tr_{\mathcal{H}_2} \Psi_{E,\Lambda_t^*}^{(\mathcal{A})} = \rho \quad \text{and} \quad tr_{\mathcal{H}_1} \Psi_{E,\Lambda_t^*}^{(\mathcal{A})} = \Lambda_t^* (\rho)$$

and the relative entropy between $\Psi_{E,\Lambda_t^*}^{(\mathcal{A})}$ and $\sigma_0 = \rho \otimes \Lambda_t^* \rho$ is obtained by

$$\begin{aligned}
&S \left(\Psi_{E,\Lambda_t^*}^{(\mathcal{A})}, \rho \otimes \Lambda_t^* \rho \right) \\
&= - \sum_{n_k \in \Delta} \lambda_{n_k} \log \lambda_{n_k} + S (\Lambda_t^* \rho) \\
&\quad + \sum_{\ell=2} \sum_{n_\ell \in J_\ell} \left(\sum_{n_i \in A_{n_\ell}} \lambda_{n_i} \right) \log \left(\sum_{n_i \in A_{n_\ell}} \lambda_{n_i} \right) .
\end{aligned}$$

Since

$$\sum_{\ell=2} \sum_{n_\ell \in J_\ell} \left(\sum_{n_i \in A_{n_\ell}} \lambda_{n_i} \right) = \sum_{n_k \in \Delta} \lambda_{n_k}$$

is held, then we have

$$\begin{aligned}
&S \left(\Psi_{E,\Lambda_t^*}^{(\mathcal{A})}, \rho \otimes \Lambda_t^* \rho \right) \\
&= - \sum_{\ell=2} \sum_{n_\ell \in J_\ell} \left(\sum_{n_i \in A_{n_\ell}} \lambda_{n_i} \right) \log \frac{\lambda_{n_i}}{\left(\sum_{n_k \in A_{n_\ell}} \lambda_{n_k} \right)} + S (\Lambda_t^* \rho) \\
&\geq S (\Lambda_t^* \rho) .
\end{aligned}$$

Therefore we get the following inequality:

$$I^{(\mathcal{A})} (\rho; \Lambda_t^*) \geq S \left(\Psi_{E,\Lambda_t^*}^{(\mathcal{A})}, \rho \otimes \Lambda_t^* (\rho) \right) \geq S (\Lambda_t^* \rho) . \qquad \square$$

It shows that the mutual entropy at time t defined by using the entangled compound state greater than the von Neumann entropy $S (\Lambda_t^* \rho)$ of the final state $\Lambda_t^* \rho$. When $t \to 0$ is held, one has the following inequality

$$I^{(\mathcal{A})}(\rho; id) \geq \lim_{t \to 0} S\left(\Psi_{E,\Lambda_t^*}^{(\mathcal{A})}, \rho \otimes \Lambda_t^* \rho\right) \geq S(\rho).$$

It means that the mutual entropy type measure defined by using the entangled compound state at initial time $t = 0$ greater than the von Neumann entropy $S(\rho)$ of the initial state ρ.

Let Λ^* be a completely positive channels Λ^* given by

$$\Lambda^*(\rho) = \sum_n V_n(\rho) V_n^*$$

satisfying

$$\sum_n V_n^* V_n = I.$$

We introduced (1)–(3) compound states in [41] and here we define a modified compound state (4):

(4) The modified compound state $\Psi_{E,\Lambda^*}^{(\mathcal{A})}$ concerned with the partitions $\mathcal{A}$ of the total index set Q with respect to the Schatten decomposition of the initial state ρ and the completely positive channel Λ^* is defined by

$$\begin{aligned}
\Psi_{E,\Lambda^*}^{(\mathcal{A})} &= \sum_n (id \otimes V_n) \Phi_E^{(\mathcal{A})} (id \otimes V_n^*) \\
&= \sum_n \sum_{\ell=2} \sum_{n_\ell \in J_\ell} \left(\sum_{n_i \in A_{n_\ell}} \sqrt{\lambda_{n_i}} \left| x_{n_i} \right\rangle \otimes V_n \left| x_{n_i} \right\rangle \right) \\
&\quad \times \left(\sum_{n_j \in A_{n_\ell}} \sqrt{\lambda_{n_j}} \left\langle x_{n_j} \right| \otimes \left\langle x_{n_j} \right| V_n^* \right) \\
&\quad + \sum_{n_k \in Q \setminus \Delta} \lambda_{n_k} E_{n_k} \otimes \Lambda^*(E_{n_k}).
\end{aligned}$$

Here we define the mutual entropy type measure as follows: For a Schatten decomposition $\sum_{n \in Q} \lambda_n E_n$ of the initial state ρ, let $\Psi_{E,\Lambda^*}^{(\mathcal{A})}$ be an entangled compound state with respect to a partition $\mathcal{A}$ of the total index set Q and the CP channel $\Lambda^*(\rho) = \sum_n V_n(\rho) V_n^*$ for any $\rho \in \mathfrak{S}(\mathcal{H}_1)$ with $\sum_n V_n^* V_n = I$ and $V_m^* V_n = \delta_{mn} V_n^* V_n$. The mutual entropy type measure $I^{(\mathcal{A})}(\rho; \Lambda^*)$ with respect to a partition $\mathcal{A}$ of the total index set Q and the CP channel Λ^* is defined by taking the supremum of the relative entropy between $\Psi_{E,\Lambda^*}^{(\mathcal{A})}$ and $\rho \otimes \Lambda^* \rho$ for all Schatten decomposition $E = \{E_n\}$ of the initial state ρ

$$I^{(\mathcal{A})}(\rho; \Lambda^*) = \sup\left\{ S\left(\Psi_{E,\Lambda^*}^{(\mathcal{A})}, \rho \otimes \Lambda^* \rho\right); \quad E = \{E_n\} \right\}.$$

Theorem 5.4. *For a Schatten decomposition $\sum_{n\in Q}\lambda_n E_n$ of the initial state ρ, let $\Psi_{E,\Lambda^*}^{(\mathcal{A})}$ be an entangled compound state with respect to a partition $\mathcal{A}$ of the total index set Q and the CP channel $\Lambda^*(\rho)=\sum_n V_n(\rho)V_n^*$ for any $\rho\in\mathfrak{S}(\mathcal{H}_1)$ with $\sum_n V_n^*V_n=I$ and $V_m^*V_n=\delta_{mn}V_n^*V_n$. It satisfies two marginal conditions*

$$tr_{\mathcal{H}_2}\Psi_{E,\Lambda^*}^{(\mathcal{A})}=\rho \quad \text{and} \quad tr_{\mathcal{H}_1}\Psi_{E,\Lambda^*}^{(\mathcal{A})}=\Lambda^*(\rho).$$

The mutual entropy type measure $I^{(\mathcal{A})}(\rho;\Lambda^*)$ with respect to the relative entropy between $\Psi_{E,\Lambda^*}^{(\mathcal{A})}$ and $\sigma_0=\rho\otimes\Lambda^*\rho$ holds the following inequality:

$$I^{(\mathcal{A})}(\rho;\Lambda^*)\geq I(\rho;\Lambda^*),$$

where $I(\rho;\Lambda^*)$ in the right-hand side is the Ohya mutual entropy.

Proof. One has

$$\begin{aligned}
&\Psi_{E,\Lambda_t^*}^{(\mathcal{A})}\\
&=\sum_n\sum_{\ell=2}\sum_{n_\ell\in J_\ell}\left\|\left|\phi_n^{(A_{n_\ell})}\right\rangle\right\|^2\frac{\left|\phi_n^{(A_{n_\ell})}\right\rangle}{\left\|\left|\phi_n^{(A_{n_\ell})}\right\rangle\right\|}\frac{\left\langle\phi_n^{(A_{n_\ell})}\right|}{\left\|\left|\phi_n^{(A_{n_\ell})}\right\rangle\right\|}\\
&\quad+\sum_n\sum_{n_k\in Q\setminus\Delta}\lambda_{n_k}\left\|V_n\left|x_{n_k}\right\rangle\right\|^2E_{n_k}\otimes\frac{V_n\left|x_{n_k}\right\rangle}{\left\|V_n\left|x_{n_k}\right\rangle\right\|}\frac{\left\langle x_{n_k}\right|V_n^*}{\left\|V_n\left|x_{n_k}\right\rangle\right\|},
\end{aligned}$$

where

$$\left|\phi_n^{(A_{n_\ell})}\right\rangle=\sum_{n_i\in A_{n_\ell}}\sqrt{\lambda_{n_i}}\left|x_{n_i}\right\rangle\otimes V_n\left|x_{n_i}\right\rangle,$$

$$\left\|\left|\phi_n^{(A_{n_\ell})}\right\rangle\right\|^2=\sum_{n_k\in A_{n_\ell}}\lambda_{n_k}\left\|V_n\left|x_{n_k}\right\rangle\right\|^2.$$

Since

$$\begin{aligned}
&tr\Psi_{E,\Lambda^*}^{(\mathcal{A})}\log\Psi_{E,\Lambda^*}^{(\mathcal{A})}\\
&=\sum_n\sum_{\ell=2}\sum_{n_\ell\in J_\ell}\left(\sum_{n_i\in A_{n_\ell}}\lambda_{n_i}\left\|V_n\left|x_{n_i}\right\rangle\right\|^2\right)\log\left(\sum_{n_j\in\Delta}\lambda_{n_j}\left\|V_n\left|x_{n_j}\right\rangle\right\|^2\right)\\
&\quad+\sum_n\sum_{n_k\in Q\setminus\Delta}\lambda_{n_k}\left\|V_n\left|x_{n_k}\right\rangle\right\|^2\log\left\|V_n\left|x_{n_k}\right\rangle\right\|^2\\
&\quad+\sum_{n_k\in Q\setminus\Delta}\lambda_{n_k}\log\lambda_{n_k},
\end{aligned}$$

under the condition

$$\sum_n (I \otimes V_n^*)(I \otimes V_n) = I \otimes \left(\sum_n V_n^* V_n \right) = I \otimes I$$

$$\text{and} \quad \Lambda^*(\rho) = \sum_n V_n \rho V_n^* \text{ for any } \rho \in \mathfrak{S}(\mathcal{H}_1)$$

$$\text{with} \sum_n V_n^* V_n = I \quad \text{and} \quad V_m^* V_n = \delta_{mn} V_n^* V_n.$$

For any $x, x' \in \mathcal{H}_1$, one has

$$\begin{aligned} &\left\langle x, tr_{\mathcal{H}_2} \Psi_{E,\Lambda^*}^{(\mathcal{A})} x' \right\rangle \\ &= \left\langle x, \left(\sum_{\ell=2} \sum_{n_\ell \in J_\ell} \sum_{n_i \in A_{n_\ell}} \lambda_{n_i} E_{n_i} + \sum_{n_k \in Q \setminus \Delta} \lambda_{n_k} E_{n_k} \right) x' \right\rangle \\ &= \langle x, \rho x' \rangle \end{aligned}$$

and for any $y, y' \in \mathcal{H}_2$, one obtains

$$\begin{aligned} &\left\langle y, tr_{\mathcal{H}_1} \Psi_{E,\Lambda^*}^{(\mathcal{A})} y' \right\rangle \\ &\quad + \sum_n \sum_i \left\langle x_i \otimes y, \sum_{n_k \in Q \setminus \Delta} \lambda_{n_k} E_{n_k} \otimes V_n E_{n_k} V_n^* x_i \otimes y' \right\rangle \\ &= \left\langle y, \left(\sum_n V_n \rho V_n^* \right) y' \right\rangle \\ &= \langle y, \Lambda^*(\rho) y' \rangle . \end{aligned}$$

Therefore $\Psi_{E,\Lambda^*}^{(\mathcal{A})}$ holds two marginal conditions

$$tr_{\mathcal{H}_2} \Psi_{E,\Lambda^*}^{(\mathcal{A})} = \rho \quad \text{and} \quad tr_{\mathcal{H}_1} \Psi_{E,\Lambda^*}^{(\mathcal{A})} = \Lambda^*(\rho).$$

$$\begin{aligned} &-tr \Psi_{E,\Lambda^*}^{(\mathcal{A})} \log \rho \otimes \Lambda^* \rho \\ &= S(\rho) + S(\Lambda^* \rho). \end{aligned}$$

Since

$$\begin{aligned} &tr \Psi_{E,\Lambda^*}^{(\mathcal{A})} \log \Psi_{E,\Lambda^*}^{(\mathcal{A})} \\ &= \sum_n \sum_{\ell=2} \sum_{n_\ell \in J_\ell} \left(\sum_{n_i \in A_{n_\ell}} \lambda_{n_i} \left\| V_n \left| x_{n_i} \right\rangle \right\|^2 \right) \log \left(\sum_{n_j \in A_{n_\ell}} \lambda_{n_j} \left\| V_n \left| x_{n_j} \right\rangle \right\|^2 \right) \\ &\quad + \sum_n \sum_{n_k \in Q \setminus \Delta} \lambda_{n_k} \left\| V_n \left| x_{n_k} \right\rangle \right\|^2 \log \lambda_{n_k} \left\| V_n \left| x_{n_k} \right\rangle \right\|^2 \end{aligned}$$

is held, then the relative entropy between $\Psi_{E,\Lambda^*}^{(\mathcal{A})}$ and $\sigma_0 = \rho \otimes \Lambda^* \rho$ is obtained by

$$
\begin{aligned}
& S\left(\Psi_{E,\Lambda^*}^{(\mathcal{A})}, \rho \otimes \Lambda^* \rho\right) \\
&= S\left(\Lambda^* \rho\right) - \sum_n \sum_{n_k \in Q} \lambda_{n_k} S\left(\Lambda^* E_{n_k}\right) \\
& \quad - \sum_n \sum_{\ell=2} \sum_{n_i \in A_{n_\ell}} \lambda_{n_i} \left\| V_n \left| x_{n_i} \right\rangle \right\|^2 \log \frac{\lambda_{n_i} \left\| V_n \left| x_{n_i} \right\rangle \right\|^2}{\left(\sum_{n_j \in A_{n_\ell}} \lambda_{n_j} \left\| V_n \left| x_{n_j} \right\rangle \right\|^2 \right)}.
\end{aligned}
$$

Thus we have the inequality

$$I^{(\mathcal{A})}\left(\rho; \Lambda^*\right) \geq I\left(\rho; \Lambda^*\right).$$

□

If $\Lambda^* = id$ is held, then we obtain the following inequality:

$$I^{(\mathcal{A})}\left(\rho; \Lambda^*\right) \geq I\left(\rho; \Lambda^*\right) = S\left(\rho\right).$$

It shows that the mutual entropy defined by using the entangled compound state $\Psi_{E,\Lambda^*}^{(\mathcal{A})}$ with respect to a partition $\mathcal{A}$ of the total index set Q, the initial state ρ and the quantum channel $\Lambda^* = id$ greater than the von Neumann entropy $S(\rho)$ of the initial state ρ.

If the above completely positive channel Λ^* has orthogonality (i.e., $E_n \perp E_m \, (n \neq m) \Rightarrow \Lambda^*(E_n) \perp \Lambda^*(E_m)$) then we have the following theorem.

Theorem 5.5. *For a Schatten decomposition of the initial state ρ, let $\Psi_{E,\Lambda^*}^{(\Delta)}$ be an entangled compound state with respect to a partition $\mathcal{A}$ of the total index set Q and the CP channel $\Lambda^*(\rho) = \sum_n V_n(\rho) V_n^*$ for any $\rho \in \mathfrak{S}(\mathcal{H}_1)$ with $\sum_n V_n^* V_n = I$ and $V_m^* V_n = \delta_{mn} V_n^* V_n$ and orthogonality (i.e., $E_n \perp E_m \, (n \neq m) \Rightarrow \Lambda^*(E_n) \perp \Lambda^*(E_m)$). It satisfies two marginal conditions*

$$tr_{\mathcal{H}_2} \Psi_{E,\Lambda^*}^{(\mathcal{A})} = \rho \quad \text{and} \quad tr_{\mathcal{H}_1} \Psi_{E,\Lambda^*}^{(\mathcal{A})} = \Lambda^*(\rho).$$

The following inequality is held:

$$I^{(\mathcal{A})}\left(\rho; \Lambda^*\right) \geq S\left(\Psi_{E,\Lambda^*}^{(\mathcal{A})}, \rho \otimes \Lambda^*(\rho)\right) \geq S(\rho).$$

Proof. One has

$$
\begin{aligned}
&\Psi_{E,\Lambda_t^*}^{(\mathcal{A})}\\
&=\sum_{n}\sum_{\ell=2}\sum_{n_\ell\in J_\ell}\left\|\left|\phi_n^{(A_{n_\ell})}\right\rangle\right\|^2\frac{\left|\phi_n^{(A_{n_\ell})}\right\rangle}{\left\|\left|\phi_n^{(A_{n_\ell})}\right\rangle\right\|}\frac{\left\langle\phi_n^{(A_{n_\ell})}\right|}{\left\|\left|\phi_n^{(A_{n_\ell})}\right\rangle\right\|}\\
&\quad+\sum_{n}\sum_{n_k\in Q\setminus\Delta}\lambda_{n_k}\left\|V_n\left|x_{n_k}\right\rangle\right\|^2 E_{n_k}\otimes\frac{V_n\left|x_{n_k}\right\rangle}{\left\|V_n\left|x_{n_k}\right\rangle\right\|}\frac{\left\langle x_{n_k}\right|V_n^*}{\left\|V_n\left|x_{n_k}\right\rangle\right\|},
\end{aligned}
$$

where

$$
\begin{aligned}
\left|\phi_n^{(A_{n_\ell})}\right\rangle &= \sum_{n_i\in A_{n_\ell}}\sqrt{\lambda_{n_i}}\left|x_{n_i}\right\rangle\otimes V_n\left|x_{n_i}\right\rangle,\\
\left\|\left|\phi_n^{(A_{n_\ell})}\right\rangle\right\|^2 &= \sum_{n_k\in A_{n_\ell}}\lambda_{n_k}\left\|V_n\left|x_{n_k}\right\rangle\right\|^2.
\end{aligned}
$$

Since

$$
\begin{aligned}
&tr\Psi_{E,\Lambda^*}^{(\mathcal{A})}\log\Psi_{E,\Lambda^*}^{(\mathcal{A})}\\
&=\sum_{n}\sum_{\ell=2}\sum_{n_\ell\in J_\ell}\left(\sum_{n_i\in A_{n_\ell}}\lambda_{n_i}\left\|V_n\left|x_{n_i}\right\rangle\right\|^2\right)\log\left(\sum_{n_j\in\Delta}\lambda_{n_j}\left\|V_n\left|x_{n_j}\right\rangle\right\|^2\right)\\
&\quad+\sum_{n}\sum_{n_k\in Q\setminus\Delta}\lambda_{n_k}\left\|V_n\left|x_{n_k}\right\rangle\right\|^2\log\left\|V_n\left|x_{n_k}\right\rangle\right\|^2\\
&\quad+\sum_{n_k\in Q\setminus\Delta}\lambda_{n_k}\log\lambda_{n_k},
\end{aligned}
$$

under the condition

$$
\sum_n\left(I\otimes V_n^*\right)\left(I\otimes V_n\right)=I\otimes\left(\sum_n V_n^*V_n\right)=I\otimes I
$$

$$
\text{and}\quad \Lambda^*(\rho)=\sum_n V_n\rho V_n^*\ \text{for any}\ \rho\in\mathfrak{S}(\mathcal{H}_1)
$$

$$
\text{with}\ \sum_n V_n^*V_n=I\quad\text{and}\quad V_m^*V_n=\delta_{mn}V_n^*V_n.
$$

For any $x, x' \in \mathcal{H}_1$, one has

$$\begin{aligned}
&\left\langle x, tr_{\mathcal{H}_2}\Psi_{E,\Lambda^*}^{(\mathcal{A})}x'\right\rangle \\
&= \left\langle x, \left(\sum_{\ell=2}\sum_{n_\ell\in J_\ell}\sum_{n_i\in A_{n_\ell}}\lambda_{n_i}E_{n_i} + \sum_{n_k\in Q\backslash\Delta}\lambda_{n_k}E_{n_k}\right)x'\right\rangle \\
&= \langle x, \rho x'\rangle
\end{aligned}$$

and for any $y, y' \in \mathcal{H}_2$, one obtains

$$\begin{aligned}
&\left\langle y, tr_{\mathcal{H}_1}\Psi_{E,\Lambda^*}^{(\mathcal{A})}y'\right\rangle \\
&= \left\langle y, \left(\sum_{n} V_n\rho V_n^*\right)y'\right\rangle \\
&= \langle y, \Lambda^*(\rho)\, y'\rangle\,.
\end{aligned}$$

Therefore $\Psi_{E,\Lambda^*}^{(\mathcal{A})}$ holds two marginal conditions

$$tr_{\mathcal{H}_2}\Psi_{E,\Lambda^*}^{(\mathcal{A})} = \rho \quad \text{and} \quad tr_{\mathcal{H}_1}\Psi_{E,\Lambda^*}^{(\mathcal{A})} = \Lambda^*(\rho)\,.$$

$$\begin{aligned}
&-tr\Psi_{E,\Lambda^*}^{(\mathcal{A})}\log\rho\otimes\Lambda^*\rho \\
&= S(\rho) + S(\Lambda^*\rho)\,.
\end{aligned}$$

The relative entropy between $\Psi_{E,\Lambda^*}^{(\mathcal{A})}$ and $\sigma_0 = \rho\otimes\Lambda^*\rho$ is obtained by

$$\begin{aligned}
&S\left(\Psi_{E,\Lambda^*}^{(\mathcal{A})}, \rho\otimes\Lambda^*\rho\right) \\
&= S(\rho) \\
&\quad - \sum_{n}\sum_{\ell=2}\sum_{n_\ell\in J_\ell}\sum_{n_i\in A_{n_\ell}}\lambda_{n_i}\left\|V_n\left|x_{n_i}\right\rangle\right\|^2\log\frac{\lambda_{n_i}\left\|V_n\left|x_{n_i}\right\rangle\right\|^2}{\left(\sum_{n_j\in A_{n_\ell}}\lambda_{n_j}\left\|V_n\left|x_{n_j}\right\rangle\right\|^2\right)} \\
&\geq S(\rho)\,.
\end{aligned}$$

Therefore we obtain the following inequality:

$$I^{(\mathcal{A})}(\rho;\Lambda^*) \geq S\left(\Psi_{E,\Lambda^*}^{(\mathcal{A})}, \rho\otimes\Lambda^*(\rho)\right) > S(\rho)\,.$$

□

It shows that the mutual entropy defined by using the entangled compound state $\Psi_{E,\Lambda^*}^{(\Delta)}$ with respect to a subset $\Delta \subset Q$ ($\#(\Delta) \geq 2$), the initial state ρ and the quantum channel Λ^* greater than the von Neumann entropy $S(\rho)$ of the initial state ρ.

If $\#(\Delta) \leq 1$ is held, then the mutual entropy type measure $I^{(1)}(\rho;\Lambda^*)$ is equals to the Ohya mutual entropy taking the von Neumann entropy of the initial state ρ

$$I^{(1)}(\rho;\Lambda^*) = I(\rho;\Lambda^*) = S(\rho).$$

If $\Delta = Q$ is held, then the mutual entropy type measure $I^{(Q)}(\rho;\Lambda^*)$ is equals to the Lindblad-Nielsen entropy taking two times of the von Neumann entropy of the initial state ρ

$$I^{(Q)}(\rho;\Lambda^*) = I_{LN}(\rho;\Lambda^*) = 2S(\rho).$$

It is shown that the mutual entropy defined using the entangled compound state $\Psi_{E,id}^{(\Delta)}$ for the subset $\Delta \subset Q$ $(\#(\Delta) \geq 2)$, the initial state ρ, and the quantum channel Λ^* greater than the von Neumann entropy $S(\rho)$ in the initial state ρ. It does not satisfy the fundamental inequality.

6. Conclusion

As mentioned above, we discuss quantum mutual entropy-type measure with entangled compound states. The mutual entropy type measure $I^{(\mathcal{A})}(\rho;\Lambda_t^*)$ at time t defined using an entangled compound state for a partition $\mathcal{A}$ of all index sets Q greater than the von Neumann entropy $S(\Lambda_t^*\rho)$ in the final state $\Lambda_t^*\rho$. Then, the mutual entropy type measure at the initial time $t = 0$, which is larger than the von Neumann entropy $S(\rho)$ in the initial state ρ. The mutual entropy type measure $I^{(\Delta)}(\rho;\Lambda^*)$ greater than $S(\rho)$. It does not satisfy the fundamental inequality. It seems reasonable to think that a quantum mutual entropy type measure defined using an entangled compound state is not useful in discussing the efficiency of information transfer from the initial system to the final system.

References

1. Shannon, C. E. A mathematical theory of communication. *Bell System Tech. J.* **1948**, *27*, 379–423 and 623–656.
2. A. Jamiołkowski, Linear transformations which preserve trace and positive semidefiniteness of operators, Rep. Math. Phys. 3, 275–278 (1972).
3. Ohya, M. Quantum ergodic channels in operator algebras. *J. Math. Anal. Appl.* **1981**, *84*, 318–327.
4. von Neumann, J. *Die Mathematischen Grundlagen der Quantenmechanik.* Springer - Berlin, 1932.
5. Umegaki, H. Conditional expectations in an operator algebra IV (entropy and information). *Kodai Math. Sem. Rep.* **1962**, *14*, 59–85.

6. Araki, H. Relative entropy of states of von Neumann Algebras. *Publ. RIMS, Kyoto Univ.* **1976**, *11*, 809–833.
7. Araki, H. Relative entropy for states of von Neumann algebras II. *Publ. RIMS, Kyoto Univ.* **1977**, *13*, 173–192.
8. Uhlmann, A. Relative entropy and the Wigner-Yanase-Dyson-Lieb concavity in interpolation theory. *Commun. Math. Phys.* **1977**, *54*, 21–32.
9. Donald, M. J. On the relative entropy. *Commun. Math. Phys.* **1985**, *105*, 13–34.
10. Urbanik, K. Joint probability distribution of observables in quantum mechanics. *Stud. Math.* **1961**, *21*, 317.
11. Ohya, M. On compound state and mutual information in quantum information theory. *IEEE Trans. Infor. Theo.* **1983**, *29*, 770–777.
12. Ohya, M. Note on quantum probability. *L. Nuo. Cimen.* **1983**, *38*, 402–404.
13. Schatten, R. *Norm Ideals of Completely Continuous Operators.* Springer - Verlag, 1970.
14. Ohya, M. Some aspects of quantum information theory and their applications to irreversible processes. *Rep. Math. Phys.* **1989**, *27*, 19–47.
15. Ohya, M.; Petz, D. *Quantum Entropy and its Use.* Springer - Verlag, 1993.
16. Holevo, A. S. Some estimates for the amount of information transmittable by a quantum communication channel (in Russian). *Prob. Pered. Infor.* **1973**, *9*, 3–11.
17. Ohya, M. Entropy Transmission in C*-dynamical systems. *J. Math. Anal. Appl.* **1984**, *100*, 222–235.
18. Muraki, N.; Ohya, M.; Petz, D. Entropies of general quantum systems. *Open Sys. Infor. Dyn.* **1992**, *1*, 43–56.
19. Ohya, M.; Watanabe, N. Construction and analysis of a mathematical model in quantum communication processes. *Scripta Thechnica, Inc., Elect. Commun. Japan* **1985**, *68*, 29–34.
20. Ohya, M. State change and entropies in quantum dynamical systems. *Springer Lect. Not. in Math.* **1985**, *1136*, 397–408.
21. Ohya, M.; Petz, D.; Watanabe, N. On capacities of quantum channels. *Prob. and Math. Stat.* **1997**, *17*, 179–196.
22. Ohya, M.; Petz, D.; Watanabe, N. Numerical computation of quantum capacity. *Inter. J. of Theor. Phys.* **1998**, *38*, 507–510.
23. Ohya, M.; Watanabe, N. A mathematical study of information transmission in quantum communication processes. *Quant. Commun. and Measur.* **1995**, *2*, 371–378.
24. Ohya, M.; Watanabe, N. A new treatment of communication processes with Gaussian Channels. *Japan J. Appl. Math.* **1986**, *3*, 197–206.
25. K.H. Fichtner, W. Freudenberg and V. Liebscher, Beam splittings and time evolutions of Boson systems, Fakultat fur Mathematik und Informatik, Math/ Inf/96/ 39, Jena, 105 (1996).
26. Ohya, M. Fundamentals of quantum mutual entropy and capacity. *Open Sys. Infor. Dyn.* **1999**, *6*, 69–78.
27. Ohya, M.; Volovich, I. V. On Quantum Entropy and its Bound. *Infi. Dimen. Anal., Quant. Prob. and Rela. Topics* **2003**, *6*, 301–310.

28. Schumacher, B. W. Sending entanglement through noisy quantum channels. *Phys. Rev. A* **1996**, *54*, 2614.
29. Ingarden, R. S.; Kossakowski, A.; Ohya, M. *Information Dynamics and Open Systems.* Kluwer, 1997.
30. Emch, G. G. Positivity of the K-entropy on non-abelian K-flows. *Z. Wahrscheinlichkeitstheory verw. Gebiete* **1974**, *29*, 241–252.
31. Ohya, M.; Volovich, I. V. *Mathematical Foundations of Quantum Information and Computation and Its Applications to Nano- and Bio-systems.* Springer, Berlin, 2011.
32. Shor, P. *The quantum channel capacity and coherent information.* Lecture Notes, MSRI Workshop on Quantum Computation, 2002.
33. Barnum, H.; Nielsen, M. A.; Schumacher, B. W. Information transmission through a noisy quantum channel. *Phys. Rev. A* **1998**, *57*, 4153–4175.
34. Bennett, C. H.; Shor, P. W.; Smolin, J. A.; Thapliyalz, A. V. Entanglement-Assisted Capacity of a Quantum Channel and the Reverse Shannon Theorem. quant-ph/0106052.
35. Schumacher, B. W.; Nielsen, M. A. Quantum data processing and error correction. *Phys. Rev. A* **1996**, *54*, 2629.
36. Ohya, M.; Watanabe, N. Comparison of mutual entropy-type measures. TUS preprint, 2003.
37. Watanabe, N. Efficiency of optical modulations with coherent states. *Springer Lect. Notes in Phys.* **1991**, *378*, 350–360.
38. Accardi, L.; Ohya, M. Compound Channels, Transition Expectations, and Liftings. *Appl. Math. Optim.* **1999**, *39*, 33–59.
39. L.B. Levitin, 1987, Information Theory for Quantum Systems. Information, Complexity and Control in Quantum Physics, A. Blaquiere, S. Diner, and G. Lochak, eds., Springer-Verlag, pp. 15–47, Wien-New York.
40. N. Watanabe, Note on entropies for compound systems, Open Systems and Information Dynamics, Vol. 22, No. 2, 1550001–1550021, 2015.
41. N. Watanabe, A Note on Complexities by Means of Quantum Compound Systems, Entropy, 2020, 22(3), 298, 1–24; https://doi.org/10.3390/e22030298.

Infinite Dimensional Analysis, Quantum Probability and Related Topics

pp. 273–283

CONDITIONAL DISTRIBUTION OF A RANDOM VARIABLE, CONDITIONED BY HIDA DISTRIBUTIONS, ON EUCLIDEAN QUANTUM FIELDS

M. W. YOSHIDA

Department of Information Systems, Kanagawa University, Japan
E-mail: washizuminoru@hotmail.com

Conditional probability distributions of a random variable, conditioned by Hida distributions, on Euclidean quantum fields are considered. An explicit form of the characteristic function of the probability distribution is derived. For a corresponding future work to apply the above results, a support property of Φ_3^4 Euclidean quantum field is discussed through the Bochoner-Minlos's Theorem.

Keywords: Hida distribution, Bochoner-Minlos's Theorem.

1. Characteristic function of conditional probability distributions

Let μ be a Borel probability measure on $\mathcal{S}' \equiv \mathcal{S}'(\mathbb{R}^d \to \mathbb{R})$, which corresponds to an Euclidean (scalar) field, e.g., Euclidean free field, Euclidean $P(\phi)_2$ field, Euclidean Φ_3^4 field, etc., where $d \in \mathbb{N}$ denotes a given space dimension. Denote by S_n, $n \in \mathbb{N}$, the corresponding Schwinger functions, and suppose that $S_{2n-1} = 0$ for $n \in \mathbb{N}$.

$$S_{2n}(\varphi_1, \dots, \varphi_{2n}) \equiv \int_{\mathcal{S}'} < \varphi_1, \phi > \cdots < \varphi_{2n}, \phi > \mu(d\phi),$$

$$\varphi_i \in \mathcal{S}(\mathbb{R}^d \to \mathbb{R}),\ i = 1, \dots, 2n.$$

For each $n \in \mathbb{N}$ let S_{2n}^c be the tempered distribution in $\mathcal{S}'(\mathbb{R}^d \to \mathbb{R})^{\otimes 2n}$, which is defined through S_{2n}, such that (*formally* it is a Schwinger function having only the *connected* Feynman graphs):

$$S_2^c(\varphi_1, \varphi_2) \equiv S_2(\varphi_1, \varphi_2),$$

and generally, for $n \in \mathbb{N}$,

$$\begin{aligned} &S^c_{2n}(\varphi_1, \cdots, \varphi_{2n}) \\ &\equiv \text{remove all the terms that are composed by } S_{2n-2} \\ &\text{from } S_{2n}(\varphi_1, \cdots, \varphi_{2n}). \end{aligned} \tag{1}$$

Let $\kappa \in C_0^\infty(\mathbb{R}^3)$, and denote its boundary by $\partial\kappa$. For a (smooth) function $z(y)$, $y \in \mathbb{R}^3$ denote the restriction of $\partial_n z(\cdot)$ to the boundary $\partial\kappa$ by $\partial_n z_{\partial\kappa}$, where ∂_n is the derivative to the direction of the transversal of $\partial\kappa$. Then, for regular $z(\cdot)$,

$$\begin{aligned} &S^c_{2k}(i\xi\kappa + \partial_n z_{\partial\kappa} \cdot \delta_{\partial\kappa}, \cdots, i\xi\kappa + \partial_n z_{\partial\kappa} \cdot \delta_{\partial\kappa}) \\ &= x_{2k,0} + \binom{2k}{1} x_{2k,1} + \cdots + \binom{2k}{2k} x_{2k,2k}, \end{aligned}$$

where

$$x_{2k,m} = S^c_{2k}((i\xi\kappa)^{\otimes(2k-m)} \otimes (\partial_n z_{\partial\kappa} \cdot \delta_{\partial\kappa})^{\otimes m}), \tag{2}$$

and $\partial_n z_{\partial\kappa} \cdot \delta_{\partial\kappa}$ is the function such that for regular f

$$\int_{\mathbb{R}^3} \partial_n z_{\partial\kappa} \cdot \delta_{\partial\kappa}\, f(x)\, dx = \int_{\partial\kappa} \partial_n z(x) f(x) dx. \tag{3}$$

Consider the characteristic function of the conditional probability distribution of the real valued random variable $\kappa(\phi)$ such that

$$f_{\kappa,z}(\xi) \equiv E^\mu \left[e^{i\xi\kappa(\cdot)} \mid z \right], \tag{4}$$

where the condition is given by the sub σ-field generated by the elements in $\mathcal{S}'$, the support of which is in

$$(\text{outside of the support of } \kappa) \cup \partial\kappa. \tag{5}$$

If the random field μ has a support in a sufficiently regular sub space of $\mathcal{S}'$, in order that (3) is well defined, then it is possible to see that

$$f_{\kappa,z}(\xi) = \frac{E^\mu\left[e^{(i\xi\kappa)+(\partial_n z_{\partial\kappa}\cdot\delta_{\partial\kappa})}\right]}{E^\mu\left[e^{(\partial_n z_{\partial\kappa}\cdot\delta_{\partial\kappa})}\right]}, \tag{6}$$

and the following lemma holds:

Lemma 1.1. *Suppose that the random field μ has a support in a sufficiently regular sub space of $\mathcal{S}'$. Then for μ-a.e. $z(\cdot)$, the characteristic function*

$f_{\kappa,z}(\xi)$ of the conditional probability distribution of the random variable $\kappa(\phi)$ is given as follows, assuming that the right hand side is well defined:

$$f_{\kappa,z}(\xi) = e^{\frac{1}{2}\{2x_{2,1}+x_{2,0}\}} \left[1 + \sum_{l=2}^{\infty} \prod_{k=2}^{l} \left(\prod_{m=0}^{2k} e^{\frac{1}{(2k)!}\binom{2k}{m} x_{2k,m}} - 1 \right) \right]$$
$$\times \left[1 + \sum_{l=2}^{\infty} \prod_{k=2}^{l} (e^{\frac{1}{(2k)!} x_{2k,2k}} - 1) \right]^{-1} . \qquad (7)$$

Remark 1.1. In case where the random field μ is a Gaussian field, then (7) is given simply by

$$f_{\kappa,z}(\xi) = e^{\frac{1}{2}\{2x_{2,1}+x_{2,0}\}}. \qquad (8)$$

2. Support property of Φ_3^4 Euclidean field

In Lemma 1.1, it is assumed that the random field μ has a support in a sufficiently regular sub space of $\mathcal{S}'$.

As a future work, in order to apply this lemma to Φ_3^4 Euclidean field, we have to certify the sub space of $\mathcal{S}'(\mathbb{R}^3 \to \mathbb{R})$ in which the support of Φ_3^4 Euclidean field measure exists. The following is the corresponding consideration for Φ_d^4, $d = 2, 3$, Euclidean field measure. In the subsequent discussions let $d = 2, 3$.

For this purpose, we recall the Bochner-Minlos's theorem (cf. e.g., Section 3.2 of [9]). Let E be a nuclear space. Suppose in particular that E is a countably Hilbert space, namely the sequence of semi-norms by which E is characterized by a sequence of Hilbert norms $\| \|_n$, $n \in \mathbb{N} \cup \{0\}$ such that $\| \|_0 < \| \|_1 < \cdots < \| \|_n < \cdots$. Let E_n be the completion of E with respect to the norm $\| \|_n$, then $E = \bigcap_{n \geq 0} E_n$ and $E_0 \supset E_1 \supset \cdots \supset E_n \supset \cdots$. Define

$$E_n^* \equiv \text{the dual space of } E_n, \text{ in order that } E_0^* = E_0 \ ,$$

then

$$E \subset \cdots \subset E_{n+1} \subset E_n \subset \cdots \subset E_0 = E_0^* \subset \cdots \subset E_n^* \subset E_{n+1}^* \subset \cdots \subset E^*.$$

Since E is a nuclear space, for any $m \in \mathbb{N} \cup \{0\}$ there exists an $n \in \mathbb{N} \cup \{0\}$, $n > m$, such that the (canonical) injection $T_m^n : E_n \to E_m$ is a trace class (nuclear class) operator. The Bochner-Minlos's theorem is given as follows:

Proposition 2.1. ***(Bochner-Minlos's Theorem)***
Let $C(\varphi)$, $\varphi \in E$, be a complex valued function on E such that

i) $C(\varphi)$ *is continuous with respect to the norm* $\|\cdot\|_m$ *for some* $m \in \mathbb{N} \cup \{0\}$*;*
ii) **(positive definiteness)** *for any* $k \in \mathbb{N}$*,*

$$\sum_{i,j=1}^{k} \bar{\alpha}_i \alpha_j C(\bar{\varphi}_i - \varphi_j) \geq 0, \qquad \forall \alpha_i \in \mathbb{C},\ \forall \varphi_i \in E,\ i = 1, \dots, k;$$

(where $\bar{\alpha}$ *means complex conjugate of* α*).*
iii) **(normalization)** $C(0) = 1$.
Then, there exists a unique Borel probability measure ν *on* E^* *such that*

$$C(\varphi) = \int_{E^*} e^{i<\phi,\varphi>} \nu(d\phi), \qquad \varphi \in E.$$

Moreover, if the (canonical) injection $T_m^n : E_n \to E_m$*,* $n > m$*, is a Hilbert-Schmidt operator, then the support of* ν *is in* E_n^**, where* $< \phi, \varphi >=\, _{E^*}< \phi, \varphi >_E$ *is the dualization between* $\phi \in E^*$ *and* $\varphi \in E$*.* □

We next prepare an adequate countably Hilbert nuclear space $\mathcal{H} \supset \mathcal{S}(\mathbb{R}^d \to \mathbb{R}) \equiv \mathcal{S}(\mathbb{R}^d)$, for a given $d \in \mathbb{N}$. Let

$$\mathcal{H}_0 \equiv \left\{ f : \|f\|_{\mathcal{H}_0} = \left((f, f)_{\mathcal{H}_0} \right)^{\frac{1}{2}} < \infty,\ f : \mathbb{R}^d \to \mathbb{R}, \text{ measurable} \right\} \supset \mathcal{S}(\mathbb{R}^d), \tag{9}$$

where

$$(f, g)_{\mathcal{H}_0} \equiv (f, g)_{L^2(\mathbb{R}^d)} = \int_{\mathbb{R}^d} f(x) g(x)\, dx. \tag{10}$$

Let

$$H \equiv (|x|^2 + 1)^{\frac{d+1}{2}} (-\Delta + 1)^{\frac{d+1}{2}} (|x|^2 + 1)^{\frac{d+1}{2}}, \tag{11}$$

$$H^{-1} \equiv (|x|^2 + 1)^{-\frac{d+1}{2}} (-\Delta + 1)^{-\frac{d+1}{2}} (|x|^2 + 1)^{-\frac{d+1}{2}}, \tag{12}$$

be the pseudo differential operators on $\mathcal{S}'(\mathbb{R}^d \to \mathbb{R}) \equiv \mathcal{S}'(\mathbb{R}^d)$ with the d-dimensional Laplace operator Δ. For each $n \in \mathbb{N}$, define

$$\mathcal{H}_n \equiv \text{the completion of } \mathcal{S}(\mathbb{R}^d) \text{ with respect to the norm } \|f\|_n,\ f \in \mathcal{S}(\mathbb{R}^d), \tag{13}$$

where $\|f\|_n^2 = (f, f)_n$ (in case where $n = 1$, to denote the $\mathcal{H}_1$ norm we use the exact notation $\|\ \|_{\mathcal{H}_1}$, in order to avoid a confusion between the notation of some L^1 or l^1 norms) with

$$(f, g)_n = (H^n f, H^n g)_{\mathcal{H}_0}, \qquad f,\ g \in \mathcal{S}(\mathbb{R}^d), \tag{14}$$

and

$$\mathcal{H}_{-n} \equiv \text{the completion of } \mathcal{S}'(\mathbb{R}^d)$$

$$\text{with respect to the norm } \|f\|_{-n}, \quad f \in \mathcal{S}'(\mathbb{R}^d), \tag{15}$$

where $\|f\|_{-n}^2 = (f, f)_{-n}$ with

$$(f, g)_{-n} = ((H^{-1})^n f, (H^{-1})^n g)_{\mathcal{H}_0}, \qquad f, g \in \mathcal{S}(\mathbb{R}^d). \tag{16}$$

Then obviously, for $f \in \mathcal{S}(\mathbb{R}^d)$,

$$\|f\|_n \le \|f\|_{n+1}, \qquad \|f\|_{-n-1} \le \|f\|_{-n}. \tag{17}$$

by taking an inductive limit and setting $\mathcal{H} = \bigcap_{n \in \mathbb{N}} \mathcal{H}_n$, we have the followings:

$$\mathcal{H} \subset \cdots \subset \mathcal{H}_{n+1} \subset \mathcal{H}_n \subset \cdots \subset \mathcal{H}_0 \subset \cdots \subset \mathcal{H}_{-n} \subset \mathcal{H}_{-n-1} \subset \cdots \subset \mathcal{H}^*. \tag{18}$$

The (topological) dual space of $\mathcal{H}_n$ is $\mathcal{H}_{-n}$, $n \in \mathbb{N}$. By the operator H^{-1} given by (12) on $\mathcal{S}(\mathbb{R}^d)$ we can define, on each $\mathcal{H}_n$, $n \in \mathbb{N}$, the bounded symmetric operators (and hence a self-adjoint operator on it)

$$(H^{-1})^k, \quad k \in \mathbb{N} \cup \{0\} \tag{19}$$

(by using the same notations as the operators on $\mathcal{S}(\mathbb{R}^d)$). Hence, the canonical injection

$$T_n^{n+k} : \mathcal{H}_{n+k} \longrightarrow \mathcal{H}_n, \qquad k, n \in \mathbb{N} \cup \{0\}, \tag{20}$$

is given by the self-adjoint operator $(\mathcal{H}^{-1})^k$ on $\mathcal{H}_n$. By a simple calculation by means of the Fourier transform, and by Young's inequality, we see that for each $n \in \mathbb{N} \cup \{0\}$, H^{-1} on $\mathcal{H}_n$ is a Hilbert-Schmidt operator and hence $(H^{-1})^2$ on $\mathcal{H}_n$ is a trace class operator.

As a rough explanation, the Euclidean Φ_d^4, $d = 2, 3$, resp., fields are probability measures on $\mathcal{S}'(\mathbb{R}^d \to \mathbb{R})$, which are probability measures having self interactions with fourth power, and are defined by adding corresponding interaction term to the Euclidean free fields on $\mathcal{S}'(\mathbb{R}^d \to \mathbb{R})$, $d = 2, 3$, resp. (cf., e.g., [10], for fields with general polynomial interactions). There are several strategies through which the Φ_d^4, $d = 2, 3$, Euclidean fields are constructed. One of a standard construction strategy is to start from (continuum) random fields on $\mathcal{S}'(\mathbb{R}^d \to \mathbb{R})$, and a procedure that start from fields on lattice $(\epsilon\mathbb{Z})^d$ (with the lattice spacing $\epsilon > 0$, which subsequently tends to 0, i.e., taking a continuum limit) is another standard strategy.

Here, we make use of the lattice formulation given by [7]. The procedure, through which the Φ_d^4 Euclidean field measures are defined, is the following: Let $d = 2, 3$. For each bounded region $\Lambda \subset \mathbb{R}^d$ and a lattice spacing $\epsilon > 0$, let

$$L_{\epsilon,\Lambda} \equiv (\epsilon\mathbb{Z})^d \cap \Lambda, \tag{21}$$

and define a family of real valued random variables $\phi \equiv \{\phi(x) : x \in L_{\epsilon,\Lambda}\}$, the probability distribution of which is given by

$$\nu_{\epsilon,\Lambda} \equiv \frac{1}{Z_{\epsilon,\Lambda}} \prod_{x \in L_{\epsilon,\Lambda}} e^{-S_{L_{\epsilon,\Lambda}}(\phi)} d\phi(x), \tag{22}$$

where $Z_{\epsilon,\Lambda}$ is the normalizing constant,

$$S_{L_{\epsilon,\Lambda}}(\phi) \equiv \frac{1}{2} \sum_{<x,y>} \epsilon^{d-2} \big(\phi(x) - \phi(y)\big)^2 + \frac{1}{2} a_\epsilon \sum_{x \in L_{\epsilon,\Lambda}} \epsilon^d \phi^2(x) + \frac{\lambda}{2} \sum_{x \in L_{\epsilon,\Lambda}} \epsilon^d \phi^4(x), \tag{23}$$

with a_ϵ a counter term depending on $\epsilon > 0$ and $d = 2, 3$, $\lambda \geq 0$ a coupling constant and $< x, y >$ denotes the nearest neighbor points in $L_{\epsilon,\Lambda}$. In [7], roughly speaking, it is shown that, for adequately small $\lambda \geq 0$, there exists a subsequence $\{\nu_{\epsilon_i,\Lambda_j}\}_{i,j \in \mathbb{N}}$ of $\{\nu_{\epsilon,\Lambda}\}_{\epsilon>0,\Lambda \subset \mathbb{R}^d}$ with $\lim_{i\to\infty} \epsilon_i = 0$ and $\lim_{j\to\infty} \Lambda_j = \mathbb{R}^d$, and a weak limit

$$\nu^* \equiv \lim_{i\to\infty} \lim_{j\to\infty} \nu_{\epsilon_i,\Lambda_j} \tag{24}$$

exists in the space of Borel probability measures on $\mathcal{S}'(\mathbb{R}^d \to \mathbb{R})$ by interpreting $\nu_{\epsilon,\Lambda}$ as a Borel probability measure on $\mathcal{S}'(\mathbb{R}^d \to \mathbb{R})$ for each $\epsilon > 0$ and $\Lambda \subset \mathbb{R}^d$. For each $d = 2, 3$, ν^* given by (24) is known as a Φ_d^4, Euclidean field measure, defined through the lattice argument.

Now, we shall certify the support of the Φ_d^4, $d = 2, 3$, Euclidean field measure given by (24).

For each $\epsilon > 0$, bounded region $\Lambda \subset \mathbb{R}^d$, and for $F(\phi)$, a polynomial in $\{\phi(x) : x \in L_{\epsilon,\Lambda}\}$, by (21) and (22) let

$$< F >_{\epsilon,\Lambda} \equiv \int F(\phi) \nu_{\epsilon,\Lambda}(d\phi), \tag{25}$$

then by [11] (cf. also Section 2 of [7]), the following limit exists:

$$< F >^{(\epsilon)} \equiv \lim_{\Lambda \uparrow \mathbb{R}^d} < F >_{\epsilon,\Lambda} . \tag{26}$$

Also, for each $\epsilon > 0$, there exists a weak limit

$$\nu_\epsilon \equiv \lim_{\Lambda \uparrow \mathbb{R}^d} \nu_{\epsilon,\Lambda}, \tag{27}$$

that is a Borel probability measure on $\mathcal{S}'(\mathbb{R}^d \to \mathbb{R})$. Let $D^{(\epsilon)}$ be the two point function of the lattice field with the lattice spacing $\epsilon > 0$ and a given mass $m_0 > 0$, such that for $x, y \in (\epsilon\mathbb{Z})^d$

$$D^{(\epsilon)}(x-y)$$
$$= (2\pi)^{-1} \int_{[-\frac{\pi}{\epsilon},\frac{\pi}{\epsilon}]^d} \left(\epsilon^{-2} \sum_{i=1}^{d} (1 - \cos \epsilon k_i) + m_0^2 \right)^{-1} e^{ik\cdot(x-y)} dk_1 \cdots dk_d,$$

which is an interpretation of the operator $(-\Delta + m_0^2)^{-1}$ in the continuous Φ_d^4 model to the lattice one. By (26), define

$$S^{(\epsilon)}(x-y) \equiv <\phi(x)\cdot\phi(y)>^{(\epsilon)}, \qquad x, y \in (\epsilon\mathbb{Z})^d, \tag{28}$$

and

$$S_n^{(\epsilon)}(x_1,\ldots,x_n) \equiv <\prod_{i=1}^{n} \phi(x_i)>^{(\epsilon)}, \quad x_i \in L_\epsilon,\ i = 1,\ldots,n,\ n \in \mathbb{N}. \tag{29}$$

From Theorem 6.1, Lemma A.1 with the formulas (A13) and (8.2) of [7] we see that there exist universal constants λ_0, K_1, K_2 such that if $0 \le \lambda \le \lambda_0$, then the following hold:

$$|||S^{(\epsilon)} - D^{(\epsilon)}|||^{(\epsilon)} \le K_1\lambda^2, \qquad \forall \epsilon > 0, \tag{30}$$

$$|||D^{(\epsilon)}|||_{l^1}^{(\epsilon)} \le K_2 m_0^{-2}, \qquad \forall \epsilon > 0, \tag{31}$$

and the *Gaussian inequality*

$$0 \le S_{2n}^{(\epsilon)}(x_1,\ldots,x_{2n}) \le \sum_{\text{pairing}} \prod S^{(\epsilon)}(x_\alpha, x_\beta),$$

$$S_{2n-1}^{(\epsilon)}(x_1,\ldots,x_{2n-1}) = 0, \quad n \in \mathbb{N}, \tag{32}$$

where

$$|||f|||^{(\epsilon)} \equiv \|f\|_{l^1}^{(\epsilon)} + \|f\|_\infty^{(\epsilon)} \equiv \epsilon^d \sum_{x\in(\epsilon\mathbb{Z})^d} |f(x)| + \sup_{x\in(\epsilon\mathbb{Z})^d} |f(x)|. \tag{33}$$

By making use of (30), (31) and (32), from Proposition 2.1 we see that the supports of the Borel probability measure ν_ϵ on $\mathcal{S}'(\mathbb{R}^d \to \mathbb{R})$, for each $\epsilon > 0$, and a weak limit of a subsequence of $\{\nu_\epsilon\}_{\epsilon>0}$, denoted by ν, are in the Hilbert space $\mathcal{H}_{-2} \subset \mathcal{S}'(\mathbb{R}^d \to \mathbb{R})$ defined by (15):

$$\nu_\epsilon \text{ and } \nu \text{ can be understood as probability measures on } \mathcal{H}_{-2}. \tag{34}$$

In fact, by the Sobolev's embedding theorem and by (13) with (11) and (14), we see that there exist some constants K_3, K_4, K_5 and the following inequality with respect to the norms, and the corresponding *continuous* embedding hold:

$$\sup_{x\in\mathbb{R}^d} \left|(|x|^2+1)^{\frac{d+1}{2}} f(x)\right| \le K_3 \left\|(|x|^2+1)^{\frac{d+1}{2}} f\right\|_{W^{d+1,2}} \tag{35}$$

$$\le K_4 \left\|(-\Delta+1)^{\frac{d+1}{2}}(|x|^2+1)^{\frac{d+1}{2}} f\right\|_{L^2(\mathbb{R}^2)} \le K_5 \|f\|_{\mathcal{H}_1},\ \forall f\in\mathcal{S}(\mathbb{R}^d\to\mathbb{R}),$$

$$\mathcal{H}_1 \hookrightarrow (|x|^2+1)^{-\frac{d+1}{2}} W^{d+1,2}(\mathbb{R}^d) \hookrightarrow (|x|^2+1)^{-\frac{d+1}{2}} C_b(\mathbb{R}^d\to\mathbb{R}),$$

where $W^{d+1,2} = W^{d+1,2}(\mathbb{R}^d)$ is the Sobolev space, the elements of which are real measurable functions having square integrable (with respect to the Lebesgue measure on $\mathbb{R}^d$) partial derivatives (in the sense of distribution) of all orders up to $d+1$, and $C_b(\mathbb{R}^d\to\mathbb{R})$ is the space of real valued continuous functions on $\mathbb{R}^d$. Then, by the Young's inequality, which is valid for both integrals and sums on lattice (cf., e.g., Appendix of [7]), from (30), (31), (32), (33) and (35), we see that there exists a constant K and

$$\begin{aligned}
&\left|\int_{\mathcal{S}'} \left({}_{\mathcal{S}'}<\phi,\varphi>_{\mathcal{S}'}\right)^2 \nu_\epsilon(d\phi)\right| \\
&= \left|\epsilon^d \sum_{x\in(\epsilon\mathbb{Z})^d} \varphi(x) \left(\epsilon^d \sum_{y\in(\epsilon\mathbb{Z})^d} S^{(\epsilon)}(x-y)\varphi(y)\right)\right| \\
&\le \left|\epsilon^d \sum_{x\in(\epsilon\mathbb{Z})^d} |\varphi(x)| \left(\epsilon^d \sum_{y\in(\epsilon\mathbb{Z})^d} S^{(\epsilon)}(x-y)|\varphi(y)|\right)\right| \\
&\le \left(\epsilon^d \sum_{x\in(\epsilon\mathbb{Z})^d} |\varphi(x)|^2\right)\cdot(K_1\lambda^2+K_2 m_0^{-2}) \\
&\le K_5^2 \left(\epsilon^2 \sum_{x\in(\epsilon\mathbb{Z})^d} \|\varphi\|^2_{\mathcal{H}_1}(|x|^2+1)^{-d-1}\right)(K_1\lambda^2+K_2 m_0^{-2}) \\
&\le K\|\varphi\|^2_{\mathcal{H}_1}, \qquad \forall\epsilon>0,\ \forall\varphi\in\mathcal{H}_1,
\end{aligned} \tag{36}$$

where $\| \ \|_{\mathcal{H}_1}$ is the $\mathcal{H}_1$ norm defined by (13), and $\mathcal{S} = \mathcal{S}(\mathbb{R}^d\to\mathbb{R})$, $\mathcal{S}' = \mathcal{S}'(\mathbb{R}^d\to\mathbb{R})$. In (36) for the inequality last but one, we used the fact, which follows from (35), such that

$$\begin{aligned}
|\varphi(x)| &\le \Big(\sup_{x'\in\mathbb{R}^d} (|x'^2+1)^{-\frac{d+1}{2}}|\varphi(x')|\Big)\cdot(|x|^2+1)^{-\frac{d+1}{2}} \\
&\le K_5\|\varphi\|_{\mathcal{H}_1}(|x|^2+1)^{-\frac{d+1}{2}},\ \forall x\in\mathbb{R}^d, \qquad \forall\varphi\in\mathcal{H}_1.
\end{aligned}$$

By (36), from the *Gaussian inequality* (32) we have

$$\begin{aligned}
&\left|\int_{\mathcal{S}'} \left({}_{\mathcal{S}'}<\phi,\varphi>_{\mathcal{S}'}\right)^{2n} \nu_\epsilon(d\phi)\right| \\
&= \left|(\epsilon^d)^{2n} \sum_{x_1,\ldots,x_{2n}\in(\epsilon\mathbb{Z})^d} \varphi(x_1)\cdots\varphi(x_{2n})\, S_{2n}^{(\epsilon)}(x_1,\ldots,x_{2n})\right| \\
&\leq (2n-1)!!K^n\|\varphi\|_{\mathcal{H}_1}^{2n}, \qquad \forall\epsilon>0,\ \forall\varphi\in\mathcal{H}_1,\ n\in\mathbb{N}. \qquad (37)
\end{aligned}$$

Now, by (37), it holds that

$$\begin{aligned}
&\left|\int_{\mathcal{S}'} e^{i_{\mathcal{S}'}<\phi,\varphi>_{\mathcal{S}'}}\nu_\epsilon(d\phi)\right| \\
&\leq \left|\sum_{n=0}^{\infty}\frac{(-1)^n}{(2n)!}\int_{\mathcal{S}'}\left({}_{\mathcal{S}'}<\phi,\varphi>_{\mathcal{S}'}\right)^{2n}\nu_\epsilon(d\phi)\right| \\
&\leq \sum_{n=0}^{\infty}\frac{(2n-1)!!}{(2n)!}K^n\|\varphi\|_{\mathcal{H}_1}^{2n} = e^{\frac{1}{2}K\|\varphi\|_{\mathcal{H}_1}}, \quad \forall\epsilon>0,\ \forall\varphi\in\mathcal{H}_1, \qquad (38)
\end{aligned}$$

(cf., (12.51) of [8] for corresponding regularity of (general) Euclidean $P(\phi)_2$, i.e., fields with polynomial interactions, measures). Denote

$$C^{(\epsilon)}(\varphi) \equiv \int_{\mathcal{S}'} e^{i_{\mathcal{S}'}<\phi,\varphi>_{\mathcal{S}'}}\nu_\epsilon(d\phi). \qquad (39)$$

Since ν_ϵ is a Borel probability measure on $\mathcal{S}'$, $C^{(\epsilon)}$ satisfies conditions ii) and iii) for C in Proposition 2.1. Moreover, by (38), since $C^{(\epsilon)}(\cdot)$ is continuous with respect to the $\mathcal{H}_1$ norm $\|\ \|_{\mathcal{H}_1}$, from i) of Proposition 2.1 with its last statement we conclude that the support of ν_ϵ is in $\mathcal{H}_{-2}$ for any $\epsilon>0$. This guarantees (34) for ν_ϵ.

To certify (34) for a weak limit of $\{\nu_\epsilon\}_{\epsilon>0}$, by (36), (37) and (32), recall that the distribution $S_k^{(\epsilon)}$, $k\in\mathbb{N}$, $\epsilon>0$, satisfies

$$S_{2n-1}^{(\epsilon)} = 0 \quad \text{and} \quad \left\|S_{2n}^{(\epsilon)}\right\|_{(\mathcal{H}_{-1})^{\otimes 2n}} \leq (2n-1)!!(K+1)^n, \qquad \forall\epsilon>0,\ \forall n\in\mathbb{N}. \qquad (40)$$

Namely, for each $n\in\mathbb{N}$, the family of distributions $\{S_{2n}^{(\epsilon)}\}_{\epsilon>0}$ forms a bounded set in the Hilbert space $(\mathcal{H}_{-1})^{\otimes 2n}$, the space of $2n$-th tensor power of $\mathcal{H}_{-1}$ defined by (15). Thus, for each $n\in\mathbb{N}$, we can take a sequence $\{\epsilon_{n,i}\}_{i\in\mathbb{N}}$, with $\epsilon_{n,i}>0$ and $\lim_{i\to\infty}\epsilon_{n,i}=0$ such that $\{S_{2n}^{(\epsilon_{n,i})}\}_{i\in\mathbb{N}}$ converges weakly (as $i\to\infty$) to some $S_{2n}\in(\mathcal{H}_{-1})^{\otimes 2n}$, that satisfies the same bound as (37). By taking subsequences, through the diagonal argument, we then see that there exists a sequence $\{\epsilon_i\}_{i\in\mathbb{N}}$ with $\epsilon_i>0$ and $\lim_{i\to\infty}\epsilon_i=0$

such that $\{S_{2n}^{\epsilon_i}\}_{i\in\mathbb{N}}$ converges weakly (as $i \to \infty$) to $S_{2n} \in (\mathcal{H}_{-1})^{\otimes 2n}$ for any $n \in \mathbb{N}$. By these, we can define a functional

$$C(\varphi) \equiv \sum_{n=0}^{\infty} \frac{(-1)^n}{(2n)!} \left\langle S_{2n}, \varphi^{\otimes 2n} \right\rangle, \tag{41}$$

that satisfies the same bound as (38):

$$|C(\varphi)| \leq e^{\frac{1}{2}K\|\varphi\|_{\mathcal{H}_1}}, \qquad \forall\varphi \in \mathcal{H}_1, \tag{42}$$

where $< S_{2n}, \varphi^{\otimes 2n} >= {}_{(\mathcal{H}_{-1})^{\otimes 2n}}\left\langle S_{2n}, \varphi^{\otimes 2n} \right\rangle_{(\mathcal{H}_1)^{\otimes 2n}}$ is the dualization between $(\mathcal{H}_{-1})^{\otimes 2n}$ and $(\mathcal{H}_1)^{\otimes 2n}$. For each $\epsilon > 0$, since $C^{(\epsilon)}$ defined by (39) satisfies the conditions i), ii), iii) of Proposition 2.1, by the construction $C(\varphi)$ defined by (41) also satisfies the same conditions. In particular, by (42), $C(\varphi)$ is continuous with respect to the $\mathcal{H}_1$ norm. Hence, from Proposition 2.1, we can conclude an existence of Borel probability measure ν on $\mathcal{H}_{-2}$ corresponding to $C(\varphi)$ defined by (41). This guarantees (34) for ν.

Theorem 2.1. *The φ_d^4, $d = 2, 3$, Euclidean field measure ν defined through the lattice arguments has support in $\mathcal{H}_{-2} \subset \mathcal{S}'(\mathbb{R}^d \to \mathbb{R})$.* □

References

1. Albeverio, S., Höegh-Krohn, R., Quasi invariant measures, symmetric diffusion processes and quantum fields. Les mèthodes mathèmatiques de la thèorie quantique des champs. (*Colloq. Internat. CNRS, No. 248, Marseille, 1975*), Èditions Centre Nat. Recherche Sci., Paris (1976), 11–59.
2. Albeverio, S., Höegh-Krohn, R., Dirichlet forms and diffusion processes on rigged Hilbert spaces. *Z. Wahrscheinlichkeitstheor. Verv. Geb.* 40 (1977), 1–57.
3. Albeverio, S., Höegh-Krohn, R., Streit, L., Energy forms, Hamiltonians, and distorted Brownian paths, *J. Mathematical Phys.* 18 (1977), 907–917.
4. Albeverio, S., Röckner, M., Classical Dirichlet forms on topological vector spaces- the construction of the associated diffusion processes, *Probab. Theory Related Fields,* 83 (1989), 405–434.
5. Albeverio, S., Röckner, M., Classical Dirichlet forms on topological vector spaces-closability and a Cameron-Martin formula, *J. Functional Analysis,* 88 (1990), 395–43.
6. Albeverio, S., Röckner, M., Stochastic differential equations in infinite dimensions: solution via Dirichletforms, *Probab. Theory Related Fields,* 89 (1991), 347–386.
7. Brydges, D., Fröhlich, J., Sokal, A., A New proof of the existence and non triviality of the continuum φ_2^4 and φ_3^4 quantum field theories, *Commn. Math. Phys.,* 91 (1983), 141–186.

8. Glimm, J., Jaffe, A., *Quantum Physics: A Functional Integral Point of View, 2nd ed.*, Springer, Berlin, 1987.
9. Hida, T., *Brownian motion,* Springer-Verlag, New York Heidelberg Berlin 1980.
10. Simon, B., *The* $P(\Phi)_2$ *Euclidean (Quantum) Field Theory,* Princeton Univ. Press, Princeton, NJ., 1974.
11. Sokal, A.D., An alternative constructive approach to φ_3^4-quantum field theory, and a possible destructive approach to φ_4^4, *Ann. Inst. Henri Poincarèm,* A 37 (1982), 317–398.
12. Treves, F., Topological vector spaces, distributions and kernels, *Academic Press, New York-London,* 1967.

www.ingramcontent.com/pod-product-compliance
Lightning Source LLC
LaVergne TN
LVHW010332220926
848484LV00004BB/888